Synthesis and Applications of Electrospun Nanofibers

Synthesis and Applications of Electrospun Nanofibers

Ramazan Asmatulu

*Department of Mechanical Engineering,
Wichita State University, Wichita, KS, United States*

Waseem S. Khan

*Department of Mechanical and Mechatronic Engineering,
Dubai Men's College Higher Colleges of Technology, UAE*

Elsevier
Radarweg 29, PO Box 211, 1000 AE Amsterdam, Netherlands
The Boulevard, Langford Lane, Kidlington, Oxford OX5 1GB, United Kingdom
50 Hampshire Street, 5th Floor, Cambridge, MA 02139, United States

British Library Cataloguing-in-Publication Data
A catalogue record for this book is available from the British Library

Library of Congress Cataloging-in-Publication Data
A catalog record for this book is available from the Library of Congress

ISBN: 978-0-12-813914-1

For Information on all Elsevier publications
visit our website at https://www.elsevier.com/books-and-journals

Publisher: Matthew Deans
Acquisition Editor: Simon Holt
Editorial Project Manager: Ana Claudia A. Garcia
Production Project Manager: Anitha Sivaraj
Cover Designer: Greg Harris

Typeset by MPS Limited, Chennai, India

Contents

Introduction to electrospun nanofibers

1

CHAPTER OUTLINE

1.1 INTRODUCTION

1.1.1 WHAT IS NANOTECHNOLOGY?

Nanotechnology is highly integrated with our society and will continue to be in the next few decades. It will have a more significant impact on our society than other technologies. The prefix nano is derived from the Greek word ναῦος or Latin word nannus, both meaning dwarf. It is adopted as an official SI prefix meaning 10^{-9} of an SI base unit. Nanotechnology is the engineering at the atomic or molecular level. It is the collective term for a wide range of technologies, processing techniques, modeling, and measurements that involve the manipulating of matter at the smallest scale (from $1-100$ nm). Nanotechnology is the study of controlling matter on these scales. According to the National Science Foundation and National Nanotechnology Initiative, nanotechnology is the ability to understand control, and manipulate matter at the level of atoms and molecules, as well as at the "supramolecular" level involving clusters of molecules, in order to create materials, systems, and devices with fundamentally new properties and functions because of their small structures. Generally, nanotechnology involves components

Synthesis and Applications of Electrospun Nanofibers. DOI: https://doi.org/10.1016/B978-0-12-813914-1.00001-8

1

and structures with nanosize, and entails developing, creating, or modifying devices, systems, and materials within that length scale. Nanotechnology is concerned with the creation of fibers, particles, and materials at nanoscale dimensions. These fibers, particles, and materials are referred to as nanofibers, nanoparticles, and nanomaterials, respectively, and they exhibit unusual and exotic properties that are not present in traditional bulk materials [1].

Nanotechnology is a multidisciplinary field which includes molecular physics, materials science, chemistry, biology, computer science, electrical engineering, and mechanical engineering. Nanotechnology is associated with design, manufacturing, characterization, modeling, and application of materials, devices, and systems at nanometer scale, by manipulating their shape and dimensions in a controlled manner. These nanoscale products and materials exhibit at least one novel or superior property due to their nanoscale size. Nanoscience is another term that is used frequently in the literature. Nanoscience is closely related to nanotechnology; however, some distinctions occur, which need to be explained. Nanoscience is the study of the fundamental principles of molecules, structures, and systems with at least one of the dimensions usually between 1 and 100 nms. These structures are known as nanostructures. Nanotechnology is the application of these nanostructures into useful nanoscale devices [2]. Nanoscience involves the study of the physical properties of materials and products at atomic, molecular, and micromolecular levels. Nanotechnology combined with nanoscience controls the matter at the nanometer scale and is involved in almost every field at the nanoscale level. Nanoscience and nanotechnology deal with the potential to see, organize, and control individual atoms and molecules for useful scientific and technological uses.

Sometimes, the laws of science may not be enough to deal with engineered nanomaterials or nanostructures. Nanomaterials possess a large surface area, a high aspect ratio, and a high surface-to-mass ratio. The unusual features of nanomaterials can significantly influence the mechanical, thermal, electrical, and other physical, chemical, physicochemical, and biological properties [3,4]. The term "nanotechnology" was first introduced by a Japanese engineer, Norio Taniguchi, in 1974. The term implied a new technology which can control materials beyond the micrometer scale [5]. The ideas and concepts of creating nanoscale machines started with a talk entitled "There's plenty of Room at the bottom" by the famous American physicist, Richard Feynman, at California Institute of Technology in 1959. In his lecture, Feynman illustrated a process in which researchers and scientists would be able to control and manipulate atoms and molecules. In the 1980s, IBM Zurich scientists invented the tunneling microscope, a landmark achievement in nanotechnology development, which allowed scientists and researchers to analyze materials at the atomic or molecular level. Recently, similar studies on nanotechnology research and development have increased globally. Research in nanotechnology is expected to continue to grow worldwide, and in the next few decades, nanotechnology and its products could have a more than $1 trillion impact on the global economy soon [1−4].

Nanotechnology has the potential to change our standard of living. The latest applications of nanotechnology include electronic components, nanopaints, storage devices, stain-free fabrics, bio- and nanosensors, and medical components. Nanotechnology is spreading into almost every field, such as energy storage and production, information technology, medical purposes, manufacturing, food and water purification, instrumentation, biomedical, DNA computers, microelectromechanical systems (MEMS), nanoelectromechanical systems (NEMS), motors, nanosensors, nanowires, nano-satellite missions, and many others, somewhat at atomic, molecular, or macromolecular scales. The basic feature of this technology is the size that makes it so feasible to be used in many different fields. The nano size of the materials provides certain advantages, such as high surface area, quantum effect, and low surface defects/lesser imperfections in the material, thereby improving the material properties. One of the fundamental aspects of nanotechnology is the creation of new materials having one of the dimensions at nanoscale. These materials, known as nanomaterials, are engineered at nanoscale have entirely different properties than their "bulk" counterpart. The commonly used engineered nanomaterials in consumer products are nanosilver, carbon nanotubes, nanosized metal oxides (ferrous oxides, titanium dioxide, and zinc oxide), silica, platinum, and gold. Other engineered nanomaterials used in consumer, medical, and industrial products are nanocarbon, cerium oxide, nickel, aluminum oxide, and the nanoclays copper oxide, iron oxide, and quantum dots.

1.1.2 WHAT IS ELECTROSPINNING?

There are various processes available to generate nanofibers. These processes include template synthesis, phase separation, and self-assembly. However, electrospinning is the simplest, most straightforward, and cheapest process of producing nano- and micro-sized fibers in a very short period of time with minimum investment. Generally, electrospinning is used to produce high-surface-area submicron and nanosized fibers. These fibers possess more exceptional physical properties (e.g., mechanical, magnetic, electrical, optical, and thermal) than their bulk-size fibers. Electrospinning is related to the principle of spinning of polymeric solutions or melt at elevated temperature in a high DC electric field. Most of the synthetic and naturally occurring polymers can be electrospun after dissolving in appropriate solvents. In conventional spinning, shearing, rheological, gravitational, inertia, and aerodynamics forces act on the fibers. However, in electrospinning only electrostatic forces are used to generate fibers. In electrospinning, the shearing forces are generated by the interaction of an applied electrostatic field with the electrical charges carried by the polymer jet rather than by the spindles utilized in conventional spinning. Electrospinning is a process in which a high voltage, and consequently a high electrostatic field, is applied to a polymeric solution or melt to generate nanofibers in a very quick time. A polymer melt or solution is held by its surface tension at the end of a capillary tube, when charges are introduced into the liquid by an electrostatic field. These charged ions move in response to the applied field towards the collector screen having opposite polarity, thereby transferring tensile forces to the polymer solution. At the tip of the capillary tube, the pendant drop takes the shape of a hemispherical drop, generally referred to as a *Taylor cone* in the presence of an electrostatic field. When the intensity of the electrostatic field overcomes the surface tension of the polymer solution, a jet is emanated from the Taylor cone, which travels linearly for some distance, called the jet length, and then experiences a whipping motion or pirouette motion, which is commonly referred to as bending instability of the electrified jet [6,7]. The bending instability makes fibers very long and reduces the fiber diameter from micron size to nanosize. Evaporation of the solvent with the occurrence of bending instability results in the formation of a charged polymer fiber which is collected as an interconnected web on the collector, placed at some distance from the capillary tube. This bending instability stretches the jet of polymeric solution thousands of times more, experiencing plastic deformation and thereby resulting in ultra-thin fibers before arriving at the collector screen. Fig. 1.1 shows the experimental setup of an electrospinning process.

The term "electrospinning" was derived from "electrostatic spinning" because of the electrostatic field utilized during the fabrication process. The use of the term "electrospinning" has increased since the 1990s. Formhals [8] patented the electrospinning process in 1934 entitled "process and apparatus for preparing artificial threads," wherein an experimental set-up was demonstrated for the generation of polymer filaments using electrical forces. Taylor also studied the electrospinning process in detail and demonstrated that the jet is ejected from the vertex of the cone (Taylor cone), formed when the electrostatic force surpasses the surface tension of the polymer solution. Electrospinning is a

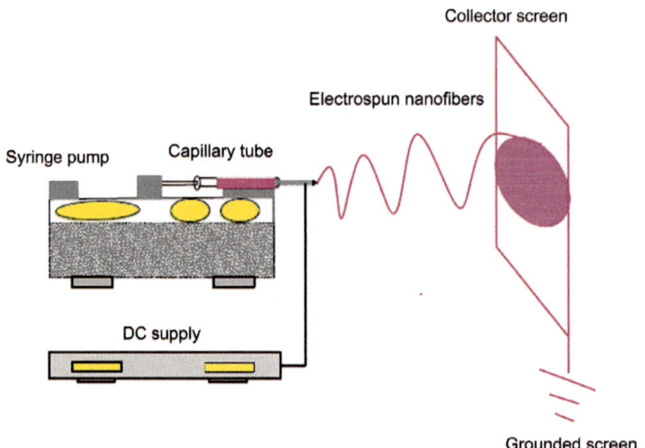

FIGURE 1.1

Setup of the electrospinning process.

relatively simpler, easier, and direct process of fabricating a nonwoven mat of polymer fibers compared to conventional methods, such as melt spinning, wet spinning, and extrusion molding with minimum initial investment and in the shortest possible time. Electrospinning is not a new method of fabricating submicron-size fibers. This technology has existed since the 1930s; however, it never gained considerable importance in the past due to low productivity and lack of interest. Nevertheless, it has getting considerable industrial importance recently owing to the exploration of excellent properties of electrospun fibers in many industrial applications. Electrospinning generally produces fibers with diameters in the range of 40−2000 nm [9]. However, fibers with even the thinner diameter can be produced from liquid crystal or other disentangled systems that can produce nanofibers down to 3 nm [9]. The smallest polymer fiber must contain at least one polymer molecule and a typical molecule has a diameter of a few tenths of a nanometer [9]. A number of different shapes and sizes of micron and nanoscale fibers can be fabricated from various classes of polymers. There are several advantages to electrospun nanofibers, including:

- Many flame-resistant polymers, such as polyetherether ketone (PEEK), polyvinyl chloride (PVC), polyacrylonitrile (PAN), and polystyrene (PS) can be electrospun.
- The surface area of nanofibers is 100−10,000 times greater than that of conventional fibers.
- The noise-absorption rate in nanofibers is expected to be exponentially higher because of the interaction of air molecules of sound waves with the fiber surfaces.
- The overall weight of materials used for many industrial applications is less.
- Nanofibers can enhance the physical properties of composites.
- Nanofibers can be electrospun on both composite and metal surfaces.
- Adhesives can be added to polymers to improve the adhesion between the fiber and the surface.
- Electrospinning is an economical and technologically mature method for bulk production for different industries.

1.1.3 CONVENTIONAL FIBER-FORMING TECHNIQUES

1.1.3.1 Solution spinning

This is one of the oldest methods for producing fibers. It was developed at the end of the 19th century. In this process, a polymer is dissolved in a solvent and drawn through a bath of nonsolvent. Fig. 1.2 shows a schematic view of a solution spinning process. When the fibers are drawn in the nonsolvent, the polymer is precipitated, forming a gel in the coagulation bath, which is then stretched by means of rotating drums. Stretching reorients the molecules of the polymer and solvent which separate out [10].

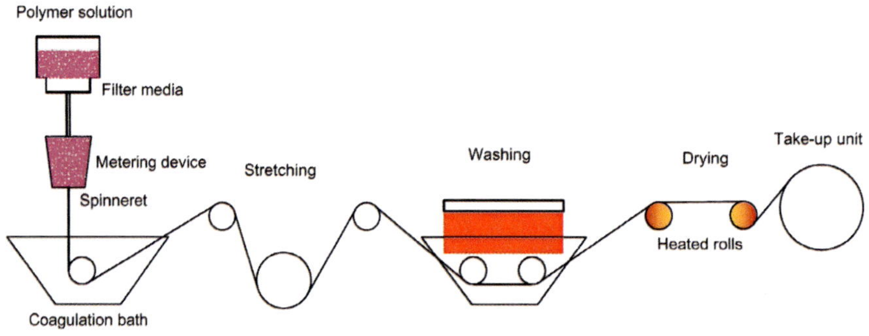

FIGURE 1.2

Schematic of the solution-spinning process.

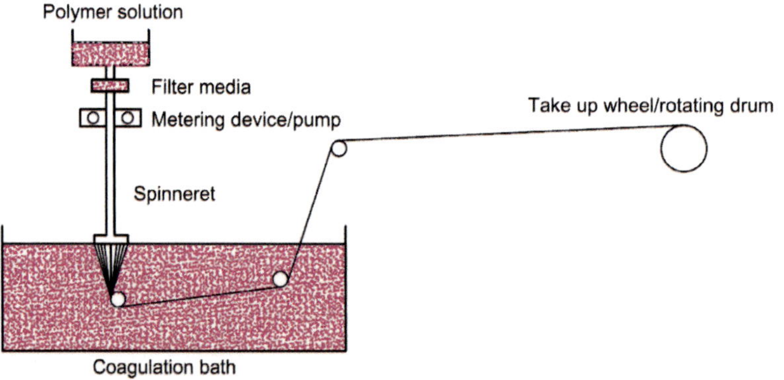

FIGURE 1.3

Schematic of the wet-spinning process.

1.1.3.2 Wet spinning

The wet-spinning process is capable of spinning a large number of fibers simultaneously since several spinnerets can be placed in a coagulation bath. In this process, a polymer is dissolved in an appropriate solvent, which is then drawn into a nonsolvent (coagulation bath) by submerging spinnerets in the coagulation bath. When the fibers come out of the bath, they precipitate and solidify. These solidified fibers are then stretched on a rotating drum. This process is used to make rayon, acrylic, modacrylic, and spandex fibers. Fig. 1.3 shows the wet-spinning process.

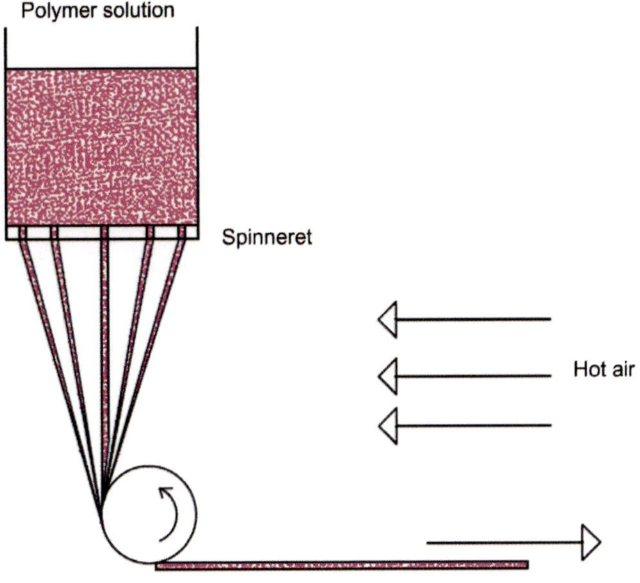

FIGURE 1.4

Schematic of the dry-spinning process.

1.1.3.3 Dry spinning

In dry spinning, the polymer is dissolved in an appropriate solvent and then the polymer solution is pumped through a spinneret (die) with a number of holes. As the polymer solution is pushed through a spinneret, it enters into a heating column, where the solvent evaporates, leaving behind dry fibers. In the heating column, steam of hot air or inert gas is used to solidify fibers and remove solvent. Fig. 1.4 shows a schematic of the dry-spinning process.

1.1.3.4 Melt spinning

Melt spinning is the most economical process of spinning due to the fact that no solvent is recovered or evaporated just like in solution spinning, and the spinning rate is fairly high. Melt spinning is the preferred method of fabricating polymeric fibers and is used extensively in the textile industry. Fig. 1.5 shows a schematic of the melt-spinning process. Melt spinning is used for polymers that can be melted easily. In this process, a viscous melt of polymer is extruded through a spinneret containing a number of holes into a chamber, where a blast of cold air or gas is directed on the surface of fibers emanating from the spinneret. As the air strikes the fibers, the fibers are solidified and collected on a take-up wheel.

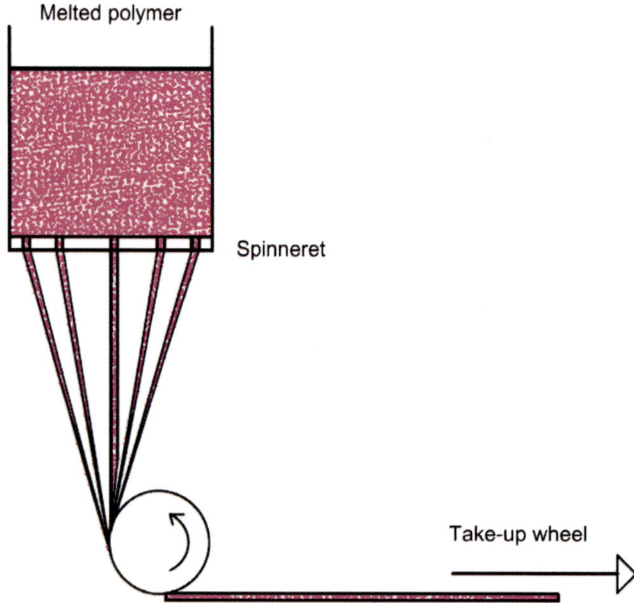

FIGURE 1.5

Schematic of the melt-spinning process.

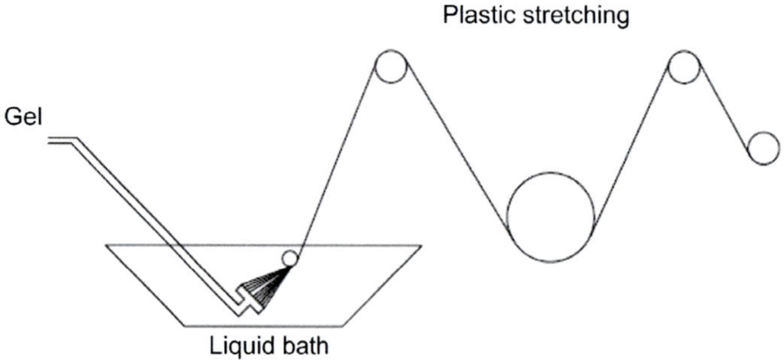

FIGURE 1.6

Schematic of the gel-spinning process.

1.1.3.5 Gel spinning

Gel spinning is an old process of fabricating polymeric fibers. This process is used to make high-strength fibers. In this process, a highly viscous polymer solution (semidiluted) is extruded through a spinneret into a liquid bath where the solution readily solidifies and forms a gel of polymer, wherein the polymer molecules are randomly aligned. The gel of polymer is then stretched on rollers, making it 100 times longer than its original length, which increases the tensile strength. Fig. 1.6 shows a schematic of the gel-spinning process.

1.1.4 NANOFIBER-FORMING TECHNIQUES

Nanofibers are fibers with a diameter in the nanometer scale. Nanofibers can be produced from almost all polymers. However, their properties and applications are different. The diameters of nanofibers depend on the type of polymers used and the method of their production. There are various methods to fabricate nanofibers, including drawing, template synthesis, phase separation, self-assembly, and electrospinning. These processes are outlined below [10,11].

1.1.4.1 Drawing

This process is based on the principle of drawing nanofibers from polymer droplet at a specific rate. This method makes long single strands of nanofibers. The biggest advantage to this process is the production of a single fiber in order to observe the properties of a single fiber and explore its applications. Some applications of single nanofibers include nano-optics, tissue engineering, nano-electronics, and so on [10,11].

1.1.4.2 Template synthesis

The template synthesis method is an effective method to synthesize an array of aligned micro-/nanofibers, nanotubes, and nanowires with controllable length and diameter. The template synthesis method utilizes a nanoporous membrane template containing pores of uniform diameter to make nanofibers/wires. Many porous materials are used as templates for the fabrication of nanofibers and nanotubes. The uniform pores allow for control of the dimensions of the nanofibers. The disadvantage of this synthesis technique is that a post-synthesis process is required to remove the template [10,11].

1.1.4.3 Phase separation

Phase separation is another method to produce nanofibers, which involves the following steps [1−3]:

- Polymer dissolution;
- Liquid−liquid phase separation;
- Gelation;
- Extraction through a solvent;
- Freeze-drying.

The homogeneous polymer solution preparation is the first step in the phase separation. The homogeneous polymer solution tends to separate into polymer-rich and polymer-lean phases, depending on the temperature. After solvent removal, the polymer-rich phase forms the matrix and polymer-lean phase forms pores. Liquid−liquid separation is generally used to form bicontinuous phase structures, whereas solid−liquid phase separation is used to form crystal structures. Gelation is the most crucial step in phase separation as it controls the morphology of nanofibers. The duration of gelation varies with the polymer concentration temperature. Low gelation temperature causes the formation of nanoscale fiber networks, while high gelation temperature causes the formation of a platelet structure. After the gelation process, gel is placed in DI water for solvent exchange. After removing the gel from DI water, the gel goes through a freezing and freeze-drying process [10,12].

1.1.4.4 Self-assembly

In a self-assembling system, the individual components interact with a predetermined surface, which causes self-organization of components into higher-order structures. Self-assembly is used to make patterns, ordered entities, and functional systems. In self-assembly, the fiber-forming substances (polymers) organize themselves into a nanoscale preferential pattern. The steps involve designing a system that takes the advantage of self-assembly for nanofabrication [13−15]. First, the interaction between the elements that makes the final system should be tailored for an appropriate response. The interaction between the elements, such as molecules, can be controlled by means of chemistry. Chemistry involves hydrogen bonding, electrostatic forces, hydrophobic−hydrophilic interaction, and van der Waals forces [13]. The second step is the determination of external parameters in order to achieve the desired result. For instance, magnetic, electrostatic, and hydrodynamics forces can be used to guide a self-assembly process towards a distinct outcome. Some of the driving forces of self-assembly are summarized below [14,15]:

- Assembly by magnetic forces;
- Assembly by hydrophobic interactions;
- Assembly by capillary forces;
- Assembly by van der Waals force;
- Assembly by electrostatic forces;
- Assembly by hydrogen and coordination bonds.

These bonds are fairly weak compared to the other strong bonds, such as covalent, ionic, and metallic bonds.

1.1.4.5 Electrospinning for nanofiber production

As defined in Section 1.2, electrospinning is a straightforward and most common process of fabricating nanofibers from polymer solution or melt. Although the concept of electrospinning has been known for almost a century, the interest in electrospinning has been spurred in the last two decades. In an electrospinning process, fibers having a diameter from 3 to 2000 nm or greater can be produced by applying high electrostatic forces to a polymer solution instead of mechanical or shearing forces. The electrical potential provides a charge to the polymer solution. Mutual charge repulsion in the polymer solution due to an applied electrostatic field causes a force (tangential force) that is directly opposite to the surface tension of the polymer solution. The electrospinning technique can produce various nanofibers in the forms of woven, nonwoven, and hollow structures.

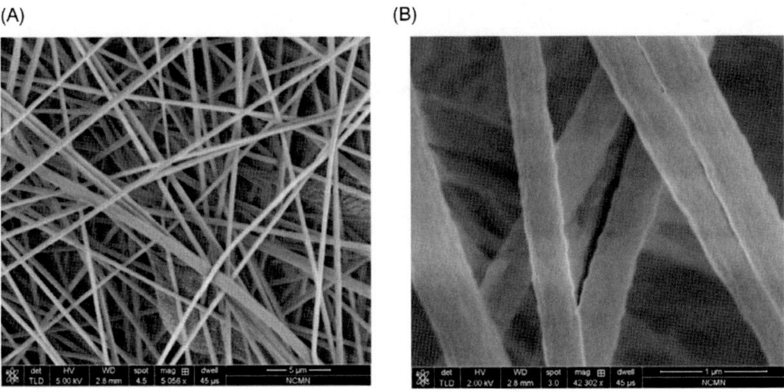

FIGURE 1.7

(A) Low and (B) high magnifications of the PAN nanofibers produced through the electrospinning process at 25 kV DC voltage, 3 ml/h pump speed, and 25 cm tip-to-collector distance [16].

Fig. 1.7 shows the PAN nanofibers produced through an electrospinning process at 25 kV DC voltage, 3 ml/h pump speed, and 25 cm tip-to-collector distance [16]. PAN powder was dissolved in DMF/acetone mixture of 90:10 ratio at 80:20 solvent concentration (80% solvent) prior to the electrospinning process.

1.1.5 NANOMATERIALS

Nanomaterials are new classes of materials having excellent physical and chemical properties. The classification of nanomaterials is based on the number of dimensions, which are in nano range (≤ 100 nm). Table 1.1 gives the different classifications of nanomaterials with regard to various applications [17]. The properties of nanomaterials are substantially different from their bulk counterparts. When the size of the bulk materials is reduced to nanoscale ranges, these materials can exhibit interesting and unusual properties. These properties include increased mechanical strength, chemical reactivity, and conductivity. In general, the silver element is opaque in the larger-scale bulk forms; nevertheless, once it is reduced to a nanosize, it becomes transparent. In the same manner, gold can become liquid when it is in nanosize at lower temperatures. Many researchers have studied this abnormal behavior of nanomaterials, and they have outlined three reasons for the abnormal behavior: quantum size effect, surface and interface effects, and characteristic length scale. Hundreds of products containing nanomaterials are already available, such as batteries, coatings, antibacterial clothing, etc. Nano innovation will soon be seen in occupational safety and health, environment, information technology, energy, transport, security, and space [18].

Table 1.1 Classification of Different Nanomaterials With Regard to Their Applications [17]

Classification	Examples
Dimension	
Three dimensions <100 nm	Particles, quantum dots, hollow spheres, etc.
Two dimensions <100 nm	Tubes, fibers, wires, platelets, etc.
One dimension <100 nm	Films, coatings, multilayer, etc.
Phase composition	
Single-phase solids	Crystalline, amorphous particles and layers, etc.
Multiphase solids	Matrix composites, coated particles, etc.
Multiphase systems	Colloids, aerogels, ferrofluids, etc.
Manufacturing process	
Gas phase reaction	Flame synthesis, condensation, CVD, PVD, etc.
Liquid phase reaction	Sol-gel, precipitation, hydrothermal processing, etc.
Mechanical procedures	Ball milling, plastic deformation, etc.

A number of nanoscale materials, such as nanotubes, nanoparticles, nanofilms, nanofibers, nanowires, and nanocomposites are usually considered to be the future generation of advanced materials for stronger military equipment, more powerful computers and satellites, faster cars and planes, and better microchips and batteries owing to the excellent mechanical, thermal, electrical, magnetic, optical, and biological properties and behaviors. Nanomaterials can also be used to make biosensors, artificial muscle, nanofiltration units, and medicines. More than 1600 nanomaterials have been found in different industrial products, including sunscreens, concretes, antibacterial cloths, car bumpers, polymeric coatings, toothpastes, wrinkle-resistant clothes, tennis rackets, and other electronic, optical, diagnostic, and sensing devices [17,19].

1.1.6 QUANTUM SIZE EFFECT

The quantum size effect dominates the behavior of materials at nanoscales, affecting the electrical, optical, and magnetic properties of materials. When materials are reduced to nanosize, they contain few atoms; therefore, the density of states in the conduction and valence bands decreases, thereby properties change significantly. Electron movement is confined, which leads to a discrete energy level. The energy of the electrons is not enough to break this confinement. As a result, exotic properties are observed, which is known as the quantum size effect [1–3]. For instance, copper element is opaque in bulk scale; however, if it is reduced to nanoscale range, it becomes transparent. In addition, platinum is an inert material, but it becomes a catalyst at nanoscale; silicone becomes a conductor at nanoscale; and aluminum becomes combustible at nanoscale. These examples can be extended significantly.

1.1.7 SURFACE AND INTERFACE EFFECTS

The physical and chemical properties of materials depend on the surface properties of the materials, whether they are in bulk form or in nanoscale form. At nanoscale, the surface area to volume ratio is exponentially increased. Surface science mainly involves the study of the physical, chemical, physicochemical, and biological properties of surfaces. However, a term called "interface" is commonly used in surface science, which emphasizes the importance of a boundary between the surface and surrounding environment. In surface science the chemical groups that are attached at the interface determine the physical, chemical, and biological properties of material interfaces. Nanomaterials have a large number of atoms on the surface, which causes a profound effect on the properties of nanomaterials [10].

Atoms that exist at the surface or interface are different from interior atoms. Atoms at the interface have high reactivity and an enhanced tendency to agglomerating and clustering. These atoms are unstable and possess high surface energy. Nanomaterials contain a large number of atoms at the surface or interface, which behave fairly different in a liquid medium. The number of atoms at the surface increases as the material is reduced in nanosize. The surface on the top possesses fewer atoms than the interior surface; thus, there are broken bonds exposed to the surface. Surface atoms are inward-directed, and the bond distance between surface atoms and subsurface atoms is smaller than interior atoms. When the material is in nanosize, this decrease in bond length between surface atoms and interior atoms becomes critical, and the lattice constant of the entire nanoparticle shows remarkable reduction, as well. Because of the large number of broken bonds on the surface of nanomaterials, they possess a high surface area and high surface energy, both of which cause them to become unstable.

1.1.8 CHARACTERISTIC LENGTH SCALE

In nanomaterials, length scale is an important parameter to characterize and classify the dimensions of nanomaterials (0D, 1D, 2D, and 3D). There are two noticeable features of nanograin materials that lead to unusual properties of nanomaterials: one is the dimension characteristic of the physical phenomenon involved, generally called characteristic length, and the other is the microstructural dimension, termed the size parameter. The range in which these two physical properties differ or coincide is of great interest to scientists and researchers. Conventional-size laws often fail to explain this, and in many cases, these laws are reversed based on the size and dimension.

1.2 CONCLUSIONS

In this first chapter, nanotechnology and the electrospinning method are generally defined and important basic information is provided. Some of the conventional fiber-forming techniques, such as solution spinning, wet spinning, dry spinning, melt spinning, and gel spinning are explained. Nanofiber-forming techniques drawing, template synthesis, phase separation, self-assembly, and electrospinning are also described and recent developments in the fields are mentioned. In addition to this information, some other nanomaterials, quantum size effect, surface and interface effects, and characteristic length scale were analyzed for the future chapters.

REFERENCES

[1] M.C. Roco, Handbook on Nanoscience, Engineering and Technology, second ed., Taylor and Francis, Boca Raton, FL, 2007.

[2] M. Ratner, D. Ratner, Nanotechnology: A Gentle Introduction to the Next Big Idea, Prentice Hall, Upper Saddle River, NJ, 2002.

[3] L. Stander, L. Theodore, Environmental implications of nanotechnology—an update, Int. J. Environ. Res. Public Health 8 (2011) 470–479.

[4] W.S. Khan., M. Ceylan, R. Asmatulu, "Effects of Nanotechnology on Global Warming," ASEE Midwest Section Conference, Rollo, MO, September 19–21, 2012, 13 pages.

[5] A. Genaidy, W. Karwowski, Nanotechnology occupational and environmental health and safety: education and research needs for an emerging interdisciplinary field of study, Human Factors Ergonomics Manuf. 16 (3) (2006) 247–253.

[6] W.S. Khan, R. Asmatulu, M.M. Elatbey, Dielectric properties of electrospun PVP and PAN nanocomposite fibers at various temperatures, J. Nanotechnol. Eng. Med. 1 (2010) 041017-1–041017-6.

[7] W.S. Khan, R. Asmatulu, Y.H. Lin, Y. Chen, J. Ho, Electrospun polyvinylpyrrolidone-based nanocomposite fibers containing $(Ni_{0.6}Zn_{0.4})$ Fe_2O_4, J. Nanotechnol. 2012 (2012). Article ID 138438, doi:10.1155/2012/1384382012, 5 pages.

[8] J.H. He, Y. Liu, L.F. Mo, Y.Q. Wan, L. Xu, Electrospun Nanofibers and Their Applications, iSmithers, 2008, UK.

[9] D.H. Reneker, I. Chun, Nanometer diameter fibers of polymer. Produced by electro-spinning, Nanotechnology 7 (3) (1996) 216–223.

[10] W.S. Khan, "Fabrication and Characterization of Polyvinylpyrrolidone and Polyacrylonitrile Electrospun nanocomposite Fibers," PhD dissertation, Department of Mechanical Engineering, Wichita State University, December, 2010.

[11] B.K. Raghavan,"Forming of Integral Webs of Nanofibers via Electrospinning, "Master's thesis, Miami University, Oxford, 2006.

[12] Z.M. Huang, Y.Z. Zhang, M. Kotaki, S. Ramakrishna, A view on polymer nanofibers by electrospinning and their applications in nanocomposite, Composite Sci. Technol. 63 (15) (2003) 2223–2253.

[13] B.A. Parviz, D. Ryan, G.M. Whitesides, Using self-assembly for the fabrication of nano-scale electronic and photonic devices, IEEE Trans. Adv. Packing 26 (3) (2003) 233–241.

[14] N. Nuraje, R. Asmatulu, G. Mul, Green Photo-Active Nanomaterials: Sustainable Energy and Environmental Remediation, RSC Publishing, Cambridge, England, 2015.

[15] R. Asmatulu, August Nanotechnology Safety, Elsevier, Amsterdam, The Netherland, 2013.

[16] M.A. Alamir, "Designing and Evaluating of Superhydrophilic Nanofiber Mats for Fog Catching in the Atmosphere," M.S. Thesis, Wichita State University, December 11, 2017.

[17] R. Asmatulu, E. Asmatulu, A. Yourdkhani, "Importance of Nanosafety in Engineering Education" ASEE Midwest Conference, Lincoln, NB, September, 2009, 8 pages.

[18] W.S. Khan, N. Hamadneh, W.A. Khan, Polymer Nanocomposites − Synthesis Techniques, Classification and Properties, One Central Press, Cambridge, UK, 2016.

[19] E. Asmatulu, J. Twomey, M. Overcash, Life cycle and nano-products: end-of-life assessment, Journal of Nanoparticles Research 14 (2012) 720.

FURTHER READING

W.S. Khan, R. Asmatulu, M. Ceylan, A. Jabbarnia, Recent progress on conventional and non-conventional electrospinning processes, Fibers Polymers 14 (2013) 1235–1247.

Historical background of the electrospinning process

2

CHAPTER OUTLINE

2.1 BRIEF HISTORICAL BACKGROUND

Electrospinning or electrostatic spinning is not a new technology for processing polymer solution or melt into micro- and nanofibers. The first report on the influence of an electrical charge on a liquid droplet dates back to the 17th century. It was reported by William Gilbert about 400 years ago that a drop of water on a flat and dry surface is drawn into a conical shape, when a piece of rubbed amber is held at a known distance [1]. Much later, in 1969, Taylor used theoretical investigations and determined that a conducting fluid can exist in equilibrium in the form of a cone under the influence of an electrostatic field only when the semivertical angle is 49.3 degrees [2]. A drop of liquid in a uniform electrical field is pulled into a spheroidal shape, which becomes unstable when the electrical field reaches a critical value of $1.62(T/r_o)^{1/2}$ and the droplet becomes 1.85 times longer than the equatorial diameter [2]. The units used in this relation are

Synthesis and Applications of Electrospun Nanofibers. DOI: https://doi.org/10.1016/B978-0-12-813914-1.00002-X

17

electrostatic units for electrical field and centimeter$-$gram$-$second (C.G.S.) units for surface tension (dyne/cm) and r_o for the initial radius. The same analysis anticipates equilibrium shapes which are longer than 1.85 times the diameter when the electrical field value is less than $1.62(T/r_o)^{1/2}$ [1]. However, when the intensity of the electrical field is high the drop does not acquire spherodial shapes, but develops pointed ends from which narrow jets emerge. In 1846, Christian Friedrich Schönbein produced highly nitrated cellulose fibrous structures [3]. Charles V. Boys also illustrated the electrospinning process and working mechanisms [3].

Lord Rayleigh reported the hydrodynamic stability mechanism of a jet of liquid with and without an applied electrical field. In 1882, he studied the instability mechanism during electrospinning and found that when the electrostatic force overcomes surface tension, liquid emanates in the form of fine jets. The first electrospinning patent was filed in the early 1900s [3]. John Zeleny attempted to mathematically model the behavior of fluids under an electrostatically charged field and the behavior of fluid droplets at the end of metal capillaries was published in 1914 [3]. Formhals disclosed at least 22 patents on electrospinning between 1931 and 1944 [3]. N.D. Rozenblum and I.V. Petryanov$-$Sokolov produced electrospun fibers in 1938 [3]. In 1934, Formhals filed his first patent describing the apparatus for producing artificial threads using electrical charges [4]. However, the method of producing artificial filaments using an electrical field had already been experimented a long time previously, but it never gained importance due to some technical concerns and applications of produced fibers, until Formhals' work [4]. Formhals described the spinning of cellulose acetate fibers with acetone solvent, in his first patent [4]. Formhals' spinning process was capable of producing threads aligned parallel on a receiving device in such a way that the threads can be unwounded easily [4]. Later, in 1940, Formhals filed another patent describing a method for producing composite fiber web from multiple polymers and fiber substrate by electrospinning polymer fibers on a moving base substrate [5]. In the early 1960s, the fundamental studies on the jet-forming mechanisms were pioneered by Taylor. Subsequently, in 1969, Taylor studied the shape of the droplet that emerged at the tip of the capillary tube, when an electrical field is applied, and showed that it was in a cone shape and the jet emanated from the vertices of the cone [1]. This conical shape of the jet was referred to as a "Taylor cone" by later researchers in subsequent literature. Taylor found that an angle of 49.3 degrees is needed to balance the surface tension of the polymer solution with the electrostatic forces.

In subsequent years, researchers shifted their focus to studying the structural morphology of nanofibers and developing the understanding of nanofiber properties' dependence on fiber morphology and process parameters. The equipment, such as scanning electron microscopy (SEM), transmission electron microscopy (TEM), X-ray diffraction (XRD), and differential scanning calorimetry (DSC), has been employed by many researchers to characterize electrospun nanofibers [6]. Baumgarten reported the acrylic microfiber via electrospinning with a diameter ranging from 500 to 1100 nm in 1971 [7]. Baumgarten electrospun polyacrylonitrile (PAN) fibers with dimethylformamide (DMF) solution and observed the dependence of fiber diameter on the viscosity of the solution [6]. Larrondo and Manley [8,9] generated polyethylene and polypropylene fibers from the melt and studied the relationship between fiber diameters and melt temperature and determined that higher melt temperature results in a reduction of fiber diameter. Their studies revealed that the fiber diameter is reduced by approximately 50% when the spinning voltage is doubled. In 1987, Hayati et al. studied the effects of an electrical field, experimental conditions, and all factors affecting the fiber stability, morphology, and atomization. They found that high conducting fluid with a high electrostatic field produced highly unstable stream that looped and whipped around in different directions. Their work also showed that this unstable jet produced fibers with much broader diameter distributions.

After a short break of around two decades, an extensive upsurge in research on electrospinning began due to the increased knowledge of the potential applications of electrospun nanofibers in different industries. Research on electrospun nanofibers spurred in the recent era is based on the pioneering work of Doshi and Reneker [10]. The real surge in the electrospinning process started in 1995, when Reneker and coworkers began publishing several papers on the electrospinning processes and their potential applications in industries. In 1995, Doshi and Reneker studied the features of electrospinning polyethylene oxide (PEO) by varying the solution concentration and applied voltage [10]. The jet diameters emanating from the Taylor cone were measured as a function of the distance from the apex of the cone, and they observed that the jet diameter decreases as the distance that jet travels from the apex of the cone increases [6]. The authors also demonstrated that the PEO solution with viscosity less than 800 centipoise was too dilute to form a stable jet for electrospinning and a solution with viscosity more than 4000 centipoise was too viscous to form fibers [6].

Reneker and Srinivasan [11] electrospun a liquid crystal system of polymer solution (p-phenyleneterephthalamide) in sulfuric acid and an electrically conducting polymer (poly (aniline)) in the same acid. Again, in 1995, Reneker and Chun [12] electrospun poly(amic acid) and poly(acrylonitrile) fibers via electrospinning and demonstrated the electrospinning of polymer melts both in air and in a vacuum. In 1996, Vancso et al. [14] electrospun poly(ethylene oxide) fibers and employed scanning probe microscopy to study their surface profiles and other morphologies. Reneker and Feng used nylon and polyimide to electrospin nanofibers via electrospinning. Jeager et al. [13] studied the thinning of PEO/water electrospun fibers and observed that the diameter of the jet decreased to 19 μm in traveling 1 cm from the tip of the capillary tube, 11 μm after traveling 2 cm, and 9 μm after traveling 3.2 cm. Deitzel et al. [16] demonstrated that an increase in applied electrical field changes the shape and surface of the droplet from which the jet emerges and the shape changes are related to bead defects. Stenoien et al. [17] manufactured a silicone polyester composite vascular graft using the electrospinning method in 1998. Comprehensive analysis of the electrospinning process by experimental characterization and assessment of fluid instabilities was performed by Hohman et al. [18], Yarin et al. [19], Warner et al. [20], and Feng [21]. Spivak and Dzenis applied nonlinear a rheological constitutive equation to the electrospinning process and analyzed the test results [22].

MacDiarmid et al. [23] fabricated a conducting electrospun mat by blending a conducting material, polyaniline, which was doped with camphorsulfonic acid along with poly(ethylene oxide). Poly(ethylene terephthalate-co-ethylene isophthalate copolymers), poly(hexyl isocyanate), cellulose, poly(3,4 ethylenedioxythiophene), and acrylonitrile-butadiene-styrene were electrospun in appropriate solvents in 2003 [24]. The number of publications on electrospinning have been increasing since 2000. New applications of electrospun fibers have been explored and innovative processes similar to electrospinning have been demonstrated by many researchers for almost two decades. Amongst the new processes, some are near-field electrospinning, needleless electrospinning, and multiple jet electrospinning, etc. Electrospinning has had major attention from scientists and researchers around the world. In 2003, more than 250 research articles were published on electrospinning. In 2004, more than 300 technical articles were published in the electrospinning field. The number of publications on electrospinning materials and processes has been gradually increasing each year. This increase in publications is due to new applications of electrospun nanofibers in a wide variety of industrial uses. The number of publications will keep on increasing and new applications of electrospun nanofibers will emerge in the near future.

2.2 PROCESS DETAILS AND EXPERIMENTAL DESIGN

Electrospinning is a unique approach of utilizing an electrostatic field to generate ultrafine fibers. Electrospinning has spurred the attention of the scientific community due to its potential of producing various fibers. Electrospun fibers have high porosity, flexibility, and surface area to volume ratio, and also a simple, straightforward, and cost-effective method to create fibers with a diameter in the range of 3 nm to 10 μm [25]. Electrospun nanofibers are of crucial importance for scientific and economic revival of all countries. Electrospinning is a process in which a high electrostatic field (mainly DC voltage) is applied to produce nanofibers with various properties. Polymeric solution or melt that has to be electrospun is forced by means of a syringe pump through a spinneret to form a pendant drop of the polymeric solution at the tip of the capillary tube (spinneret). The tip of the capillary is connected to an electrode and the other end of the electrode is connected to a high DC supply. The electrical forces then draw this pendant drop into a hemispherical shape, commonly known as a Taylor cone [26]. If the viscosity of the polymeric solution and surface tension are accurately tuned, the varicose break-up can be avoided, and a stable jet originates from the Taylor cone. A Taylor cone is formed, when equilibrium between electrostatic forces and surface tension exists. As soon as the applied electrostatic field increases and surpasses surface tension, an elongated cone is formed and a jet is originated. The jet originated from the Taylor cone travels linearly for some distance (generally 1−2 cm), also known as the jet length, and then experiences a whipping or spiral motion, commonly termed as the bending instability of the electrified jet of fiber. The bending instability makes fibers very long and thin due to the plastic deformation. The bending instability stretches the jet thousands of times more than its original size, thus experiencing a large amount of plastic deformation and thereby resulting in ultrafine fibers before arriving at the metallic screen (collector). During the jet's flight from the capillary tube to the collector screen, most of the solvent evaporates and the jet is finally collected on a grounded collector screen. Fig. 2.1 shows the experimental set-up of an electrospinning process.

The electrospinning process is similar to the drawing process except for the use of electrical forces rather than mechanical, magnetic, or shearing forces. A syringe pump controls the flow rate and the distance between the spinneret and collector screen is appropriately tuned, so that the jet has enough distance to travel in order to facilitate plastic deformation. Throughout the electrospinning process, the liquid meniscus of the liquid jet originating from the spinneret has a stress of the order of γ/r, where "γ" is the surface tension of the polymeric solution, and "r" is the radius of the meniscus. The stress induced due to an applied field (Maxwell stress) can be calculated using the following equation [27]:

$$\sigma = (\varepsilon V^2)/H^2 \tag{2.1}$$

where "ε" is the permittivity, "V" is the applied electrostatic field (spinning voltage), and "H" is the distance between the capillary and the collection screen.

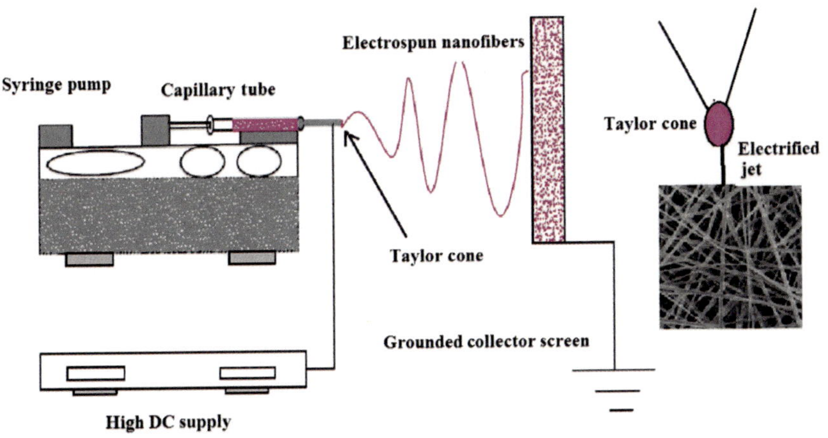

FIGURE 2.1

Experimental setup of the electrospinning process.

As far as the other forces, such as hydrostatic pressure, viscoelastic, and inertia are concerned, their effects can be neglected in comparison with the high-electrostatic field. By balancing two stresses, one can estimate the critical voltage (V_C) that must be overcome to cause electrospinning [27];

$$V_C = \sqrt{\frac{\gamma H^2}{r\varepsilon}} \tag{2.2}$$

Almost all polymeric solutions can be electrospun between 7 and 30 kV after dissolving in appropriate solvents. However, to produce fibers in nanosize, the applied electrostatic field should be higher. When an electrostatic field is applied to the polymeric solution, charge is induced in the polymeric solution. As the intensity of the electrostatic field increases, more charges are induced in the polymeric solution and mutual charge repulsion creates longitudinal stresses. These longitudinal stresses overcome the surface tension of the polymeric solution to form a Taylor cone from which an electrified jet originates, which experiences bending instability before collecting on the collector screen in the form of a fibrous mat. When an external electrostatic field is applied to the polymeric solution, the positive and negative ions move in opposite directions in response to the applied field.

Negative ions primarily move towards the positive electrode of the system. The difference in the number of positive and negative ions is known as the excess charge, or simply charge. For those polymeric solutions that are insulating, salt can be added in the polymeric solution, which dissociates into equal numbers of positive and negative ions and thus ionic conductivity increases to facilitate electrospinning. When the electrostatic field is applied, the solution is charged and migration of ions through the solution occurs, thus forming an electrified jet, which carries an excess charge [28]. The jet originating from the spinneret first follows a straight path (jet length), and then undergoes a sequence of unstable bending back and forth with growing amplitude in three dimensions. The jet loop grows longer and thinner before being collected on a collector screen placed some distance from the spinneret. The loop diameter and circumference increase continuously as the jet leaves the spinneret. The cycle of bending instability can be described in three steps as outlined below [28]:

- A smooth and straight segment of the jet can develop an array of bends or experience curvilinear motion during the start of electrospinning process.
- The segment of the jet in each bend can elongate, and the array of bends can become a series of spiraling loops or whipping loops with increasing diameters.
- As soon as the perimeter of the loops increases, the diameter of the fibrous jet can grow smaller. After having the first step established on a small scale, the next cycle of bending instability starts forming.

The cycle of bending instability repeats itself several times and then more cycles form and grow before the jet diameter is reduced continuously, thereby

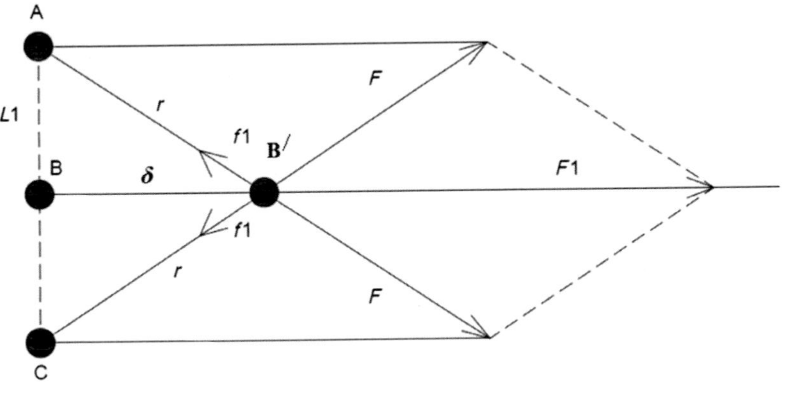

FIGURE 2.2

Illustration of bending instability during the electrospinning process.

creating nanofibers (or micron and submicron fibers based on the conditions). The solvent evaporates due to natural convection process leaving a nanofibrous mat on the collector screen. In a coordinate system with rectilinear electrified jet, and the electrical charges can be treated as a static system of charges interacting with each other, according to Coulomb's law. Such a system can also be unstable, based on the Earnshaw's theorem [28].

> *Collection of point charges cannot be maintained in a static equilibrium solely by the electrostatic interactions of the point charges.*

To explain the instability mechanism, consider three point charges having same polarity (e) and initially in a straight path at A, B, and C, as shown in Fig. 2.2 [28].

Two coulomb forces with magnitude $(F = e^2/r^2)$ push charge B (in the middle) from opposite directions, which causes B to move B' by a distance δ. A force with magnitude $F_1 = 2F\cos\theta = 2(e^2/r^3)\delta$ acts on charge B and moves in the direction of perturbation (growing amplitude of bending instability). The growth of perturbations is mainly directed by the following equation [28]:

$$m.\frac{d\delta^2}{dt^2} = 2e^2.\frac{\delta}{L_1^3} \tag{2.3}$$

where "m" is the mass of the jet. The solution to this equation, $\delta = \delta_0 \exp\{[2e^2/L_1^3 m]^{1/2} t\}$, shows that small bending perturbations follow an exponential law. This increase in perturbations is in response to an electrostatic field. The electrostatic energy decreases as e^2/r, and then the perturbations grow. The jet becomes unstable after traveling linearly for some distance due to stress relaxation. When stress relaxation occur, the jet deviates from the linear path and a number of instabilities occur, as the jet travels towards the collector screen, as shown in Fig. 2.3 [29].

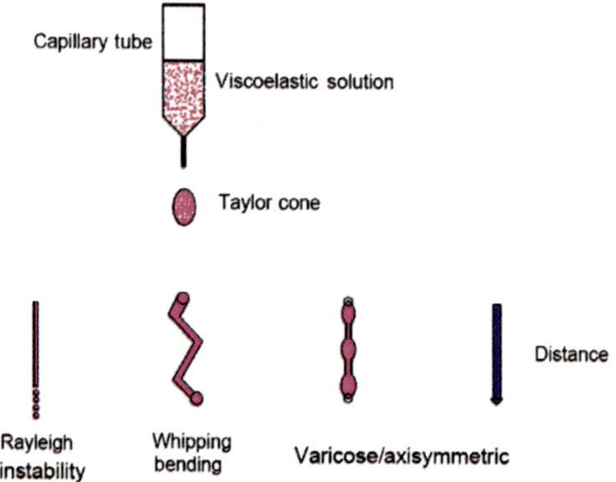

FIGURE 2.3

Schematic of various types of instabilities emerging from a Taylor cone.

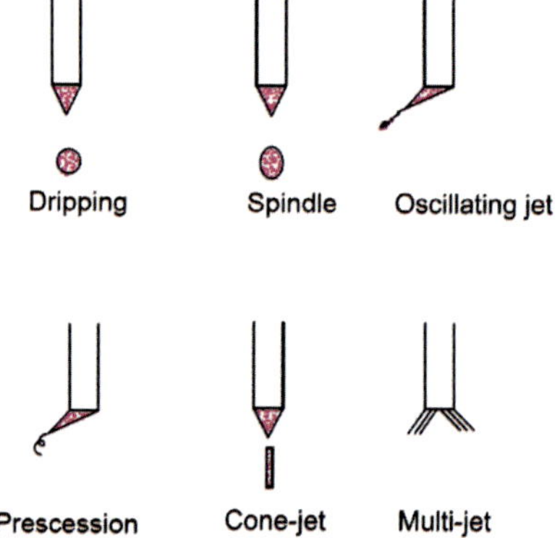

FIGURE 2.4

Schematic of various modes of charged jets emerging from a Taylor cone [29].

The instabilities occur after stress relaxation. The jet experiences a large reduction in the cross-sectional area, and spiral motion takes place, which is commonly known as the bending/whipping instability of the jet. Consequently, the jet loop forms by itself, as can be seen in Fig. 2.3. This type of instability is termed varicose instability, in which the radius of the jet is modulated but the centerline remains pretty straight [29]. In Rayleigh instability, a continuous cylindrical column of liquid breaks down into spherical droplets. This type of instability can be repressed by a high electrical field. Depending upon the applied electrostatic field and process parameters, a number of jets can emanate from a Taylor cone, as shown in Fig. 2.4 [29]. In the dipping mode, the spherical droplets originate from the Taylor cone, which is also the starting point of the fibers. In the spindle mode, the jet is extended and then breaks up into small droplets. In oscillating

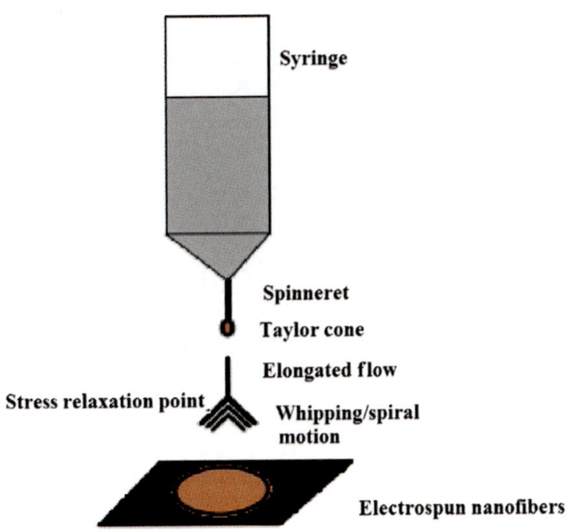

FIGURE 2.5

Schematic of various regions in electrospinning process [29].

mode, the jet twists around and droplets are expelled forward. In the precession mode, the initially formed jet loops up before breaking into small droplets/objects. In multiple jet mode, many jets emanate from the spinneret simultaneously due to the viscosity effect. Generally, electrospinning is carried out in cone-jet mode.

The electrospinning jet has three regions, as shown in Fig. 2.5: (a) the base, where the jet emerges or initiates from the Taylor cone; (b) another region just after the base, where the jet stretches and accelerates, the length of the jet increases and the diameter decreases; and (c) the last region, where the jet experiences plastic deformation and undergoes whipping and spiral motion and is finally collected on a collector screen in the fibrous form [29].

2.3 SHAPE OF FIBERS PRODUCED BY ELECTROSPINNING

A large variety of fibers can be produced by an electrospinning process. The cross-section of fibers via an electrospinning process largely depends on the polymeric solutions. Round fibers, bead fibers, branched fibers, flat ribbons fibers, ribbons with other shapes, and split fibers can be produced by various polymeric solutions. The cross-section fibers have also been produced by some polymeric solutions. By appropriately controlling the process parameters, fibers with different sizes and shapes can be fabricated. Scanning electron microscopy (SEM) is usually employed to analyze the fiber size (diameter), uniformity, and morphology. If the processing parameters are not appropriately controlled, instead of nanofibers, micron- or submicron-sized fibers can be produced. Among the

FIGURE 2.6

SEM images showing uniform-diameter PAN nanofibers incorporated with PEG polymers.

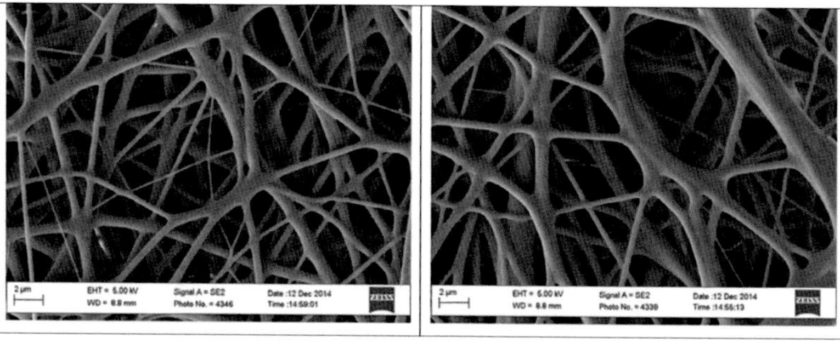

FIGURE 2.7

SEM images showing branched PVP nanofibers.

processing parameters, the polymeric solution concentration is the most critical parameter. As the solution concentration increases, the fibers produced have larger diameters.

Demir et al. [30] demonstrated that a strong relationship exists between fiber diameter and polymeric solution concentration. According to their study, the fiber diameter is proportional to the cube of the solution concentration. If the concentration is too high, a second jet of smaller fibers may also be produced with the primary fibers, when the solution concentration exceeds a certain threshold value. The spinning voltage (electrostatic field) also influences the fiber diameter. Generally, fiber diameter decreases as the spinning voltage increases. Figs. 2.6–2.8 show SEM images of some of the electrospun fibers [31,32].

The presence of pores and beads is a common problem in electrospinning [32]. The presence of beads also depends on the processing parameters. As the spinning voltage is increased, the jet velocity increases subsequently and solution

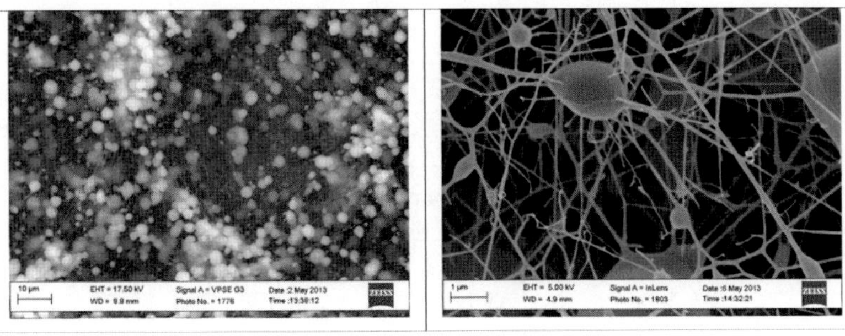

FIGURE 2.8

SEM images showing beaded PCL nanofibers.

is ejected faster from the spinneret, resulting in a lower volume of droplets as it reaches the tip of the spinneret. The Taylor cone becomes unstable, so this leads to the formation of beads. At high solution viscosity (solution concentration), the balance between electrostatic forces and surface tension is disturbed, resulting in bead formations. However, highly concentrated solutions have been shown to produce fibers with fewer beads. The morphology and shape of beads changes as the solution concentration increases. The beads becomes spindle-like from spherical, when the solution concentration increases [32].

2.4 ELECTROSPINNING PARAMETERS

The electrospinning parameters are very important for understanding the electrospinning process and the conversion of polymeric solution into fibers. These parameters are broadly divided into two categories: system parameters and process parameters. These parameters affect the fiber morphologies and by proper controlling these parameters, fibers with desirable morphologies and diameters can be produced. These parameters are mentioned below [15,28].

- System parameters:
  - Polymer, molecular weight, and solvent types and polarities;
  - Viscosity, conductivity, crosslinking, and surface tension of polymers.
- Process parameters:
  - Electric potential, flow rate, and polymer concentration;
  - Distance between capillary and collection screen;
  - Temperature, humidity, and air velocity effects in the chamber.

2.4.1 SYSTEM PARAMETERS

Most polymers can be electrospun in different solvents as pure or after blending with other polymers and organic and inorganic inclusions. In general, the polymer for electrospinning should have moderate molecular weight in order to avoid bead formations on the surfaces of the fibers. If the molecular weight is low, pores, microparticles, and beads are formed. On the other hand, if the molecular weight is too high, electrospinning would be impossible or beaded structures can be formed. The solubility of polymer in an appropriate solvent is an important issue. The polymer should be soluble in an appropriate solvent in order to form a single-phase homogeneous solution. Viscosity plays a critical role in electrospinning. It has a marked influence on fiber diameter. A high viscosity usually results in a large fiber diameter. Pores and beaded structures are less likely to form in the structure of the fibers when the viscosity is too high. If the viscosity is low, the jet will collapse into many droplets, resulting in splashing of the jets [15].

The solution concentration also plays an important role in the fiber-forming process. At low concentration, electrospraying occurs due to low viscosity and high surface tension of the solution. If the concentration is a little higher, fibers surrounded by beads and pores are formed [33]. However, if the concentration is appropriate, smooth and orderly fibers can be formed. Generally, solution viscosity is tuned by adjusting the solution concentration. Different solvents have different surface tensions. However, low surface tension is suitable for electrospinning because high electrical conductivity may adversely affect the electrostatic force formations on the polymeric solutions for further electrospinning. A low surface tension solvent needs less electrostatic field for electrospinning. The fiber diameter is primarily associated with the viscosity of the electrospinning solution by the following equation [34]:

$$d = 19.49\eta^{0.43} \tag{2.4}$$

where η is the viscosity of the solution and d is the diameter of the fiber (nm).

The electrical conductivity also influences the diameters of the micro- and nanofibers. Generally, fibers with small diameters can be created using the high electrical conductivity solution. Solution conductivity depends on the polymer types and solvent structures [33]. Generally, natural polymers are polyelectrolytic in nature. Therefore, in natural polymers the ions increase the charge-carrying capacity of the polymer jet under an electrical field subjected to high surface tension, thereby resulting in poor fiber formation in contrast to synthetic polymers [33]. Moreover, the electrical conductivity of the polymer solution can be increased by adding diluted acids, bases, and ionic salts, such as NaCl, H_2SO_4, HCl, NaOH, KOH, $FeCl_3$, and KH_2PO_4 [15,33].

2.4.2 PROCESS PARAMETERS

Applied voltage: In electrospinning, applied (spinning) voltage is another critical factor. When the applied voltage reaches its threshold value, a charged jet emanates from the Taylor cone. The jet diameter gets thinner and thinner as it advances to the collector screen due to evaporation of solvent and plastic deformation under the electrostatic forces. The fiber diameter decreases as the applied voltage increases. The fiber diameter is related to the applied potential (other parameters constant) by the following relation [35]:

$$d \sim V^{-1/2} \tag{2.5}$$

where "V" is the electrical potential. Hendricks [35] determined the minimum potential as:

$$V = 300\sqrt{20\pi\gamma r} \tag{2.6}$$

where "r" is the radius of the cone and "γ" is the surface tension of the polymeric solution. Taylor established a similar relation for the critical potential as indicated below [35]:

$$V_c^2 = 4H^2/L^2 \left(\ln\frac{2L}{R} - 3/2 \right)(0.117\pi\gamma R) \tag{2.7}$$

where "V_c" is the critical voltage, "H" is the separation between the capillary and the ground, "L" is the length of the capillary, "R" is the radius of the capillary, and "γ" is the surface tension of the liquid.

Flow rate: The flow rate of the polymer solution is another important parameter in electrospinning. The morphology and diameter of fibers depend on the flow rate. When the flow rate is high, the fibers are surrounded by pores and beads owing to the shorter drying time in reaching the collector screen and low stretching electrostatic forces. Generally, a low flow rate is recommended for producing thinner fibers due to the fact that the jet will have enough time to polarize and elongate.

Distance between the collector and capillary: The distance between the collector screen and capillary tip (spinneret) also influences the fiber morphology and diameter. If the distance is too short, fibers will not have enough time to solidify and plastic stretching of the jet will be limited, resulting in large-diameter fibers. However, if the distance is too long, beaded structures will form. An optimum distance is generally recommended for producing smooth fibers without pores and beads. The electrospinning process is commonly performed at room temperature under normal atmospheric conditions. Some studies have shown that increasing the temperature favors reducing the fiber diameter. Some researchers use dehumidifiers to control the humidity and produce fibers in nanoscale. High airflow will increase the evaporation rate due to heat convection and will help in reducing fiber diameter.

2.5 FABRICATIONS OF ALIGNED ELECTROSPUN NANOFIBERS

Electrospinning has been recognized as the simplest and most straightforward method of producing fibers in the nanometer to micron diameter range. Electrospinning relies on electrostatic forces, which cause whipping or chaotic motion in the electrified jet, a characteristic feature of electrospinning, resulting in randomly oriented fiber structures in the form of nonwoven nanofiber mats or webs. The fibers produced via electrospinning are usually in nonwoven form, which limits their uses to a small number of applications, such as wound dressing, filtration, health monitoring, sensors, and tissue engineering. In some of the applications well aligned and ordered architectures of fibers are required. When fibers are produced in a single continuous form or uniaxial bundles form, their uses can expand into unlimited applications. Nevertheless, this is a tough task, since the electrified jet follows a three-dimensional trajectory due to bending instability, rather than a single straight line. Significant efforts have been made by many research groups to produce continuous long aligned fibers with ordered architecture, such as rotating disk electrospinning, needleless electrospinning, nearfield electrospinning, and electrospinning with a rotating disk and translating spinneret. These processes are outlined below [15,34].

2.5.1 ELECTROSPINNING WITH ROTATING DRUM

A continuous thin film of fibers and mats with uniform orientation and ordered architecture can be produced by rotating a drum (target) at high speed. Fig. 2.9 shows a schematic view of the electrospinning process with a rotating drum. The electrified jet flows from the Taylor cone to the rotating drum (or disk) and experiences bending instability and is finally collected on a rotating drum. The

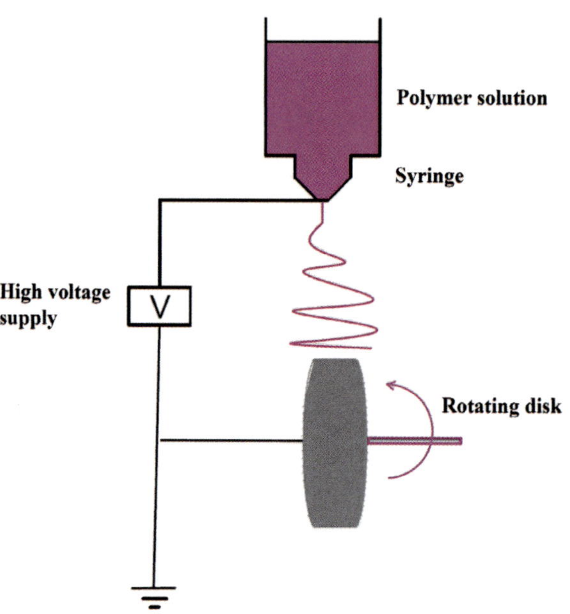

FIGURE 2.9

Schematic view of the electrospinning process with a rotating disk.

high-speed drum will reorient the nonaligned fibers into aligned fibers in the same directions. Thus, continuous thin films of nanofibers with uniform orientation can be produced in this way.

2.5.2 NEEDLELESS ELECTROSPINNING

Electrospinning has been marred by low productivity and nonuniformity. Therefore, efforts have been made to improve its productivity by increasing the number of needles (multiple needle setup) and by placing air jackets to improve the solution flow rate [36]. Needleless electrospinning addresses the issue of low productivity and nonuniformity in fiber orientation by replacing the needle with a wide-open bath of polymer solution and a copper spiral coil (electrode) inside the polymer solution. The experimental setup is depicted in Fig. 2.10. The copper coil is used as a fiber generator under electrostatic forces. The bath is filled with polymer solution and a coil is rotated inside the solution at 40 rpm [36]. The polymeric solution is charged by a high-power DC voltage supply by inserting the electrode into the solution. A rotating drum covered with an aluminum coil, placed at some distance above the solution bath, is used to collect fibers. Since the rotation of the copper coil in the solution is not too high, the viscoelastic property of the solution can also help in the formation of an evenly distributed solution layer on each spiral [36]. When the coil and polymer solution are charged, a number of jets are initiated from the copper coil. As soon as the voltage is increased, more and more jets are initiated from the coil and deposited on the surface of the rotating drum. Needleless electrospinning produces uniformly oriented fibers with high productivity.

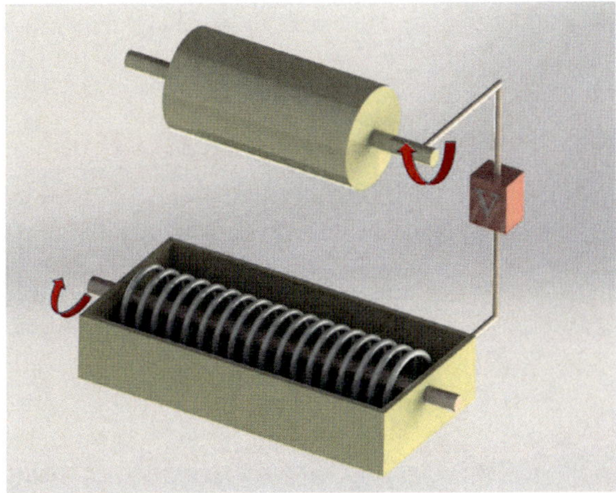

FIGURE 2.10

Schematic view of the needleless electrospinning process.

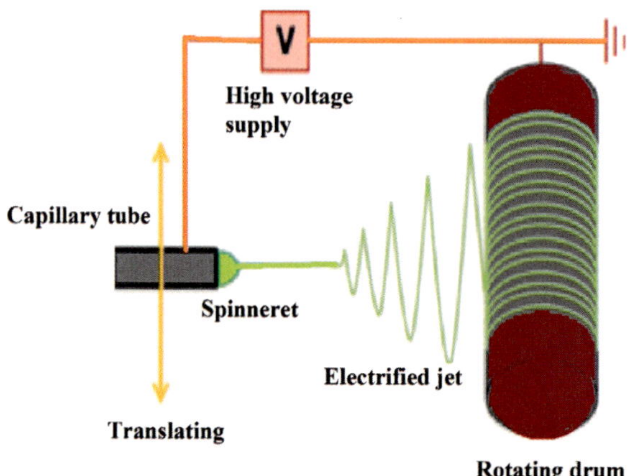

FIGURE 2.11

Schematic view of electrospinning with a rotating drum and translating spinneret.

2.5.3 ROTATING DRUM AND TRANSLATING SPINNERET

A continuous and uniformly oriented film of fibers can be produced by rotating the drum (collector) and translating the spinneret, as shown in Fig. 2.11. The main features of this process are the rotational speed of the drum and the translational movement of the spinneret. These parameters can be adjusted by experimental conditions and experience. If these parameters are appropriately adjusted, uniformly oriented fiber mat can be produced. This process has the advantage of higher production and better orientation and alignment of fibers.

2.5.4 ELECTROSPINNING WITH ROTATING ELECTRODES

Electrospinning with rotating electrodes is another process of addressing the problems with low productivity and uniformity in fiber orientation. In this process, strings of electrodes are rotated in a bath of polymer solution and a collector screen is placed at some distance above it. The electrodes are generally connected to the positive terminal of the power supply and the collector screen is either grounded or connected to the negative terminal of the power supply. Instead of using multiple spinnerets, the strings of electrode perform the same function with better control and ease in fabrication. This process is economical and faster than conventional electrospinning. Fig. 2.12 shows a schematic of electrospinning with a rotating string of electrodes. In addition to these studies, a number of other studies were conducted to produce various nanofibers for different industrial applications [37–46].

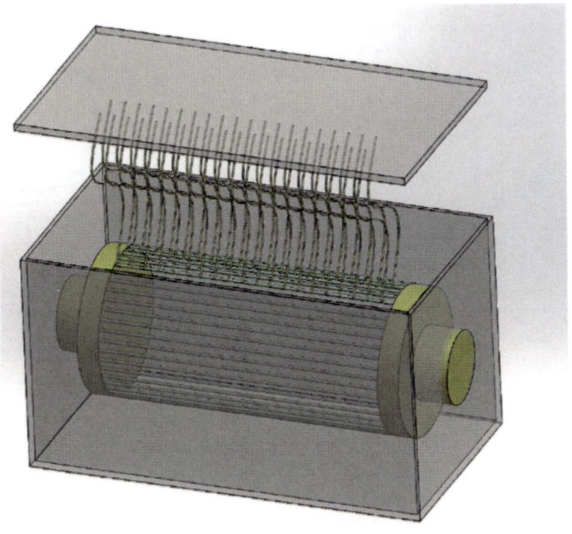

FIGURE 2.12

Schematic of electrospinning with a rotating string of electrodes.

2.6 RECENT DEVELOPMENTS IN ELECTROSPINNING METHODS

2.6.1 ELECTROSPINNING WITHOUT BENDING INSTABILITY

Among the various studies that were conducted to produce continuous and uniformly oriented fibers, electrospinning without bending instability is one of them. High charge density on the jet during electrospinning causes extensive plastic deformation, resulting in ultrafine fiber diameter and leads to an unstable and whipping motion of the jet, which is generally termed bending instability [47]. The bending instability causes fibers to be randomly oriented and nonaligned. The current setup used in electrospinning relies on the whipping motion of a jet to produce nanofibers. A literature survey revealed that limited work has been dedicated towards fiber orientation and alignment. However, numerous investigations have been directed towards analytical and experimental methodologies in order to study the fundamental chemistry and physics of electrospinning for improvement and control, such as the effect of viscosity, applied voltage, and distance between the capillary and target screen [47].

For almost all industrial applications, the fibers are expected to be uniformly oriented in the form of an ordered array rather than nonwoven random mats/films. By applying a low electrostatic field near the tip of the needle, and adjusting the mass flow rate to low in the syringe pump and placing an electrode plate close to the electrospinning emitter, a uniform electrostatic field can be created and

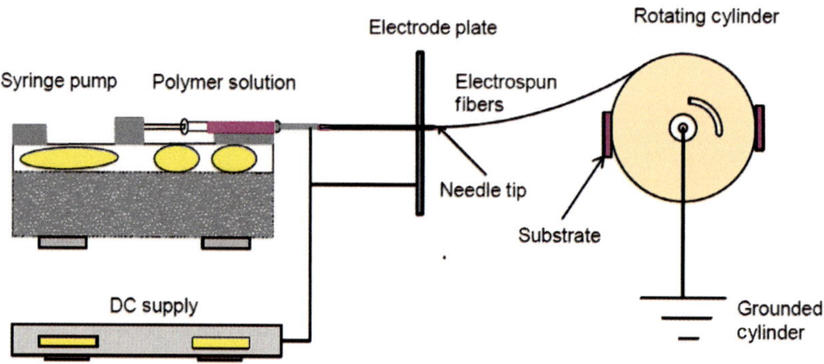

FIGURE 2.13

Schematic of electrospinning without bending instability.

bending instability can be suppressed. This adjustment in electrospinning can produce continuous uniformly oriented nanofibers. Fig. 2.13 shows a schematic of electrospinning without bending instability. The electrode plate is connected to the same DC supply as the polymer solution in the syringe pump.

2.6.2 NEAR-FIELD ELECTROSPINNING

Conventional electrospinning produces nonwoven and nonuniformly oriented fibers that have limited applications in many industries. However, for most industrial applications the fibers should be uniformly oriented in a regular pattern [48]. The thinning of fibers in electrospinning depends on bending instability, which causes random orientation and nonalignment. Near-field electrospinning (NFES) is an effective process to produce small-scale objects and also for better fiber alignment and control [48]. Fig. 2.14 illustrates the schematic setup of NFES. In NFES, the electrode-to-collector distance is minimized in order to utilize the stable region of jets emanating from the orifice for controllable deposition. Typically, a distance of 500 µm to 3 mm is used [49]. A solid tungsten spinneret of 25 µm is used to obtain nanosized fibers [49]. The key feature in NFES is to reduce the size of the polymer jet emanating from the Taylor cone by reducing the applied voltage below the critical value [50]. Theoretically, this can be accomplished by reducing the applied voltage and mass flow rate of the polymer solution after the onset of electrospinning [50]. A subcritical voltage of approximately 1.4 kV is used to charge the polymer solution [48]. This low voltage causes a meniscus of polymeric solution to form at the tip of the capillary tube or needle, without inducing electrospinning. Mechanical drawing is applied by using a tungsten probe with a 1 µm tip diameter to poke inside the meniscus [50]. The tungsten probe is quickly pulled away from the meniscus to initiate a continuous electrospinning process [50]. Some other studies have illustrated that instead of using one probe, an array of sharp tips with diameters of 20 µm can be inserted

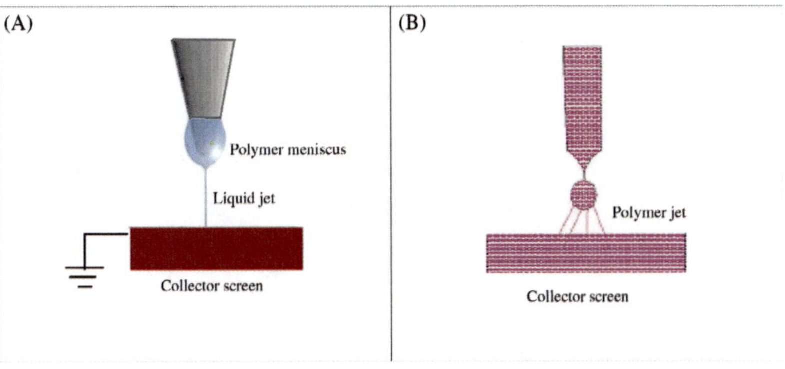

FIGURE 2.14

Schematic representation of the NFES process. (A) NFES with 1 μm tip diameter; (B) NEFS with an array of sharp tips having 20 μm diameters.

inside the meniscus and quickly removed, resulting in numerous jets emanating from the meniscus (Fig. 2.14B) [48]. The distance between two adjacent tips should be more than 50 μm to ensure effective emanation from the meniscus for better fiber formation. The repulsive forces between adjacent jets impede them from integrating with each other. NFES results in a highly orientated uniform-fiber web [48].

2.6.3 NEW PROGRESS IN SPINNING TECHNOLOGY

Conventional electrospinning is lacerated by low productivity and uniformity of fibers, a neoteric process has been developed, which is appropriately termed as forcespinning in order to address issues related to low productivity and alignment of fibers. This process utilizes centrifugal force to fabricate nanofibers, rather than high electrostatic force, which is used in conventional electrospinning. The forcespinning process uses either polymer solution or solid polymer material that are to be spun into nanofibrous form [34]. In forcespinning, the electrostatic force is replaced by centrifugal force. The combination of centrifugal force with various configurations of spinnerets makes forcespinning a versatile process that addresses many of the limitations in the existing process, including high applied electrostatic field and the polymer solution is typically dielectric. These significant changes ultimately boost the material selection criteria by allowing both conductive and nonconductive polymers to be spun into nanofibers. Moreover, numerous solid materials can be spun by dissolving them in a solvent and there is no need to recover the solvent. The forcespinning process generally uses either solution or solid materials that are spun into micro- and nanofibers depending on the centrifugal forces (G-forces) and other system and process parameters [34]. In this process, a number of organic and inorganic materials can be melted and spun at the same time. The parameters that influence the geometry and morphology of fibers are

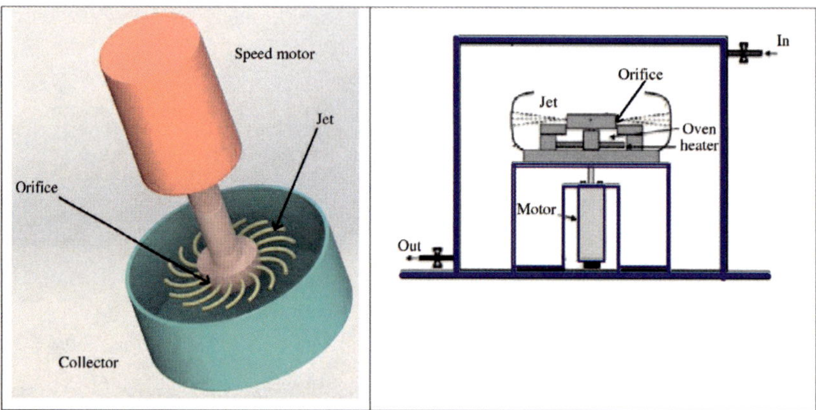

FIGURE 2.15

Schematic view of the forcespinning process.

rotational speed of the spinneret, geometry of the orifice, temperature, and the collection system of fibers [34]. In forcespinning, the polymer solution or melt is forced through the orifices of the spinneret by G-forces that cause a substantial amount of stretching on the fibers. The G-forces cause polymer solution or melt to emanate from all orifices and be stretched into a continuous web of ultrafine fibers. The main components of the forcespinning systems are the spinneret, collection system, heating source, environmental system, control system, speed motor, and brake [34]. Fig. 2.15 shows different arrangements of the forcespinning process.

2.7 CONCLUSIONS

In this chapter, historical backgrounds of electrospinning methods are generally defined and important basic information about electrospinning and their processes are provided. In this section, some of the system parameters (e.g., polymer and solvent types and polarities, molecular weight, crosslinking, viscosity, conductivity, and surface tension of polymers) and process parameters (e.g., electric potential, polymer concentration, flow rate, distance between capillary and collection screen, humidity, temperature, and air velocity effects in the chamber) of electrospinning methods were analyzed in detail. Also, fabrications of aligned electrospun nanofibers using electrospinning with a rotating drum, needleless electrospinning, rotating drum and translating spinneret, and electrospinning with rotating electrodes were studied in this chapter. Some advancements in electrospinning, such as electrospinning without bending instability and near-field electrospinning have been illustrated. A neoteric process called forcespinning has also been demonstrated, which uses centrifugal forces, rather than electrostatic forces as in electrospinning. Forcespinning addresses the issues of low productivity and fiber alignment.

REFERENCES

[1]   G. Taylor, Electrically driven jets, Proc. R. Soci. Lond. Ser. A Math. Phys. Sci. 313 (1515) (1969) 453–475.

[2]   G.I. Taylor, Disintegration of water drops in electric field, Proc. R. Soc. Lond. Ser. A Math. Phys. Sci. 280 (1964) 383–397.

[3]   N. Tucker, J.J. Stanger, M.P. Staiger, H. Razzaq, K. Hofman, The history of the science and technology of electrospinning from 1600 to 1995, J. Eng. Fibers Fabrics 7 (2012) 63–73.

[4]   A. Formhals, Process and apparatus for preparing artificial threads, US Patent 1975504A, 1934.

[5]   A. Formhals, Artificial threads and method of producing same, US Patent 2187306A, 1940.

[6]   T. Subbiah, G.S. Bhat, R.W. Tock, S. Parameswaran, S.S. Ramkumar, Electrospinning of nanofibers, J. Appl. Polymer Sci. 96 (2) (2005) 557–569.

[7]   P.K. Baumgarten, Electrostatic spinning of acrylic microfibers, J. Colloid Interface Sci. 36 (1971) 71–79.

[8]   L. Larrondo, J.R. Manley, Electrostatic fiber spinning from polymer melts. I. Experimental observations on fiber formation and properties, J. Polymer Phys. 19 (6) (1981) 909–920.

[9]   L. Larrondo, J.R. Manley, Electrostatic fiber spinning from polymer melts. II. Examination of the flow field in an electrically driven jet, J. Polymer Phys. 19 (6) (1981) 921–932.

[10] J. Doshi, D.J. Reneker, Electrospinning process and applications of electrospun fibers, J. Electrostat. 35 (2-3) (1995) 151–160.

[11] G. Srinivasan, D.H. Renekar, Structure and morphology of small diameter electrospun aramid fibers, Polym. Int. (1995) 195–201.

[12] I. Chun, Effect of Molecular Characteristic on Strength of Elastomer, Ph.D. dissertation, Department of Polymer Science, University of Akron, Akron, OH, 1997.

[13] R. Jaeger, M.M. Bergshoef, C.M. Battle, I.B. Martin, H. Schönherr, G.J. Vanso, Electrospinning of ultra-thin polymer fibers, Macromol. Symp. 127 (1998) 141–150.

[14] G.J. Vancso, H. Schonherr, D. Snetivy, Morphology, chain packing, and conformation in uniaxially oriented polymers studied by scanning force microscopy ACS symposium series: scanning probe microscopy of polymers, in: Ratner B.D., Tsukruk V. V. (Eds.), American Chemical Society, Washington, DC, USA, 1998, pp. 67–93.

[15] Deleted in review.

[16] J.M. Deitzel, J. Kleinmeyer, D. Harris, N.C.B. Tan, The effect of processing variable on the morphology of electrospun nanofibers and textile, Polymer 42 (2001) 261–272.

[17] M.D. Stenoien, W.J. Drasler, R.J. Scott, M.L. Jenson, Method of making a silicone composite vascular graft, US Patent. No. 5840240A, 1998.

[18] M.M. Hohman, M. Shin, G. Rutledge, P.M. Brenner, Electrospinning and electrically forced jets. II. Applications, Phys. Fluids 13 (8) (2001) 2221–2236.

[19] A.L. Yarin, S. Koombhongse, D.H. Reneker, Bending instability in electrospinning of nanofibers, J. Appl. Phys. 89 (5) (2001) 3018–3026.

[20] S.B. Warner, A. Buer, M. Grimler, S.C. Ugbolue, G.C. Rutledge, Y.M. Shin, "A Fundamental Investigation of the Formation and Properties of Electrospun Fibers," National Textile Center Annual Report: November 1999.

[21] J.J. Feng, The stretching of an electrified non-Newtonian jet: a model for electrospinning, Phys. Fluids 14 (11) (2002) 3912–3926.

[22] A.F. Spivak, Y.A. Dzenis, Asymptotic decay of radius of a weakly conductive viscous jet in an external electric field, Appl. Phys. Lett. 73 (21) (1998) 3067–3069.

[23] D. Norris, M.M. Shaker, F.K. Ko, A.G. MacDiarmid, Electrostatic fabrication of ultrafine conducting fibers: polyaniline/polyethylene oxide blends, Synth. Met. 114 (2) (2000) 109–114.

[24] W. Kataphinan, Electrospinning and Potential Applications, Ph.D. dissertation, Dept. of Polymer Science, University of Akron, Akron, OH, 2004.

[25] D.H. Reneker, I. Chun, Nanometre diameter fibres of polymer, produced by electrospinning, Nanotechnology 7 (3) (1996) 216–223.

[26] G. Taylor, Electrically driven jets, Proc. R. Soc. Lond. A Math. Phys. Eng. Sci. 313 (1515) (1969) 453–475.

[27] L.Y. Yeo, J.R. Fried, Electrospinning carbon nanotubes polymer composite nanofibers, J. Exp. Nanosci. 1 (2006) 177–209.

[28] D.H. Renekar, Bending instability of electrically charged liquid jets of polymer Solutions in electrospinning, J. Appl. Phys. 87 (9) (2000) 4531–4546.

[29] C.M. Hsu, Electrospinning of poly (ε-Caprolactone), Master's Thesis, Worcester Polytechnic Institute, 2003.

[30] M.M. Demir, I. Yilgor, E. Yilgor, B. Eman, Electrospinning of polyurethylane fibers, Polymer 43 (11) (2002) 3303–3309.

[31] M.A. Alamir, "Designing and Evaluating of Superhydrophilic Nanofiber Mats for Fog Catching in the Atmosphere," M.S. Thesis, Wichita State University, December 11, 2017.

[32] M. Ceylan, "Synthesis and Characterization of Electrospun Nanofibers for Advanced Drug Delivery and Cell Culturing," Ph.D. Dissertation, Wichita State University, April, 2014.

[33] Z. Li, C. Wang, Effects of working parameters on electrospinning, One-Dimensional Nanostructures. Springer Briefs in Materials, Springer, Berlin, Heidelberg, 2013.

[34] W.S. Khan, R. Asmatulu, M. Ceylan, A. Jabbarnia, Recent progress on conventional and non-conventional electrospinning processes, Fibers Polymers 14 (2013) 1235–1247.

[35] J.I. Huan, H.E. Yu-Qin, J.Y. Yu, Application of vibrating technology to polymer electrospinning, Int. J. Nonlinear Sci. Numer. Simul. 5 (3) (2004) 253–262.

[36] X. Wang, H. Niu, X. Wang, T. Lin, Needleless electrospinning of uniform Nanofibers using spiral coil spinnerets, J. Nanomater. (2012) 19–27.

[37] R. Asmatulu, M.A. Shinde, A. Alharbi, I.M. Alarifi, Integrating graphene and C_{60} into TiO_2 nanofibers via electrospinning process for the enhanced energy conversion efficiencies, Macromol. Symp. 365 (2016) 128–139.

[38] S.M. Hughes, A. Pham, K.H. Nguyen, R. Asmatulu, Training undergraduate engineering students on biodegradable PCL nanofibers through electrospinning process, Trans. Tech. STEM Educ. 1 (2016) 19–25.

[39] W.S. Khan, R. Asmatulu, S. Davuluri, V.K. Dandin, Enhancing the physical properties of recycled polystyrenes incorporated with nanoscale inclusions via electrospinning process, J. Mater. Sci. Technol. 30 (2014) 854–859.

[40] P. Gupta, R. Asmatulu, G. Wilkes, R.O. Claus, Superparamagnetic flexible substrates based on submicron electrospun Estane® fibers containing MnZnFe-Ni nanoparticles, J. Appl. Polymer Sci. 100 (2006) 4935–4942.

[41] W.S. Khan, M. Ceylan, A. Jabbarnia, L. Saeednia, R. Asmatulu, Structural investigations of electrospun PAN nanofibers incorporated with various nanoscale inclusions, J. Thermal Eng. 3 (2017) 1375−1390.

[42] A. Alharbi, I.M. Alarifi, W.S. Khan, A. Swindle, R. Asmatulu, Synthesis and characterization of electrospun polyacrylonitrile/graphene nanofibers embedded with $SrTiO_3$/NiO nanoparticles for water splitting, J. Nanosci. Nanotechnol. 17 (2017) 1−9.

[43] J. Jiang, M. Ceylan, T. Jai, L. Yao, R. Asmatulu, S.Y. Yang, Poly-ε-caprolactone electrospun nanofiber mesh as a gene delivery tool, AIMS Bioeng. 3 (2016) 528−537.

[44] I.M. Alarifi, W.S. Khan, A.K.M. Rahman, Y. Kostogorova-Beller, R. Asmatulu, Synthesis, analysis and simulation of carbonized electrospun nanofibers infused carbon prepreg composites for improved mechanical and thermal properties, Fibers Polymers 17 (2016) 1449−1455.

[45] A. Alharbi, I.M. Alarifi, W.S. Khan, R. Asmatulu, Synthesis and analysis of electrospun $SrTiO_3$ nanofibers with NiO_x nanoparticles shells as photocatalysts for water splitting, Macromol. Symp. 365 (2016) 246−257.

[46] A. Alharbi, I.M. Alarifi, W.S. Khan, R. Asmatulu, Highly hydrophilic electrospun polyacrylonitrile/polyvinypyrrolidone nanofibers incorporated with gentamicin as filter medium for dam water and wastewater treatment, J. Membr. Sep. Technol. 5 (2016) 38−56.

[47] P. Kiselev, J.R. Liompart, Highly aligned electrospun nanofibers by elimination of the whipping motion, J. Appl. Polymer Sci. 125 (3) (2012) 2433−2441.

[48] J. Pu, Y.D. Jiang, High Efficiency Electrospinning for deposition of orderly nanofiber, Mater. Res. Innov. 15 (4) (2011) 290−293.

[49] D. Sun, C. Chang, S. Li, L. Lin, Near-field electrospinning, Nano Lett. 6 (4) (2006) 839−842.

[50] C. Chang, K. Limkrailassiri, L. Lin, Continuous near-field electrospinning for large area deposition of orderly nanofiber patterns, Appl. Phys. Lett. 93 (2008) 123111−123114.

Electrospun nanofibers for drug delivery

3

CHAPTER OUTLINE

3.1 ELECTROSPUN NANOFIBERS FOR DRUG DELIVERY

3.1.1 CANCER AND TREATMENT

Cancer, also known as malignancy, is an abnormal growth of cells in almost all parts of the body. It was reported that more than 100 different cancer types can form in the body, including breast, prostate, skin, colon, brain, lung, liver, kidney, bone, leukemia, and lymphoma [1]. Normal cells grow and divide to form new cells based on the need of the body. After a certain lifetime, old and damaged cells die and new cells spontaneously form as a replacement. In all types of cancers, some of the cells in the body start continuously dividing to form tumors, and then spreading to surrounding organs and tissues [2]. These are called malignant tumors that can break off and travel to distances in the body through the blood or lymph system (carrying clear fluid) to form new tumors. The major signs and symptoms of cancer include abnormal bleeding, lumps, unexplained weight loss, changes in bowel movements, and prolonged coughing. Many cancers cells form solid tumors of certain sizes; however, some of them (e.g., leukemia) usually do not form solid structures. Unlike malignant cancer tumors, benign type cancer tumors can become larger, but do not spread to other parts of the body. They can be surgically removed and don't form new tumors; however, a benign tumor in the brain can be dangerous [1,2].

Malignant tumor cells can significantly affect the surrounding normal cells to form blood vessels that can provide enough nutrients, oxygen, and other necessary

Synthesis and Applications of Electrospun Nanofibers. DOI: https://doi.org/10.1016/B978-0-12-813914-1.00003-1
© 2019 Elsevier Inc. All rights reserved.

supports for cancer cell growth. They can also remove the waste products formed during cancer cell growth. Through cancer cell growth, these cells can evade the immune systems of the body, organs, and tissues to make the body weaker against pathogens and other diseases [1]. Even though the immune system usually protects the body against abnormal cells and other diseases, some of the malignant cells cannot be detected easily and grow to form larger tumors.

Cancer can also be a genetic disease, caused by changes in genes that control growth and division. The primary risk factors for cancer include tobacco, radiation, sex, age, race, mineral and vitamin deficiencies, alcohol, infections, and hormone level disorders. Based on the type and stage of the cancer, this disease can be treated using chemotherapy, radiation therapy, biological therapy, hormone therapy, and surgery; of these, chemotherapy is one of the most selected up-to-date methods. Chemotherapy is also a necessary step to treat the tumor before surgery, and to prevent recurrence and metastasis after surgery. However, maintaining effective concentrations of chemo-agents at a local tumor site without broadly killing remote normal cells remains an unsolved task, and needs to be addressed [1].

It was estimated that approximately 90.5 million people had cancer in 2015, of whom 14.1 million were new cases (including skin cancer other than melanoma) [3]. This disease caused about 8.8 million deaths, which is approximately 15.7% of the total deaths globally. Mainly males had lung cancer, prostate cancer, colorectal cancer, and stomach cancer, while females had breast cancer, colorectal cancer, lung cancer, and cervical cancer. Children usually have different cancers, including acute lymphoblastic leukemia and brain tumors, except in Africa where non-Hodgkin lymphoma occurs more often than other part of the world.

The total cost of cancer globally was about $1.16 trillion in 2010 [1−3].

Recently, researchers have been developing new targeted drug-delivery systems using various external forces, including magnetic fields, electric fields, ultrasound, light, hydrogel, temperature, and mechanical forces [4−12]. In these systems, the drug is localized at a specific targeted area by internal or external forces and then activated. Recently, magnetic particles carrying drug molecules were developed to target the drugs to specific sites in the body by external magnetic fields. Shortly after concentration in the targeted region, the drug molecules were gradually released, thus improving their therapeutic efficiency, while lowering the collateral toxic side effects on healthy cells or tissues [8,13−23].

In the traditional drug-delivery systems, the body is completely exposed by the drug molecules, and can cause some damage to normal cells. In recent studies, these adverse effects of drug molecules drastically eliminate using targeted drug-delivery systems. For example, using chitosan- or gelatin-based hydrogels combined with crosslinking agents, chemotherapeutic drugs, albumin, peptide, and fluorescence labeling materials, the drug molecules can be concentrated at the cancer site for a longer period of time [4,19−23]. A number of cancer drugs, such as taxotere, carboplatin, fluorouracil (5-FU), cytoxan, herceptin, goserelin, and doxorubicin, have been developed and used to treat breast cancers for many years [24].

Also, albumin within the drug-delivery system interacts with highly metabolic cancerous tissue to keep the drug and peptide molecules internally [25,26].

3.1.2 **ELECTROSPINNING OF NANOFIBERS**

Electrospinning is a relatively easy and direct method of fabricating a nonwoven mat of polymeric nanofibers compared to conventional methods, such as melt spinning [27], wet spinning [28], and extrusion molding [29]. It offers a distinct advantage of forming fibers in the micro- to nanometer range with a high surface-area-to-volume ratio compared to conventional fibers. Electrospinning utilizes a high DC electric field on the surface of a polymeric solution to overcome surface tension to produce a very slim charged jet. AC voltage is also considered in the electrospinning/electrospraying process for nanofiber manufacturing [30]. When the electric field is applied, mutual charge repulsion induces longitudinal stresses. When the intensity of the electrostatic force is increased beyond certain limits, the hemispherical surface of the solution at the tip of the capillary elongates to form a unique structure, called a Taylor cone [30]. The jet first extends in a straight path for some distance, called the jet length, and then instability occurs by bending the jet into a looping path that results in a series of spiral motions [27]. In order to minimize these bending (or whipping) instabilities, the jet undergoes a large amount of plastic stretching that consequently reduces the fiber diameter to a nanoscale [31,32]. Finally, the solvent completely evaporates, and then nanofibers are collected on a grounded screen placed at some distance from the capillary, as shown in Fig. 3.1 [31].

Generally, the diameter of electrospun nanofibers is in the range of 50−500 nm, but they can be as small as 3 nm by changing the system and process parameters [27]. Understanding and optimizing these parameters would result in nanoscale fibers with minimum bead formation. The system parameters include

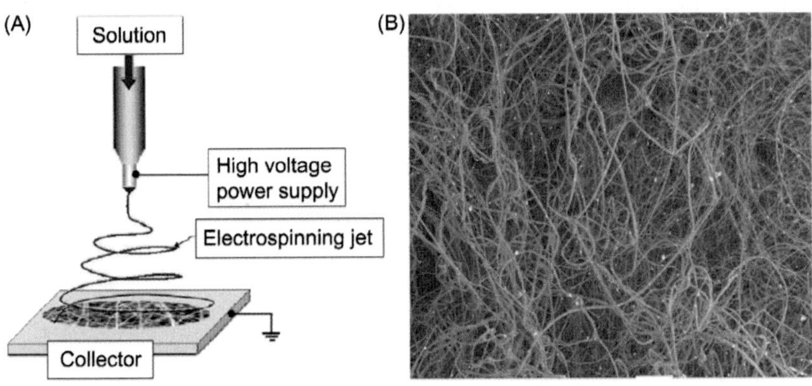

FIGURE 3.1

(A) Schematic view of the electrospinning process and (B) resulting nanofibers.

polymer and solvent types and structures, viscosity, ionic conductivity, charge-ability, and surface tension, whereas the process parameters include concentration, electric potential, flow rate, distance between capillary and collection screen, temperature, humidity, air velocity, nanoparticle inclusion, and grounded target rotation [27].

Because the electrospun nanofibers have a large surface area per unit mass, these fibers can be used in various fields: filtration and separation of micron-, submicron-, and nanosize organic, inorganic, and biological particles; HF antenna fabrication; light-weight, colorful, and invisible fabric productions; transistor; solar and hydrogen energy; and biomedical applications, such as wound dressings, tissue engineering scaffolds, and artificial blood vessels. Other promising areas of electrospun nanofibers include nanocomposite fabrications to improve crack resistance and reduce interior aircraft noise [28].

3.1.3 NANOFIBERS FOR DRUG DELIVERY

Electrospinning is a highly flexible fiber production method from various materials to form fibers down to a few nanometers level. The high surface area of fibers with functional properties makes them suitable nanomaterials for a number of different cancer treatments. These nanofibers can be embedded with other nanomaterials (e.g., Fe_3O_4), polymers, and cancer drugs to kill cancer cells locally because of their greater sensitivity to heat, pH, pressure, and light. Some electrospun nanofibers produced and utilized for cancer treatments are analyzed in detail here.

Chen et al. summarized the electrospun nanofibers for cancer diagnosis and therapy [33]. It was emphasized that progress in nanotechnology provided novel opportunities in nanomedicine and other biomedical studies because of the outstanding properties of nanofibers, including extremely large surface area, porosity, high loading capacity, high encapsulation efficiency, ease of adjustment and functionalization, low cost, and other great benefits. Nanofibers have been proposed in different cancer diagnoses and therapies (e.g., targeted cancer cell capture, ultrasensitive sensing systems for cancer detection, pH-dependent property changes, and functional and smart anticancer drug-delivery systems). The authors concluded that proper design of nanofibers, surface modifications, and adding therapeutic and sensing elements into the fibers will improve the biosensing ability for cancer biomarker detections, pre- and post-cancer treatments, and implantable drug-delivery systems for cancer therapies [33].

Chen et al. mentioned the emerging roles of electrospun nanofibers in cancer studies and the recent development of electrospun nanofibers in cancer treatments [34]. The study talked about the brief information on the importance of nanofibers that is vital for cancer research, including the control of release kinetics, incorporation of drugs, drug dissolution, alignment of nanofibers, and their applications in different biomedical fields. The study also discussed major roles of electrospun nanofibers in cancer research, such as targeted chemotherapy, localized treatments, combinatorial therapy, cancer cell capture and detection, behavior of cancer cells, a 3D cancer model with nanofibers, and cancer metastasis. The primary challenges and future trends for design, fabrication, analysis, characterization, and applications of electrospun nanofibers in diagnostics and therapeutics of different types of cancer were also analyzed [34].

Kim et al. reported the smart hyperthermia nanofibers incorporated with magnetic nanoparticles and an anticancer drug (doxorubicin) that have a heat-sensitive drug-release mechanism for cancer study [35]. The system has "on and off" switching during an alternating magnetic field for induction of skin cancer apoptosis. During the crosslinking process, the electrospun nanofiber mats displays switchable changes based on the swelling ratio since the local heat from the incorporated magnetic nanoparticles encourages the deswelling of polymer structures in the nanocomposite fibers. The test study indicated that about 70% of cancer cells were killed within 5 min of application of an alternative magnetic field in the presence of the electrospun nanofibers with magnetic nanoparticles and anticancer drug because of the double effects of heat and drug. The combination of hyperthermia material and anticancer drugs in nanofibers will be a new avenue in cancer treatment [35].

Aggarwall et al. mentioned the synthesis and analysis of drug-loaded nanofibers for the treatment of cancer [36]. The authors investigated the therapeutic efficacy of a chemotherapy drug (cisplatin) mixed with poly-caprolactone/chitosan prior to electrospinning for local chemotherapy of cervical cancers in mice. The experimental studies showed that electrospun nanofibers provided a drug-release pattern of up to one month, which might be sufficient time to kill local cancer cells. Intravaginal administration of the cancer drug-loaded nanofibers provided lesser cell viability when compared to the plain drug, indicating that more cancerous cells died during the course of treatment. The authors concluded that this novel process offered a favorable approach for the targeted delivery of the anticancer drug against the cervical cancer [36].

Saha et al. studied electrospun fibrous scaffolds for stimulating cancer cell alignment and epithelial−mesenchymal transition [37]. In this work, electrospun fibrous scaffolds were created with random and aligned fiber orientations in order to mimic the three-dimensional structure of the natural extracellular matrix. The study indicated that the elasticity and structure of the extracellular matrix changed the behavior of breast cancer cells. It was also reported that the cancer cells cultured on the surface of electrospun fiber scaffolds showed elongated spindle-like morphology in the aligned fibers; however, they did not show the same behavior on random fibers, which were mostly flat stellar shapes. It was also suggested that the cancer cells underwent epithelial−mesenchymal transitions. Overall, this study concluded that a topographical cue could play a significant role in tumor growth.

Soares et al. reported the electrospun nanofibers embedded with multiwall carbon nanotubes (MWCNTs) or gold nanoparticles for immunosensors of pancreatic cancer [38]. The authors fabricated the immunosensors using electrospun nanofiber mats made of polyamide 6 and poly(allylamine hydrochloride) because their 3D structure could be appropriate for the immobilization of antibodies to detect the pancreatic cancer biomarker (CA19-9) in these sensors. The impedance spectroscopy results confirmed that these functional sensing platforms could detect CA19-9 with a detection limit of 1.84 and 1.57 U/mL using the nanofibers with MWCNTs and gold nanoparticles, correspondingly. The biomarker absorption mechanism was related to the Langmuir—Freundlich absorption process. The sensitivity and selectivity of the immunosensors were analyzed using blood serum from patients and found that this approach was promising for a simple and effective diagnosis of pancreatic cancer at early stages [38].

Nelson et al. reported the enhanced breast cancer cell migration on biomimetic electrospun nanofibers [39]. The metastatic cancer cells can usually escape from surgical locations and colonize distant parts of organs and tissues to quickly jeopardize patient life. Thus, highly sensitive and accurate in vitro tools are necessary to allow rapid, precise, and innovative antimetastatic drug screening for those patients. It was suggested that electrospun aligned nanofibers could provide the migration potentials of cancer cells. They can also achieve a long-term goal for in vitro platform technology to investigate the efficacy of novel antimetastatic compounds [39].

Ramalingam et al. synthesized and analyzed the electrospun curcumin-loaded poly(2-hydroxyethyl methacrylate) p(HEMA) nanofibers as biomaterials for multidrug-resistant organisms [40]. It was stated that electrospun nanofibers promoted cell growth and attachment on the surfaces of nanofiber mats and retained their original morphology during experimentation. Constant and precise release pattern of plant-based curcumin against infectious diseases was observed for multidrug-resistant organisms, such as meticillin-resistant *Staphylococcus aureus* (MRSA). These antibacterial activities of curcumin proved to be an alternate herbal compound for antibiotics. The test results indicated that curcumin-loaded nanofibers could be a viable approach for wound healing of infectious diseases and an excellent drug-delivery vessel for other disease treatments [40].

Magnetite nanoparticles (10 nm in diameter) at concentrations of 0, 1, 5, 10, 20, and 30 wt.% were dispersed in DMF (90%) and sonicated for 30 minutes. *Polyacrylonitrile (PAN) powder* was added to the previous dispersion while stirring for 2 h. The dispersion was placed in a syringe with a capillary with an inside diameter of 0.5 mm, which was connected to the DC voltage via a platinum wire. The electrospinning tests were conducted at 1 mL/h pump speed, 20 kV DC voltage, and 15 cm tip-to-collector distance. Fig. 3.2 shows scanning electron microscope (SEM) images of the electrospun nanocomposite fibers as a function of magnetite concentrations [41]. It was reported that adding magnetic nanoparticles into polymeric solutions steadily increased the diameters of nanocomposite fibers. For instance, the average diameter of fibers at 0 wt.% magnetic nanoparticles is about 390 nm, whereas the average diameter of 30% magnetic nanoparticles was approximately 1.1 μm. These functional nanocomposite fibers may be utilized for biomedical, energy, filtration, and separation and defense applications [41—43].

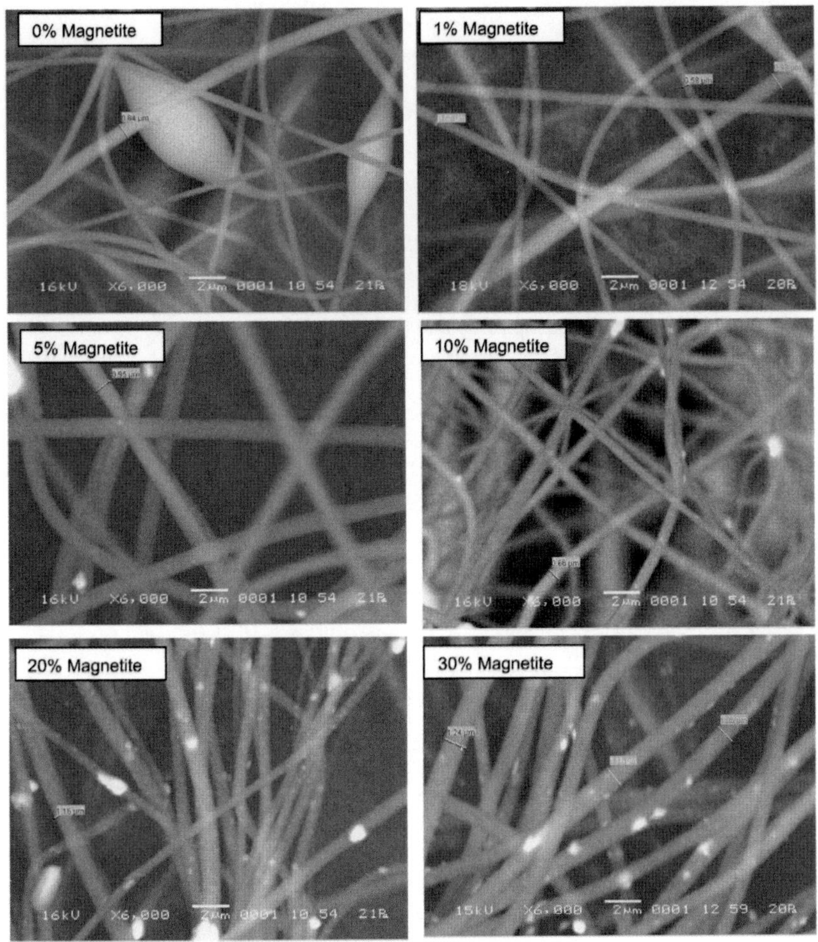

FIGURE 3.2

SEM images showing the nanocomposite fiber diameters and morphologies as a function of magnetite concentrations [41].

Son et al. reported the health benefits of different electrospun nanofibers in drug-delivery systems for potential recovery purposes [44]. It was stated that various medical drugs, including peptide, antibodies, enzymes, proteins, amino acids, and other small drug molecules could be added to the surfaces of nanofibers based on the selected drugs and their applications. Hydrophilic and hydrophobic drugs can be dissolved in many polymers using appropriate organic solvents prior to the electrospinning process. The active molecules can be physically or chemically attached to the surface of the fibers or mixed with the polymers. It was estimated that chemical absorptions on the nanofiber mesh usually promote cell differentiation and proliferation. Surface coating and embedding the drugs with nanofibers will change the drug-release regime from burst release to slow and continuous release for a longer time, where patients benefit more in the second case. Also, electrospun nanofibers can be indirectly loaded with microspheres, nanoparticles, and hydrogels with multidrug delivery systems for different release rates for various treatments [44].

Hu et al. stated that electrospun nanofibers could be great tools for various drug-delivery applications [45]. A number of different polymeric materials, such as synthetic, natural, and hybrid after dissolving in organic solvents were successfully electrospun to produce micron- and nanoscale fibers. During the fabrication process, electrospun nanofibers can be tailored to provide some specific features for drug loading, cell attachment, bio-mineralization, filtering body liquid, and mass transfer. Different drugs, such as antibiotics, anticancer agents, proteins, DNA, living cells, RNA, and various growth factors can be easily incorporated with the electrospun fibers. The authors also mentioned the recent development in electrospinning techniques and their application in various drug-delivery systems [45].

Goyal et al. reported about recent developments in nanoparticles and nanofibers for topical drug delivery for different biomedical applications [46]. The authors specified that developments in nanotechnology in general, and nanoparticles and nanofibers in particular, made huge impacts on drug-delivery systems from surface treatment to inside of the body. The proposed approaches drastically improved the drug concentration on the skin or inside the body in order to increase drug flux in the area of interest. Delivering both hydrophobic and hydrophilic drugs by those nanomaterials will control the drug release rates for a prolonged period of time. Clinical studies are needed for the new nanoproduct developments for different disease treatments worldwide [46].

Tipduangta et al. reported on electrospun blend nanofibers with tunable drug-delivery properties and their role on transformative phase separation and controlled drug release rates [47]. Blending of polymers at appropriate ratios will change the physical and chemical structures of the produced nanofibers, so the tuneability of the nanofibers was simply based on the selected materials and methods for different drug-delivery applications. This approach will provide a lot of benefits for future developments of controlled drug release formulations in which the drug release rate can be modified by blending the ratios of selected polymers. It is important that phase separation between the polymers should not be major concerns for the success of the approach; otherwise, appropriate solvents, temperatures, pH, and light sources must be considered during the blending and electrospinning process. Electrospinning usually provides micro- and nanoscale miscibility between the polymers throughout the stretching and solvent evaporation processes. The micro- and nanoscale miscibility will affect the burst release rates of nanofibers [e.g., burst release from polyvinylpyrrolidone (PVP)-rich phases and a slower and continuous release from hypromellose acetate succinate (HPMCAS)-rich phases]. Through the alteration of PVP and HPMCAS, drug release rates of nanofibers can be adjusted for different patients [47–52].

Li et al. studied nanofiber support of oligodendrocyte precursor cell growths on a neuron-free model for myelination of nerve damages [51]. It was stated that electrospun nanofiber mats could act as tools for therapeutic cell delivery to improve the axonal myelination rates of oligodendrocytes using PCL and gelatin copolymer. SEM images exhibited that PCL fibers and PCL-gelatin nanofibers were 731 ± 198.9 nm and 801.2 ± 188.4 nm, respectively. The elemental compounds of the fibers were analyzed by energy-dispersive X-ray spectroscopy to identify the gelatin component in the fibers, which also decreased the water contact angle of the electrospun fibers. The test results confirmed that both PCL nanofibers and PCL + gelatin copolymer nanofibers accelerated oligodendrocyte precursor cell (OPC) growth and differentiation (Fig. 3.3). It was also stated that differentiated oligodendrocyte cells sustained the processes along the nanofibers and ensheathed the nanofibers [51]. However, the PCL and gelatin nanofibers had better performance at the same conditions.

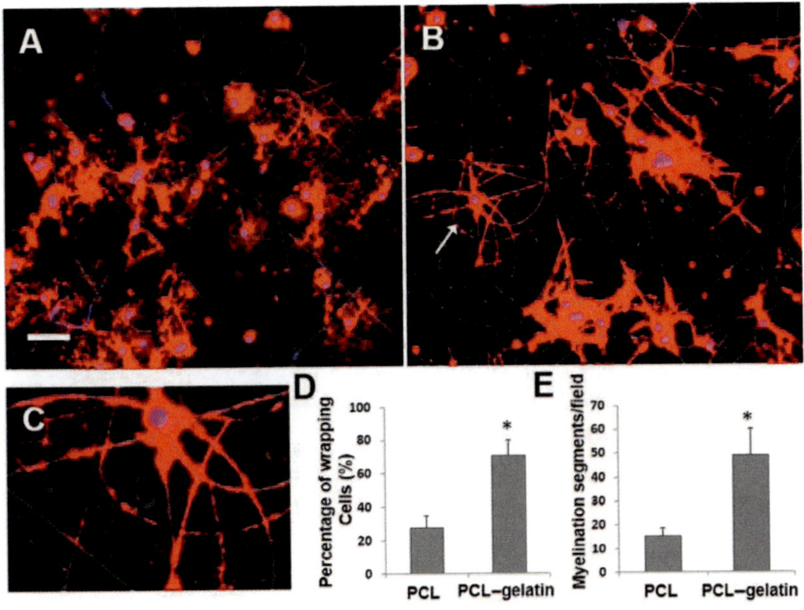

FIGURE 3.3

The differentiated OPCs ensheath nanofibers. (A) Differentiated OPCs ensheath PCL fibers. (B) Differentiated OPCs ensheath PCL-gelatin fibers. Scale bar: 100 μm. (C) Magnified image of myelinated PCL-gelatin nanofibers, as indicated by the arrow in (B). Cells were labeled with anti-MBP antibody. (D) Analysis of percentage of cells wrapping nanofibers (* indicates significant difference compared with differentiated OPCs myelinate PCL nanofibers, $P < 0.01$). (E) Analysis of percentage of myelinated nanofiber segments [51].

Ceylan et al. studied the effects of antibacterial PCL nanofibers on the growth of Gram-positive and Gram-negative bacteria for drug-delivery purposes [49]. The antibacterial agent (gentamicin) at 0, 2.5, 5, and 10 wt.% was incorporated with PCL powders and dissolved in a solvent before the electrospinning process. The produced nanofibers of different layers (1, 2, and 4 layers) were used against the growth of Gram-negative and Gram-positive bacteria, such as *Escherichia coli*, *Salmonella* sp., and *Staphylococcus epidermidis*. The SEM images showed that the diameters of the produced nanofibers were 50 and 200 nm. The test results indicated that gentamicin-loaded nanofibers released antibacterial agent during the tests and prohibited bacterial growth at different inhibition zones. Fig. 3.4 shows the antibacterial test results of PCL fibers loaded with different concentrations of gentamicin. The authors stated that the present study could improve the antibacterial properties of new drug-delivery systems for different biomedical applications: scaffolding; drug, DNA, and protein delivery; and wound dressing [49,53–57].

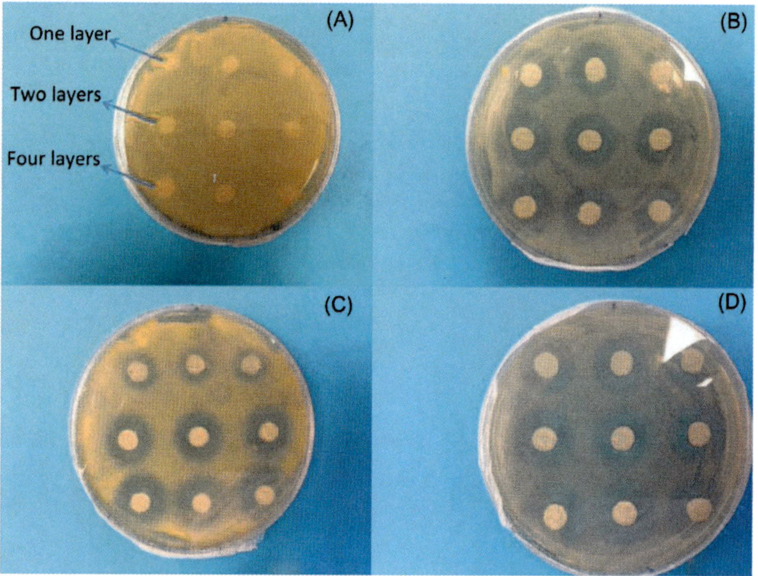

FIGURE 3.4

Photographs showing the antibacterial test results of PCL fibers loaded with (A) 0 wt. %, (B) 2.5 wt. %, (C) 5 wt. %, and (D) 10 wt. % gentamicin after 7 days of in vitro tests. Top row samples have one layer, while middle row samples two layers, and bottom row samples four layers of the nanofibers.

3.2 ELECTROSPUN NANOFIBERS FOR GENE DELIVERY

Lee et al. investigated the electrospun nanofibers as multipurpose interfaces for efficient gene-delivery tools for different disease treatments and healing [48]. The authors intended to integrate genes into electrospun nanofibers as delivery tools to enhance the potential of gene therapy for several biomedical applications, such as cancer therapy, stem cell therapy, scaffolding, and other tissue engineering. Because of the ease of production, large surface-to-volume ratios, versatile choices of numerous materials and producing many structures with different physical and chemical properties, electrospun nanofibers possess many advantages in gene delivery and other delivery applications. When properly tailored, electrospun nanofibers can spatially and temporally release gene vectors and enhance gene-delivery efficiency more than many other techniques [49−51]. The authors also discussed the powerful characteristics of electrospun nanofibers, functional properties, as well as spatial interfaces that are capable of stimulating controlled and efficient gene-delivery processes for many patients [48].

In order to direct axon regeneration in spinal cord injury (SCI) treatment, Nguyen et al. studied the 3D aligned nanofibers of hydrogel scaffolds for controlled nonviral drug/gene delivery [52]. Traffic and other accidents, sickness, and birth defects can cause SCI for many patients, which is one of the persistent neurological dysfunctions unresolved to date. The current techniques, including direct drug administration, do not effectively respond to healing the SCI owing to the rapid drug clearance in the body. The authors proposed a sustained nonviral delivery system of proteins and nucleic acids (e.g., small noncoding RNAs) using functional 3D structured nanofibers. During the experimental studies, rat spinal cord was chosen as a hemiincision model to evaluate the efficacy of the proposed scaffold design. The test results indicated that the aligned axon regeneration was noticed within a week timeframe, without any excessive inflammation in or near the tissues [52].

Jiang et al. reported poly-ε-caprolactone (PCL) electrospun nanofiber meshes as gene-delivery tools for post-cancer treatment [50]. PCL is a well-known biodegradable polymer mainly used for biomedical engineering, such as tissue scaffolds, and gene-, drug-, and protein-delivery vehicles. PCL nanofibers were produced using 25 kV DC voltage, 20 μL/min flow rate, and 25 cm needle-to-collector distance. Fig. 3.5 shows the SEM image of the PCL-nanofiber mesh and the mechanical property of the six-layer nanofiber film. The objective of this study was to develop a gene-tethering PCL-nanofiber mesh system as a gene-delivery tool after the removal of primary bone tumors, and eliminate tumor recurrence in the future. Prior to the electrospinning process, a nonviral plasmid vector was incorporated into PCL solution to produce multilayer nanomeshes. A test study showed that the plasmid-linked electrospun nanofiber mesh effectively released the gene marker and consequently expressed the transgene products at transcriptional and translational levels [50].

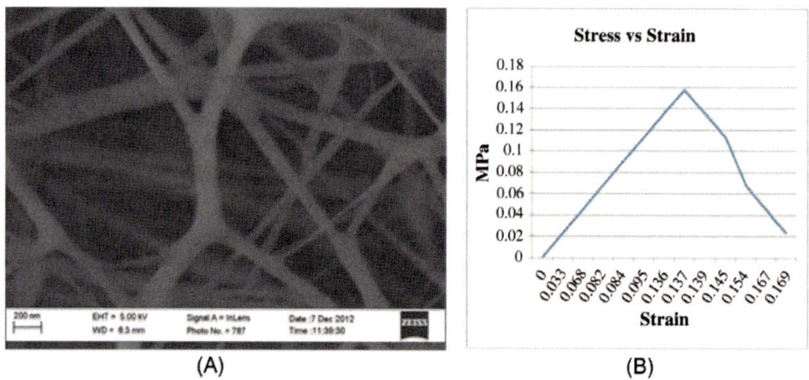

FIGURE 3.5

(A) SEM image of the PCL-nanofiber mesh and (B) mechanical property of the six-layer nanofiber films.

Ramalingam et al. mentioned in the study of "nanofiber composites for biomedical applications" that electrospun nanocomposite fibers could be utilized in gene-delivery studies for tissue engineering and regenerative medicine because of the superior physical, chemical, physicochemical, and biological properties of the prepared nanofibers [58]. It was stated that viral-based vectors could be loaded with genes of different types to achieve higher gene-delivery efficiency with an extended period of time. Because of the solvents utilized during the electrospinning process, some of the genes preserved from the sources could be adversely affected during the gene-delivery process. More studies may be conducted to eliminate the harsh electrospinning processes prior to the in vivo and in vitro studies [58].

Sultanova et al. investigated the coaxial electrospun PCL/polyvinyl alcohol (PVA)-chitosan nanofibers for novel nonviral gene-delivery scaffold applications [59]. It was stated that resembling the morphology and structure of the extracellular matrix (ECM) was the main objective for nanofiber-based tissue engineering processes due to the porosity and fibrous structure of those nanomaterials. Because nanofibers have unique structures, controllable membrane thickness, tissue-like elasticity, and large surface area to volume ratio, they are mainly used for ECM mimicking for tissue engineering. The electrospun nanofibers of PCL and PVA-chitosan polymers were produced using a coaxial electrospinning technique in which PCL was the shell solution, while PVA-chitosan blend was the core solution. Prior to the cell adhesion tests, the surfaces of the prepared nanofibers were made hydrophilic through O_2 plasma processing. The authors intended to use the nanofibers with plasmid loadings for gene transfection efficiencies of the scaffolds with and without plasma treatment [59].

Liang et al. reported the in vitro nonviral gene delivery process with nanofibrous scaffolds for transfection of genes [60]. They reported that the extracellular and intracellular barriers usually inhibit the nonviral gene vectors during the effective transfection process. It was reported that a core-shelled structure was created by invoking solvent-induced condensation of β-galactosidase in DMF with 6% of buffer solution and subsequent encapsulation of the gene in a triblock copolymer in the same solvent. The prepared solution was electrospun while protecting the gene and DNA to form nanofibrous nonwoven scaffolds. Finally, the bioactive plasmid could be released from the nanofiber scaffolds with a controlled release rate [60].

Zhang et al. reported on the Oxford Academic coaxial electrospun fibrous scaffold with DNA molecules for substrate-mediated gene-delivery purposes [61]. Electrospun nanofibers can be a robust platform for localized gene transfection of target cells. The authors proposed that a two-step approach could immobilize DNA molecules onto electrospun nanofibers for active gene delivery. In the first step, nonviral gene vectors consisting of polyethylene glycol (PEG)-modified polyethylenimine (PEI) could be associated with the scaffolds, while in the second step, the target DNA could be transferred onto the surface of produced nanofibers through an electrostatic interaction between DNA and PEI−PEG structures. The test results revealed that the proposed gene-delivery system offered high transfection efficiency; for example, approximately 65% of human embryonic kidney 293 cells and 40% of mesenchymal stem cells were transfected using green fluorescent gene. It was also mentioned that PEG-modified PEI improved the biocompatibility and advanced the transfection efficiency, which may be useful for local gene delivery in different tissue engineering applications [61].

Xie et al. fabricated the core-shell PEI/pBMP2-(poly(D,L-lactic-co-glycolic acid)) electrospun scaffolds for gene delivery using a stem cell process [62]. Tissue engineering for bone regeneration is one of the most commonly considered subjects, because the scaffolds incorporated with osteogenic factors advance the therapeutic effects drastically. The core-shell scaffolds of the coaxial electrospinning process were used to prepare for the controlled gene delivery to the human periodontal ligament stem cells (hPDLSCs). During the electrospinning process, pBMP2 was encapsulated with PEI phase (core structure) and PLGA was used to regulate pBMP2 release (shell structure). After the testing and characterization steps, the gene release rate was analyzed. The test results indicated that the pBMP2 release rate was considerably higher in the first few days and then stayed in the steady-state form. During the experiment, PEI/pBMP2 displayed high transfection efficiency, as well. Coaxial electrospun nanofibers exhibited a longer gene-delivery period when compared to the single axial electrospinning process. These results may be useful for prolonging the time of gene delivery [62].

3.3 ELECTROSPUN NANOFIBERS FOR DNA DELIVERY

Encapsulating and controlling DNA and its release rates at a known site are essential steps in achieving tissue regeneration and gene therapy for many patients. There are some requirements for control therapy, including targeting, stability, functionality, and minimum degradation, since naked DNA structures are susceptible to degrade easily because of the environmental factors that can cause a decrease in function and control of the release rate. Electrospinning and other encapsulation processes of DNA can drastically enhance the viability, stability, and transfection efficiency of the DNA molecules. Controlling hydrophobicity, hydrophilicity, anion and cation ratios, gelation, binding agents, and pH and temperature changes can successfully encapsulate DNA complexes. For instance, negatively charged nitrogen of polyethylene imine (PEI) polymer surrounds the other negatively charged phosphate backbone of DNA to prevent the degradation process [60−64].

In order to solve the problems associated with viral DNA delivery systems (e.g., toxicity, safety, and fate of transfected cells), new research studies have been conducted to develop nonviral means of delivering biological entities in vivo [60]. Usually, nonviral approaches consist of cationic lipids formulated into liposomes with DNA molecules, composite formation of polymers with DNA, and collagen-based DNA hydrogels. Recent studies have indicated that new biomaterials could serve as scaffolds for tissue engineering, and DNA- and drug-delivery purposes [60−62]. Although the nonviral gene delivery systems compromise enhanced safety when compared to the viral systems, their transfection efficiencies are moderately low. It was reported that nonviral gene delivery could have some major barriers, such as uptake of DNA by the surrounding system, degradation of the DNA in plasma, lack of translocation to the nucleus, insufficient characterization, and lysosomal degradation of the DNA during the process. To be able to address some of these issues, the design formulation must be considered for the fundamental understanding between DNA and surrounding solutions with various chemical species [60].

Yunfei et al. studied the controlled dual delivery of angiogenin and curcumin by electrospun nanofibers for skin regeneration [63]. A novel drug and plasmid DNA dual-delivery system was developed via an electrospinning method using polyethyleneimine-carboxymethyl chitosan/pDNA-angiogenin (ANG) nanoparticles, poly(D,L-lactic-co-glycolic acid) (PLGA), cellulose nanocrystals (CNCs), and curcumin (Cur). The in vitro test studies exhibited that the Cur and ANG was properly loaded with electrospun nanofibers, and based on the release tests, nearly 90% of the Cur was released in 6 days, while the ANG release was about 20 days. The cell culture tests indicated that the prepared nanofibers had excellent biocompatibility for biomedical studies. In order to validate the in vitro studies, in vivo studies were also conducted on the prepared nanofibers (PLGA/CNC/Cur/pDNA-ANG) after transplanting into the infected full-thickness burn wounds. A number of biopsy samples were taken for immunohistochemistry, histology, real-time quantitative PCR, immunofluorescence, and Western blotting analyses, and the test results indicated that the composite nanofibers not only prohibited local infection but also stimulated skin regeneration on the burn areas [63].

Ceylan et al. synthesized and evaluated electrospun PCL-plasmid DNA nanofibers for post-cancer treatments [56]. PCL nanofibers associated with plasmid

DNA were produced to investigate the DNA release from the PCL nanofibers. The DNA drastically improved green fluorescent protein (EGFP) with cytomegalovirus (CMV) promoter (PCMVb-GFP), which was initially amplified with *Escherichia coli*. The polymeric materials utilized in the present study were all biodegradable and biocompatible and have been used for different tissue engineering purposes, including DNA-, drug-, gene-, peptide-, and protein-delivery vehicles and tissue scaffolding. SEM studies indicated that the average diameter of nanofibers was around 100 nm. Cytotoxicity tests were conducted on L-929 cells, and exhibited that cell viability values for up to 7 days were exceeding 80%. Fig. 3.6 reveals the cytotoxicity results of PCL-plasmid DNA nanofibers as a function of time and dilution ratio with viability ratios. The tests results showed that PCMVb-GFP plasmid-linked electrospun nanofibers released double-stranded DNA for up to 7 days. At the beginning, there was a burst release of about 1.8 ng/mL for about 15 minutes. For the remaining time, the release was almost constant (release of 0.575 ng/mL). The authors concluded that this novel method

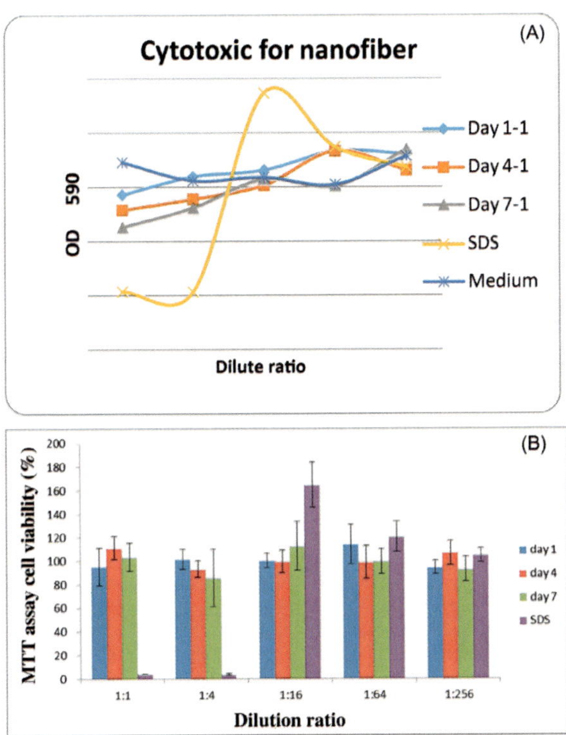

FIGURE 3.6

(A) The cytotoxicity results for PCL-plasmid DNA nanofibers as a function of time and dilution ratio, and (B) cell viability ratios.

could be an ideal approach for various biomedical applications (e.g., scaffolding, cancer treatment, tissue engineering, and DNA, gene and drug delivery) [56].

Hayenga et al. studied the incorporation of DNA particles into chitosan nanofibers for localized tissue regeneration [64]. In this study, DNA was mixed with cationic PEI polymer and then negatively charged phosphate DNA backbone was surrounded through a charge neutralization process. The prepared DNA was incorporated with chitosan and polyethylene oxide-based electrospun nanofibers as DNA delivery vehicles for tissue scaffoldings. The DNA-based nanofibers have high surface area, porosity, and flexibility that can be useful for cell interaction and tissue regeneration in many disease treatments. The encapsulated DNA could be potentially used for controlled DNA-delivery purposes [64].

Peckys et al. studied the immobilization and release strategies of double-stranded DNA (dsDNA) using carbon nanofiber arrays and a self-assembled monolayer (SAM) process [65]. In this study, the authors immobilized dsDNA molecules on the vertically aligned carbon nanofibers and analyzed the releasing mechanisms of the DNA. Prior to the immobilization step, the carbon nanofibers were coated with gold film using an SAM process to covalently bind the molecules to each other. Furthermore, modified nanofiber arrays with DNA were utilized to transfect the Chinese hamster lung epithelial cells. This study would potentially enable the transport of ligands into the cell nucleus without using extracellular matrix and cytosolic degradation. These statistically significant test results proved that end-specifically and covalently SAM-bound DNA could be expressed in cells properly [65].

Castro-Smirnov et al. investigated the cellular uptake pathways of sepiolite nanofibers as nanocarriers and their DNA transfection improvements [66]. As is known, sepiolite is a natural silicate with nanofibrous structures, which can be a suitable material for biomedical applications. Analyzing the properties (perfectly stable and intrinsic fluorescence features) of sepiolite as nanocarriers could open up new avenues for DNA transfection. The intrinsic fluorescence ability would allow to follow the DNA pathways into the cell structures through fluorescence microscopes. The test studies showed that sepiolite could be spontaneously internalized into the mammalian cells through different paths. This nanostructured silicate could be a great option for a future DNA-delivery system [66].

3.4 CONCLUSIONS

In this chapter (electrospun nanofibers for drug delivery), three major topics were analyzed in detail. In the first, the fabrication, structure, and properties of electrospun nanofibers were studied and their major applications for drug delivery were reported. Various electrospun nanofibers could be developed and used for many drug-delivery applications. In the second part of the chapter, electrospun nanofibers and their tailoring abilities were investigated for a number of gene-delivery purposes. Especially, it was indicated that gene delivery could be a major breakthrough for post-cancer treatments. In the final part of the chapter, various electrospun nanofibers incorporated with different DNA molecules as drug-delivery tools were reported for different aspects of biomedical applications.

REFERENCES

[1] R.W. Ruddon, Cancer Biology, Fourth ed., Oxford University Press, London, 2007.

[2] P. Price, K. Sikora, Treatment of Cancer, Sixth ed., CRC Press, New York, 2014.

[3] https://en.wikipedia.org/wiki/Cancer [accessed 1.4.18].

[4] F. Abedin, R. Asmatulu, R. Anwar, S.Y. Yang, Albumin-based nanocomposite drug carriers with dual chemo-agents for targeted breast cancer treatment, J. Biomater. Appl. (2015). Available from: https://doi.org/10.1177/0885328215569614.

[5] R. Asmatulu, R.O. Claus, J. Riffle, "Targeting Magnetic Nanoparticles in High Magnetic Fields for Drug Delivery Purposes," Mat. Res. Soc. Symp. Proc. (MRS 2004 Spring Meeting — San Francisco) Vol. 820, O3.8.1—O3.8.6.

[6] R. Asmatulu, M.A. Zalich, R.O. Claus, J. Riffle, Synthesis, characterization and targeting of biodegradable magnetic nanocomposite particles by external magnetic fields, J. Magn. Magn. Mater. 292C (2005) 108—119.

[7] R. Asmatulu, H.E. Misak, S.Y. Yang, P.H. Wooley, "Composite Magnetic Nanoparticle Drug Delivery System," Application Number: 61392018, October 11, 2010 (provisional).

[8] R. Asmatulu, A. Fakhari, H.L. Wamocha, H.Y. Chu, Y.Y. Chen, M.E. Elsayed, et al. Synthesizing drug-carrying magnetic nanocomposite particles for targeted drug delivery. J. Nanotechnol. Vol., 2009, Article ID 238536, 6 pages.

[9] R. Asmatulu, B. Cooper, H. Misak, "Fabrication of Ferrofluids at Controlled pH Values for Biomedical Applications," SAMPE, Fall Technical Conference 2008, Memphis, TN, September 8—11, 2008, pp. 1—8.

[10] R. Asmatulu, H. Misak, B. Cooper, W. Khan, "Nanotechnology and Relevant Technologies Lab Development at WSU," 2008 ASEE Midwest Regional Conference, September 18—19, 2008, Tulsa, OK, pp. 1—8.

[11] R. Asmatulu, A. Fakhari, H.L. Wamocha, H.H. Hamdeh, J.C. Ho, "Fabrication of Magnetic Nanocomposite Spheres for Targeted Drug Delivery," 2008 ASME International Mechanical Engineering Congress and Exposition, October 31—November 6, 2008, Boston, pp. 1—4.

[12] H.L. Wamocha, R. Asmatulu, M.M. El-Tabey, H. Misak, J.S. Gopu, B. Cooper, et al., "Increasing the Efficiency of Pharmaceutical Drugs for Magnetic Targeted Drug Delivery," SAMPE Fall Technical Conference, Wichita, October 19—22, 2009, 1—7.

[13] R. Asmatulu, B. Geist, W. Spillman, R.O. Claus, Dielectric constant and breakdown field studies of electrostatic self-assembled materials, Smart Struct. Mater. 14 (2005) 1493—1500.

[14] P. Gupta, R. Asmatulu, G. Wilkes, R.O. Claus, Superparamagnetic flexible substrates based on submicron electrospun Estane® fibers containing MnZnFe-Ni nanoparticles, J. Appl. Polym. Sci. 100 (2006) 4935—4942.

[15] W.B. Spillman Jr., R. Asmatulu, C.F. Jullian, B. Geist, R.O. Claus, J.L. Robertson, Preliminary dielectric measurement and analysis protocol for determining the melting temperature and binding energy of short sequences of DNA in solution, Biotechnol. J. 3 (2008) 252–263.

[16] R. Asmatulu, B. Zhang, N. Nuraje, A ferrofluid guided system for the rapid separation of the non-magnetic particles in a microfluidic device, J. Nanosci. Nanotechnol. 10 (5) (2010) 6383–6387.

[17] W.S. Khan, R. Asmatulu, M.M. El-Tabey, Dielectric properties of electrospun PVP and PAN nanocomposite fibers, J. Nanotechnol. Eng. Med. 1 (2010) 6.

[18] R. Asmatulu, M. Ceylan, N. Nuraje, Study of superhydrophobic electrospun nanocomposite fibers for energy systems, Langmuir 27 (2) (2011) 504–507.

[19] H.L. Wamocha, H.E. Misak, Z. Song, H.Y. Chu, Y.Y. Chen, R. Asmatulu, et al., Cytotoxicity of release products from magnetic nanocomposites in targeted drug delivery, J. Biomater. Appl. (2011) 1–7.

[20] R. Asmatulu, Toxicity of nanomaterials and recent developments in lung disease Chapter 6, in: P. Zobic (Ed.), in *Bronchitis*, InTec, 2011, pp. 95–108.

[21] R. Asmatulu, H.E. Misak, S.Y. Yang, P.H. Wooley "Composite Magnetic Nanoparticle Drug Delivery System," U.S. Patent No. 20120265001, October 18, 2012.

[22] H. Misak, B. Cooper, J. Gopu, K.-P. Man, N. Zacharias, P. Wooley, et al., Skin cancer treatment by albumin/5-Fu loaded magnetic nanocomposite spheres in a mouse model, J. Biotechnol. 164 (2013) 130–136.

[23] H.L. Wamocha, H.E. Misak, Z. Song, H.Y. Chu, Y.Y. Chen, R. Asmatulu, et al., Cytotoxicity of release products from magnetic nanocomposites in targeted drug delivery, J. Biomater. Appl. 11 (2013) 661–667.

[24] Y. Hirshaut, P.I. Pressman, J. Brody, Breast Cancer: The Complete Guide, Fifth ed., Bantam, New York, 2008.

[25] A. Wunder, U.M.L. Ernst, H.K. Stelzer, J. Funk, E. Neumann, G. Stehle, et al., Albumin-based drug delivery as novel therapeutic approach for rheumatoid arthritis, J. Immunol. 170 (2003) 4793–4801.

[26] A. Wunder, G. Stehle, H. Sinn, H.H. Schrenk, D. Hoff-Biederbeck, F. Bader, et al., Enhanced albumin uptake by rat tumors, Int. J. Oncol 11 (1997) 497–507.

[27] E. Lenk, "Melt spinning apparatus," US Patent: 4704077, November 3, 1987.

[28] S. Ramakrishna, *An Introduction to Electrospinning and Nanofibe*rs, World Scientific, Singapore, 2005.

[29] K. Cantor, Blown Film Extrusion, an Introduction - Book Review, Hanser, Munich, 2006.

[30] G. Taylor, Electrically driven jets, Proc. R. Soc. Lond. A 313 (1969) 453–475.

[31] W.E. Teo, S. Ramakrishna, A review on electrospinning design and nanofiber assemblies, Nanotechnology 17 (2006) R89–R106.

[32] Y. Gogotsi, Nanomaterials Handbook, CRC Press, Boca Raton, 2006.

[33] Z. Chen, Z. Chen, A. Zhang, J. Hu, X. Wang, Z. Yang, Electrospun nanofibers for cancer diagnosis and therapy, Biomater. Sci. 4 (2016) 922–932.

[34] S. Chen, S.K. Boda, S.K. Batra, X. Li, J. Xie, Emerging roles of electrospun nanofibers in cancer research, Adv. Healthcare Mater. (2017). Available from: https://doi.org/10.1002/adhm.201701024.

[35] Y.J. Kim, M. Ebara, T. Aoyagi, A smart hyperthermia nanofiber with switchable drug release for inducing cancer apoptosis, Adv. Funct. Mater. 23 (2013) 5753–5761.

[36] U. Aggarwall, A.K. Goyal, G. Rath, Development and characterization of the cisplatin loaded nanofibers for the treatment of cervical cancer, Mater. Sci. Eng. C Mater. Biol. Appl. 75 (2017) 125–132.

[37] S. Saha, X. Duan, L. Wu, P.K. Lo, H. Chen, Q. Wang, Electrospun fibrous scaffolds promote breast cancer cell alignment and epithelial-mesenchymal transition, Langmuir 28 (2012) 2028–2034.

[38] J.C. Soares, L.E.O. Iwaki, A.C. Soares, V.C. Rodrigues, M.E. Melendez, J.H.T.G. Fegnani, et al., Immunosensor for pancreatic cancer based on electrospun nanofibers coated with carbon nanotubes or gold nanoparticles, ACS Omega 10 (2017) 6975–6983.

[39] M.T. Nelson, A. Short, S.L. Cole, A.C. Gross, J. Winter, T.D. Eubank, et al., Preferential, enhanced breast cancer cell migration on biomimetic electrospun nanofiber cell highways, BMC Cancer 14 (2014) 825–835.

[40] N. Ramalingam, T.S. Natarajan, S. Rajiv, Preparation and characterization of electrospun curcumin loaded poly(2-hydroxyethyl methacrylate) nanofiber--a biomaterial for multidrug resistant organisms, J. Biomed. Mater. Res. Part A 103 (2015) 16–24.

[41] R. Asmatulu, W.S. Khan, K.D. Nguyen, M.B. Yildirim, Synthesizing magnetic nanocomposite fibers for undergraduate nanotechnology education, Int. J. Mech. Eng. Educ. 38 (2010) 196–203.

[42] R. Asmatulu, August Nanotechnology Safety, Elsevier, Amsterdam, The Nederland, 2013.

[43] N. Nuraje, R. Asmatulu, G. Mul, Green Photo-Active Nanomaterials: Sustainable Energy and Environmental Remediation, RSC Publishing, Cambridge, England, 2015.

[44] Y.J. Son, W.J. Kim, H.S. Yoo, Therapeutic applications of electrospun nanofibers for drug delivery systems, Arch. Pharm. Res. 37 (2014) 69–78.

[45] X. Hu, S. Liu, G. Zhou, Y. Huang, Z. Xie, X. Jing, Electrospinning of polymeric nanofibers for drug delivery applications, J. Controlled Release 185 (2014) 12–21.

[46] R. Goyal, L.K. Macri, H.M. Kaplan, J. Kohn, Nanoparticles and nanofibers for topical drug delivery, J. Control Release 240 (2016) 77–92.

[47] P. Tipduangta, P. Belton, L. Fabian, L.Y. Wang, H. Tang, M. Eddleston, et al., Electrospun polymer blend nanofibers for tunable drug delivery: the role of transformative phase separation on controlling the release rate, Mol. Pharmaceut. 13 (2016) 25–39.

[48] S. Lee, G. Jin, J.H. Jang, Electrospun nanofibers as versatile interfaces for efficient gene delivery, J. Biol. Eng. 8 (2014) 30–40.

[49] M. Ceylan, S.Y. Yang, R. Asmatulu, Effects of gentamicin-loaded PCL nanofibers on growth of gram positive and gram negative bacteria, Int. J. Appl. Microbiol. Biotechnol. Res. 5 (2017) 40–51.

[50] J. Jiang, M. Ceylan, T. Jai, L. Yao, R. Asmatulu, S.Y. Yang, Poly-ε-caprolactone electrospun nanofiber mesh as a gene delivery tool, AIMS Bioeng. 3 (2016) 528–537.

[51] Y. Li, M. Ceylan, B. Sherstha, H. Wang, Q.R. Lu, R. Asmatulu, et al., Nanofibers support oligodendrocyte precursor cell growth and function as a neuron-free model for myelination study, Biomacromolecules 15 (2014) 319–326.

[52] L.H. Nguyen, M. Gao, J. Lin, W. Wu, J. Wang, S.Y. Chew, Three-dimensional aligned nanofibers − hydrogel scaffolds for controlled non-viral drug / gene delivery to direct axon regeneration in spinal cord injury treatment, Sci. Reports 7 (2017) 42212.

[53] F. Abedin, "Magnetic and Albumin Targeted Drug Delivery for Breast Cancer Treatment," M.S. Thesis, Wichita State University, July, 2011.

[54] D.P. Ha, "Synthesis and Characterization of Graphene Nanoflakes-based Polycaprolactone Bionanocomposites," M.S. Project, Wichita State University, December, 2011.

[55] S. Patrick, "Fabrication and Characterization of Antibacterial Polycaprolactone and Natural Hydroxyapatite Nanofibers for Bone Tissue Scaffolds," M.S. Thesis, Wichita State University, May, 2013.

[56] M. Ceylan, "Synthesis and Characterization of Electrospun Nanofibers for Advanced Drug Delivery and Cell Culturing," Ph.D. Dissertation, Wichita State University, April, 2014.

[57] S.M. Hughes, "Electrospun PCL Nanofibers Incorporated with Natural and Synthetic Calcium Hydroxyapatite Particles for Human Tooth Regeneration," M.S. Thesis, Wichita State University, April 27, 2016.

[58] M. Ramalingam, S. Ramakrishna, Nanofiber Composites for Biomedical Applications, Technology and Engineering, New York, 2017.

[59] Z. Sultanova, G. Kabay, G. Kaleli, M. Mutlu "Coaxial electrospun PCL/PVA-chitosan nanofibers: a novel non-viral gene delivery scaffold," Plasma Sciences (ICOPS), 2015 IEEE International Conference, Antalya, Turkey. 10.1109/PLASMA.2015.7179972.

[60] D. Liang, Y.K. Luu, K. Kim, B.S. Hsiao, M. Hadjiargyrou, B. Chu "In vitro non-viral gene delivery with nanofibrous scaffolds," **Dehai Liang***To whom correspondence should be addressed. Benjamin Chu, Department of Chemistry, Stony Brook University, Stony Brook, NY 11794-3400, USA. Tel: +631 632 7928; Fax: +631 632 6518; Email: bchu@notes.cc.sunysb.edu **Search for other works by this author on**: Nucleic Acids Research, 200, Vol. 33, 170-180.

[61] J. Zhang, Y. Duan, D. Wei, L. Wang, H. Wang, Z. Gu, et al., Oxford Academic Co-electrospun fibrous scaffold−adsorbed DNA for substrate-mediated gene delivery, J. Biomed. Mater. Res. Part A 96 (2011) 212−220.

[62] Q. Xie, L. Jia, H. Xu, X. Hu, W. Wang, J. Jia, Fabrication of core-shell PEI/pBMP2-PLGA electrospun scaffold for gene delivery to periodontal ligament stem cells, Stem Cells Int. 2016 (2016) 11.

[63] M. Yunfei, G. Rui, Z. Yi, X. Wei, C. Biao, Z. Yuanming, Controlled dual delivery of angiogenin and curcumin by electrospun nanofibers for skin regeneration, Tissue Eng. Part A 23 (2017) 597−608.

[64] J. Hayenga, A. Copper, N. Bhattari, M. Zhang, Incorporation of DNA particles into chitosan nanofibers for tissue regeneration, J. Undergraduate Res. Bioeng. 10 (2010) 26−30.

[65] D.B. Peckys, A.V. Melechko, M.L. Simpson, T.E. McKnight, Immobilization and release strategies for DNA delivery using carbon nanofiber arrays and self-assembled monolayers, Nanotechnology 20 (2009) 10.

[66] F.A. Castro-Smirnov, J. Ayache, J.R. Bertrand, E. Dardillac, E. LeCam, O. Pietrement, et al., Cellular uptake pathways of sepiolite nanofibers and DNA transfection improvement, Sci. Reports 7 (2017) 5586.

FURTHER READING

E. Embrey, Breast Cancer: A Nurses Journey from Diagnosis Through Reconstruction, Xulon Press, Longwood, Florida, 2012.

K.K Hunt, G.L. Robb, E. Strom, and N.T. Uneu, "Breast Cancer- 2nd Edition" M.D. Andersons, Cancer Care Series, 2007.

C. Andersson, B.M. Iresjo, K. Lundholm, Identification of tissue sites for increased albumin degradation in sarcoma-bearing mice, J. Surg. Res 50 (1991) 156–162.

P.C. Gotzsche, M. Nielsen, "Screening for Breast Cancer with Mammography (Review)," John Wiley & Sons, The Cochrane Collaboration, The Cochrane Library, Issue 4, 2009.

E.A. Rakha, J.S. Reis-Filho, I.O. Ellis, Basal-like breast cancer: a critical review, J. Clin. Oncol. 26 (15) (2008) 2568–2581 (May 20).B. Cheema, A.C. Gaul, K. Lane, M.A.F. Singh, Progressive resistance training in breast cancer: a systematic review of clinical trials, Breast Cancer and Treat 109 (2008) 9–26.

Electrospun nanofibers for textiles

4

CHAPTER OUTLINE

4.1 SUPERHYDROPHOBIC ELECTROSPUN NANOFIBERS FOR NONWETTABLE SURFACES

4.1.1 JET FORMATION IN ELECTROSPINNING

Electrospinning is a micro- and nanoscale fiber manufacturing technique under applied DC electric voltage using polymeric solutions and polymer melts at higher temperatures. This technique has both electro-spraying and conventional dry spinning processes to produce woven and nonwoven fibers without the use of coagulation chemistry or other processing [1]. Because of those conditions, the electrospinning process will be mostly suitable for the production of fibers using large and complex molecules. In the electrospinning process, some solvents (e.g., polar, nonpolar, and blends of both) are properly selected to dissolve the polymeric materials; however, in some applications (e.g., biomedical and sensor), solvents may not be considered during nanofiber fabrication. In this case, electrospinning from the molten polymers can be considered to ensure that no toxic solvents can be delivered into the final products for biomedical purposes [1−4].

As specified, the initial stage of the electrospinning process, where a liquid polymer jet is drawn out from the meniscus, is theoretically simple. Yeo et al. described that the liquid meniscus from the capillary outlet in the shape of a spherical cap could have a capillary stress of an order of γ/r, where γ is the

Synthesis and Applications of Electrospun Nanofibers. DOI: https://doi.org/10.1016/B978-0-12-813914-1.00004-3

surface tension of the solution, and r is the principal curvature of the meniscus [5–8]. The capillary stress acts to restore any interfacial deformation during the spinning process. In order to start the electrospinning process, electrostatic stress must be larger than capillary stress and surface tension of the solution to form micro- and nanoscale fibers. Therefore, the critical voltage, V_c, that must be exceeded prior to a jet formation can be estimated at the meniscus tip as:

$$V_c \sim \sqrt{\frac{\gamma H^2}{\varepsilon r}} \tag{4.1}$$

where H is electrode separation and ε is permittivity [3,8].

In the late 1960s, Taylor was the first to analyze the relationship between meniscus shape and applied DC field [9–12], and determined that an equilibrium conical shape was formed at the tip, as shown in Fig. 4.1. In a weak polarization, the solution of the Laplace equation for the conical shape in an axisymmetric spherical coordinate system (r, θ) presents electrostatic potentials for liquid φ_l and gas φ_g [8]:

$$\varphi_l(r, \theta) = A_n r^n P_n[\cos \theta], \quad \text{for} \quad \theta_0 \geq \theta \geq 0 \tag{4.2}$$

and

$$\varphi_g(r, \theta) = B_n r^n P_n[\cos(\pi - \theta)] = B_n r^n P_n[- \cos \theta], \quad \text{for} \quad \pi \geq \theta \geq \theta_0 \tag{4.3}$$

where $P_n[x]$ is the Legendre function of the first kind, and A_n and B_n are constants. The equations with interfacial boundary conditions and continuity of potential and electric stresses across the interface then lead to:

$$\varepsilon_g P_n[\cos \theta_0]P_n'[- \cos \theta_0] + \varepsilon_l P_n[- \cos \theta_0]P_n'[\cos \theta_0] = 0 \tag{4.4}$$

which has a solution $0 < n < 1$ for given values of $\varepsilon_l/\varepsilon_g$ and θ_0. Since this results in an interfacial electric field, E, scaling $1/r^{n-1}$, it then becomes clear from the normal stress jump across the interface:

$$\Delta p = \gamma \kappa + \frac{\varepsilon_0(\varepsilon_l - \varepsilon_g)}{2} \left(\frac{\varepsilon_l}{\varepsilon_g} E_n^2 - E_t^2 \right) \tag{4.5}$$

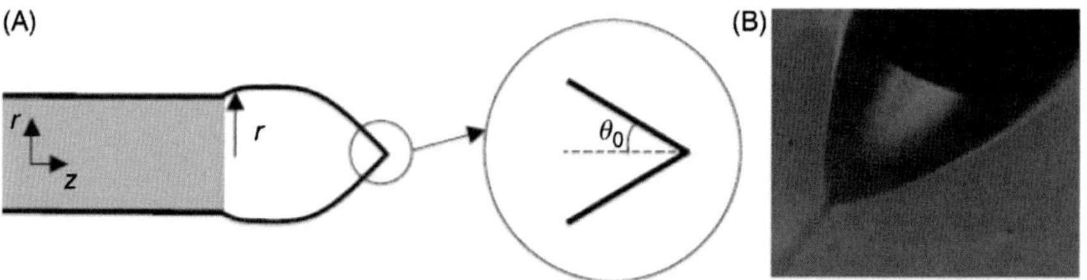

FIGURE 4.1

(A) Schematic view of electrospinning meniscus with conical end, and (B) the formation of a Taylor cone from which a liquid jet emanates prior to the bending instability development.

where Δ_p is the pressure difference across the gas–liquid interface, E_n and E_t are the normal and tangential components of the electric field, respectively, and ε_0 is the permittivity of free space. It is determined that the solution cone angle θ_0 is 49.3 degrees for a perfectly conducting drop.

The bending instability of a polymeric jet induced by electrical and aerodynamic forces that plays a fundamental role in the electrospinning process. The main difference between electrical and aerodynamical force-induced instabilities is in the nature of the destabilizing force [12,13]. In the theory of aerodynamic driven jets, the force acting on the jets is directly associated with the Bernoulli equation, which enhances perturbations and makes them grow. The aerodynamic bending force, F_{aer}, per jet length, $d\xi$, in the case of small bending perturbations is:

$$F_{\mathrm{aer}} = -\rho_a V_0^2 \pi a_0^2 IkInd\xi \tag{4.6}$$

where ρ_a is the air density, V_0 is the jet velocity, k is the curvature of jet axis at a given point, and a_0 is the jet cross-sectional radius that does not vary for small perturbations. In the case of electrical force, F_{el}, the following equation is given:

$$F_{el} = -e^2 \ln\left(\frac{L}{a}\right) IkInd\xi \tag{4.7}$$

where e is charge, and L is the length of the selected element. The aerodynamic bending may also be directly used in the case of electrically driven bending instabilities, if one replaces the factor ρV_0^2 by $e_0^2 \ln(L/a_0)/\pi a_0^2$ [13]. When the surface tension component is added to the system, the magnitude of the net force, F, acting on the jet element is equal to:

$$F = \pi a\gamma IkInd\xi \tag{4.8}$$

Therefore, the total net force acting on the jet element is given by the sum of the electric and surface tension forces [3]:

$$dF = IkInd\xi\left[\pi a\gamma - e^2\ln\frac{L}{a}\right] \tag{4.9}$$

After the jet formation, micro- and nanoscale fibers will start forming in the later stages, such as bending instabilities, fiber starching, and traveling to the collector screen. After the drying process on the collector screen for 6–8 h, the electrospun nanofibers are analyzed for further applications.

4.1.2 ELECTROSPINNING OF NANOFIBERS

Electrospun nanofibers with high aspect ratio, large surface area to volume ratio, thermal and chemical stability, high porosity, high directional strengths, good biocompatibility and biodegradability, excellent structural morphology, and flexibility have been appealing nanomaterials for various industries [14–19]. Because of these properties, electrospun nanofibers could been considered in many biological and other industrial applications, such as tissue engineering, textiles, cosmetics, wound dressing, molecular filtration, drug delivery, fuel cells, sensor devices, and many more [20−24].

Asmatulu et al. studied the hydrophilic electrospun nanofiber membranes for the filtration of micro- and nanosize suspended particles for clean water production from waste and lake water sources [25]. In this study, poly vinyl chloride (PVC) incorporated with 4 wt.% poly vinyl pyrrolidone (PVP) were dissolved in dimethylacetamide (DMAc) at 85% solvent and 15% polymer ratios on a hot plate at 65°C for about 4 h. Electrospinning tests were performed at 25 KV DC voltage, 30 cm separation distance, and 2 ml/h pump speed. Fig. 4.2 shows the scanning electron microscopy (SEM) images of the electrospun PVC nanofibers incorporated with 4 wt.% PVP at low and high magnifications. The test results indicated that these nanomembranes were highly effective in filtering micro- and nanoscale organic and inorganic materials [25].

Jabbarnia et al. synthesized and characterized electrospun polyvinylidene fluoride (PVDF)-based polymeric separators for supercapacitor applications [26−28]. Supercapacitors are new generations of storage devices to effectively store the charge with a high capacity and long life cycle when compared to other devices for storing energy. In this study, PVDF/PVP nanofibers incorporated with different weight percentages of carbon black nanoparticles were synthesized using an electrospinning process. SEM images indicated that the diameters of the nanofibers are between 100 and 200 nm. The other characterization tests, such as Fourier transform infrared spectroscopy (FTIR), Raman spectroscopy, capacitance bridge, electrochemical impedance spectroscopy, water contact angle, and X-ray

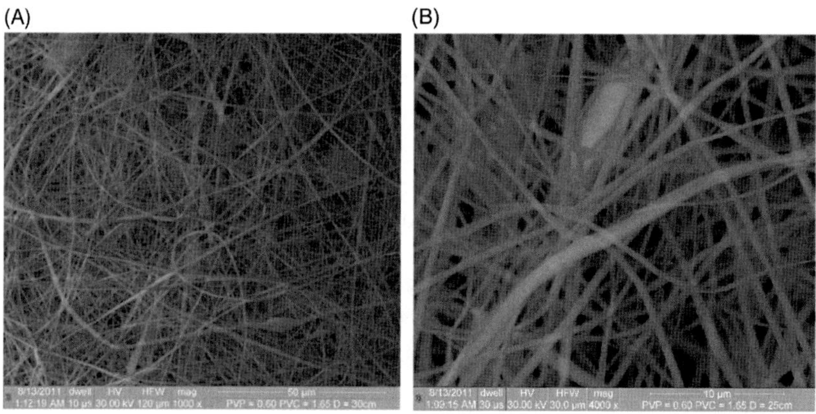

FIGURE 4.2

SEM images showing the electrospun PVC nanofibers incorporated with 4 wt.% PVP at (A) low and (B) high magnifications produced at 25 KV DC voltage, 30 cm separation distance, and 2 ml/h pump speed.

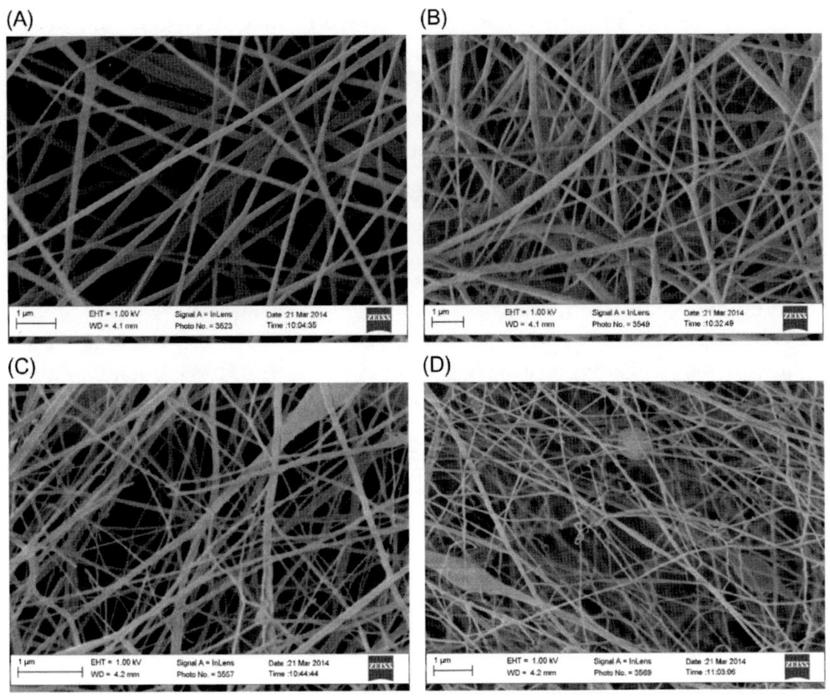

FIGURE 4.3

SEM images of electrospun PVDF/PVP nanocomposite fibers incorporated with (A) 0, (B) 1, (C) 2, and (D) 4 wt.% carbon black nanoparticles.

diffractometry (XRD) were also used to investigate the microstructures, crystal-line phases, and chemical properties of the nanocomposite fibers.

The experimental studies indicated that the ionic conductivity, dielectric constant, and wetting properties of electrospun nanofibers were considerably enhanced with the nanoscale inclusion; however, electric conductivity was not changed much, which is good to eliminate a short circuit during the electric charges and discharge. This study also showed that the other physical and chemical properties of the porous nanocomposite fibers were increased, which may be suitable for supercapacitor applications. Fig. 4.3 shows SEM images of electrospun PVDF/PVP nanofibers incorporated with 0, 1, 2, and 4 wt.% carbon black nanoparticles. The nanofibers were produced using 80:20 PVdF/PVP (total PVP is 4 wt.% in PVDF to make nanofiber wettable) and DMAC/acetone solvent ratio at 25 KV DC voltage, 25 cm separation distance, and 2 ml/h pump speed [26−28]. Adding inorganic nanoparticles into the electrospun nanofibers will greatly increase the fire retardancy of the battery membrane separator and eliminate battery fires.

In recent studies, PAN was dissolved in dimethylformamide (DMF), and different weight percentages of PVP and gentamicin sulfate powders were added to the previous solution to fabricate nanofiber-based membranes via the electrospinning process. Fig. 4.4 shows the SEM micrograms of a PAN nanomembrane used in filtration studies [21]. Antibacterial gentamicin was added to remove bacteria and prevent biofouling, whereas PVP was added to make the surface of the membrane hydrophilic to improve the filtration rate and efficiency of the nanofilter. Two water samples were chosen for the filtration processes: dam water and city wastewater. For the dam water sample, turbidity, pH, TDS, Ca^{++}, Mg^{++}, sulfates, nitrates, fluoride, chloride, alkalinity, and silica were reduced to +3.64, 89.6, 6.52, 10.5, 9.96, 5.16, 17, 19.5, 6.63, 1.43, and 63.5%, respectively. The total coliforms and *Escherichia coli* content were reduced to 4.1 and 0 MPN/100 ml, respectively, with PAN containing 10 wt.% PVP and 5 wt.% gentamicin. For the wastewater sample, pH, turbidity, TDS, TSS, BODs, phosphate, ammonia, oil-greases, and DO were reduced to +3.62, 79, 6.33, 84, 68, 1.70, 15.8, 0, and 6%, respectively. The total coliforms and *E. coli* content were also lowered to 980 and 1119.9 MPN/100 ml, respectively, with PAN containing 10 wt.% PVP and 5 wt.% gentamicin. The morphology and dimensions of the nanofibers were

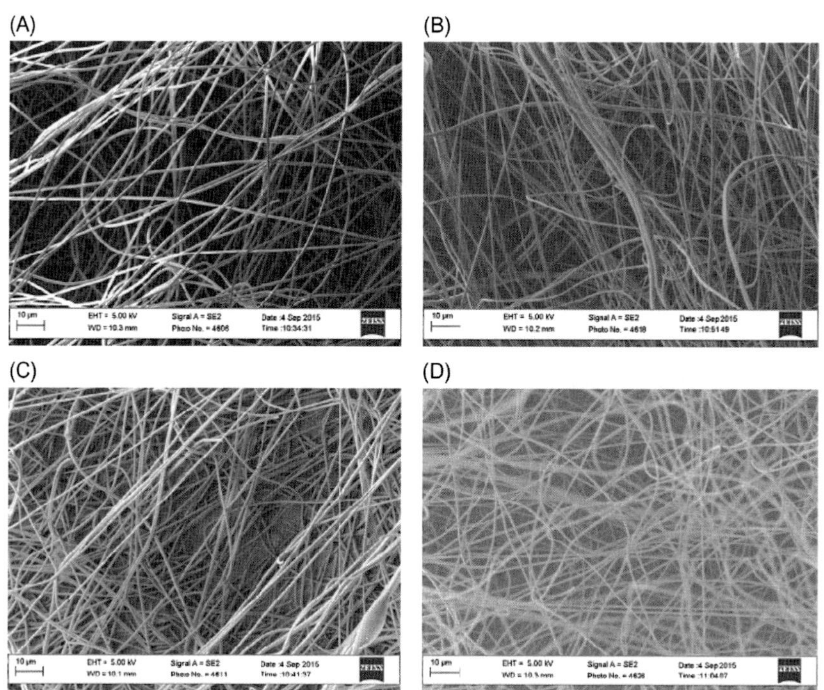

FIGURE 4.4

SEM images of electrospun nanofibers before filtration: (A) PAN + 0 wt.% PVP + 0 wt% gentamicin, (B) PAN + 0 wt.% PVP + 5 wt.% gentamicin, and (C) PAN + 5 wt.% PVP + 5 wt.% gentamicin, and (D) PAN + 10 wt.% PVP + 5 wt.% gentamicin.

observed using a scanning electron microscope (SEM). Both SEM and microscopic images of the nanomembrane before and after filtration proved that electrospun PAN nanofibers had superior water filtration performance.

4.1.3 THEORY OF SUPERHYDROPHOBICITY

Superhydrophobicity means the surface of a substrate is extremely hydrophobic with micro- and nanoscale asperities (roughness) and waxy (hydrophobic) textures/structures, and water contact angle values are usually between 150 and 180 degrees [29]. This term also refers to the lotus effect, which means the surface cannot be wetted by water molecules and many other liquids at normal conditions [30], that is, water droplets impacting the superhydrophobic surface bounce back like rubber balls [31]. The superhydrophobic properties of a surface depend mainly on surface energy and morphology/asperities. It has been shown that the chemical composition of a surface substantially influences its energy level. Surface molecules are bonded with fewer molecules than those in the interior, so there is more energy at the surface. When a drop of liquid is placed on a solid surface, an equilibrium energy state is reached between three phases of the interface [32−37]. Fig. 4.5 shows the schematic views of different superhydrophilic, hydrophilic, hydrophobic, and superhydrophobic surfaces [16]. Here, γ_{SL} is the solid−liquid interfacial energy, γ_{LV} is the liquid−vapor interfacial energy, and γ_{SV} is the solid−air interfacial energy. The interface is with air, but this is the same with any gas. The angle θ_0 is the characteristic angle achieved between the liquid surface and solid surface when a state of equilibrium is reached. The equilibrium contact angle relates to surface energy and contact angle [34]:

$$\cos \theta_o = \frac{(\gamma_{SA} - \gamma_{SL})}{\gamma_{LA}} \qquad [4.10]$$

which is valid only when a smooth flat surface is involved. The equilibrium contact angle θ_0 is a measure of the surface wettability. Materials with a contact angle less than 90 degrees are said to be hydrophilic, and materials with a contact angle greater than 90 degrees are said to be hydrophobic. A water contact angle

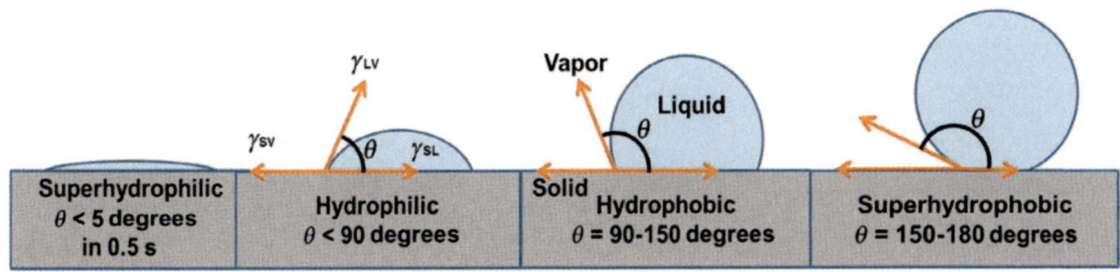

FIGURE 4.5

Schematic views of different superhydrophilic, hydrophilic, hydrophobic, and superhydrophobic surfaces [16].

on a solid greater than 150 degrees is considered superhydrophobic. When the contact angle is less than 10 degrees, the surface is said to be superhydrophilic, which does not have the time variation as indicted in Fig. 4.4 [16,34]. In some sources, superhydrophilicity is associated with the water contact angle, which is less than 5 degrees in 0.5 s [16,34−37].

The contact angle is an indication of superhydrophobicity or superhydrophilicity, but the angle at which a droplet rolls off a surface when tilted (referred to as the sliding angle) is related to the surface energy that must be overcome for the initiation of movement. A droplet that is pinned to the surface as the result of local adhesion energy, as in the case of hydrophilicity, will not start rolling without energy input, that is, mechanical energy, due to its adhesion to the surface at the line of contact. Even superhydrophobic surfaces have some adhesion at the contact line, which can, if higher enough, result in hysteresis between the advancing and receding contact angles. Contact angle hysteresis is an indication of the tendency of a droplet to bounce and bead when dropped on the surface, and the ability of a droplet to roll off the surface easily. This affects the self-cleaning property of the surface.

Two models describe the interface with a roughened surface. The Wenzel model, where the surface roughness allows the droplet to stay in contact with the surface between the asperities, thereby increases the surface area as follows [35,36]:

$$\cos \theta_w = r\frac{(\gamma_{SA} - \gamma_{SL})}{\gamma_{LA}} = r \cos\theta_o \qquad [4.11]$$

where θ_w is the Wenzel model angle, which would be greater than that of the Young's equation because r is greater than 1. The value r, a roughness factor used to adjust for the roughness of the surface and results in an increase in surface area, is as follows:

$$r = \frac{A_1}{A_2} \qquad [4.12]$$

where A_1 is the actual roughened surface area, and A_2 is the planar surface area. In the Cassie−Baxter model, the asperities are high enough to prevent the droplet from touching the surfaces between them, thus resulting in the droplet being suspended in air trapped between the rough surfaces. Therefore, the droplet is now resting on a heterogeneous surface comprised of a fraction of air and a fraction of solid surface. Cassie and Baxter developed the following equation for the equilibrium contact angle (θ_{CB}) on a porous solid using surface energy equations and Young's equation [37].

$$\cos \theta_{CB} = f_1 \cos \theta_o - f_2 \qquad [4.13]$$

where f_1 and f_2 are area functions of materials 1 and 2 ($f_1 + f_2 = 1$), respectively. The Wenzel state is a sticky surface due to the increased surface area in contact with the droplet. The Cassie−Baxter state has a low solid surface contact with the water droplet, which creates a slippery hydrophobic surface due to low surface energy. The Cassie−Baxter model combined with a micro-/nanostructured

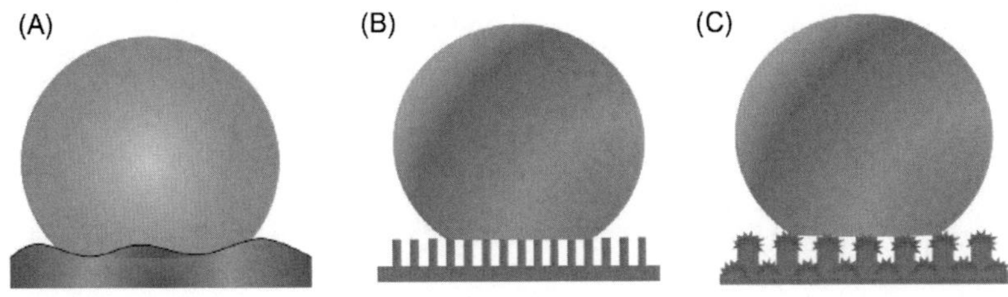

FIGURE 4.6

Various droplet-surface contact states: (A) Wenzel state, where droplet contacts the surface between asperities; (B) Cassie—Baxter state, where droplet does not make contact with surface between asperities; and (C) micro-/nanostructured hierarchical morphology to make surfaces superhydrophobic.

hierarchical morphology allows for the achievement of a superhydrophobic surface with low hysteresis and a low sliding angle. Fig. 4.6 shows the droplet-surface contact states on different surfaces for Wenzel and Cassie—Baxter states/models [33—36].

4.1.4 SUPERHYDROPHOBIC NANOFIBERS

Lately, superhydrophobic materials and devices have been gaining a lot of attention worldwide because of the nonadhesive and nonwetting features of the surfaces for a number of different multifunctional applications, including self-cleaning, anti-icing, anticorrosion, low hydrodynamic friction, and templates for directed self-assembly of nanomaterials for microelectromechanical systems (MEMS), nanoelectromechanical systems (NEMS), and microfluidic and nanofluidic materials and devices [16]. It was stated that the superhydrophobic surfaces show amazing high-contact angles and low contact-angle hysteresis between advancing and receding contact angles of the water droplets. This phenomenon can be seen in nature frequently, such as the superhydrophobic properties of many insect wings, bugs, plants, and some leafs due to their surface structure and chemistry. The water droplets are instantly repelled from or roll off the superhydrophobic surfaces [14,16,17,38—41].

Asmatulu et al. studied the state-of-the-art scientific and technological developments of electrospun nanofibers, superhydrophobicity, and their applications in membranes, deicing, self-cleaning, moisture ingression, responsive smart materials, and other related fields [14,16,17,38—40]. Self-cleaning of superhydrophobic surfaces consists of the right combinations of surface morphology and hydrophobic entities to create extremely high water contact angle values on those surfaces and drive water droplets away from the surfaces [39]. The water droplets with a water contact angle between 150 and 180 degrees simply roll off the surface, taking dirt, particles, and other contaminants way from the surface through gravity at lower angles. In their studies, the authors briefly explained the theory of the superhydrophobic self-cleaning approach, and the basic principles of the electrospinning process. In these studies, the authors also mentioned various electrospun nanofibers and compared the test results with each other in terms of the superhydrophobic properties and their logical, scientific, and industrial applications [14,16,17,38−41].

Surface wettability plays a vital role in daily human activities and industrial requirements. As is stated, the surface energy and roughness are the dominating parameters for determining the surface wettability of the substrates. A substance with a lowest surface energy of 6.7 mJ/m^2 (with closed-hexagonal-packed-CF$_3$ groups) gives a water contact angle of 120 degrees, while after the surface treatment, the surfaces of hydrophilic materials can offer water contact angles over 150 degrees [14,16]. It is stated that the CF$_3$ group provides the lowest surface energy, and when functionalized onto a flat surface, the water contact angle can be enhanced up to 120 degrees. Using the fluorinated and nonfluorinated polymeric compounds, various forms of superhydrophobic electrospun nanofibers can be fabricated [16].

Zheng et al. investigated the lotus-leaf-like microsphere/nanofiber composite film made from polystyrene (PS) using the electrohydrodynamics method and reported the superhydrophobic structures of surfaces [42]. In those studies, the authors used the block copolymer poly(styrene-b-dimethylsiloxane) to produced electrospun nanofibers with diameters of 40−150 nm and a water contact angle of 163 degrees. The surface hydrophobicity mainly comes from the surface-enrichment process due to siloxane groups on the surfaces and precise control of surface roughness of electrospun nanofibers.

Kang et al. studied the effect of solvents on the superhydrophobicity of the nanofiber surfaces [43]. If the polystyrene powder was dissolved in tetrahydrofuran and chloroform prior to the electrospinning process, water contact angle values of the nanofiber surfaces were between 138 and 139 degrees, respectively; however, the same polymer dissolved in N,N-dimethylformamide solvent provided a water contact angle value of over 154 degrees, which indicates that solvents make a huge impact on surface hydrophobicity and wettability [43]. Acatay et al. stated that the fibrous films produced through electrospraying low-molecular-weight polymer (acrylonitrile-co-α, α-dimethyl-isopropenylbenzyl isocyanate) incorporated with polymeric nanofibers and perfluorinated linear diol provided a water contact angle of 167 degrees, which is considered highly superhydrophobic and wettability is almost impossible [44].

Recently, a number of different fluorinate agents that can provide superhydrophobic surfaces have also been studied for different industrial and scientific purposes. Even though fluorinate agents may not be soluble in the electrospinning solvents, there are some new studies in this field to enhance the solubility of fluorinated polymers by combining the chains with soluble polymers. They have mostly hybrid structures to preform dual performance during the applications.

Agarwal et al. reported the possibility of changing the superhydrophobicity by modifying the surface morphology and using pentafluorostyrene (PFS)-fluorinated copolymers with polystyrene [45]. The hydrophobicity of PFS is enabled through the fluorine atoms on the aromatic ring structures of PFS polymers, and adding 30 wt.% of PFS into PS solutions and dissolving them into THF:DMF (tetrahydrofuran:dimethylformamide) (1:1 v/v) solvent mixture provided superhydrophobic nanofiber surfaces (over 160 degrees). The droplets of water could stay on the fiber surface and instantly roll off after changing the angle or moving it slightly. The same study also reported that the superhydrophobic property was not directly related to the molecular weight of the polymers [45].

Acatay et al. reported that surfaces of nonhydrophobic materials could be modified by the fluorinating agents to make those surfaces superhydrophobic [44]. The authors stated that thermoset polymer through the reactions of acrylonitrile (AN) and dimethyl meta-isopropenylbenzyl isocyanate (TMI), and crosslinking the resultant poly(AN-co-TMI) with a perfluorinated linear diol (fluorolink-D) exhibited the highest water contact angle of 166.7 degrees and lowest water roll-off angle of 4.3 degrees. DMF in solution was the key factor in controlling the topology of the electrospun film, while the reorientation of fluorine groups for a longer time provided the lowest surface energy [44]. In other words, as long as there is enough micro- and nanoscale roughness on the surfaces of the materials, adding hydrophobizing agents will take surfaces superhydrophobic or near superhydrophobic levels.

Singh et al. reported the superhydrophobic and very high hydrophobic surface of nonwoven electrospun nanofiber mats using poly[bis(2,2,2-trifluoroethoxy) phosphazene] fluorinated polymers [46]. During the experimental studies, concentrations of polymeric solutions were changed by varying the solvent concentrations to prepare numerous nanofibers with different diameters (as low as 52 nm), which directly affected the surface morphologies. The polymer utilized in this study had an inorganic backbone and fluorinated group. The hydrophobizing polymer in the prepared solutions along the textures of the nanofiber mats had a huge impact on the surface hydrophobic properties of the nanofiber materials (water contact angles of the nanofiber surfaces were between 135 and 159 degrees) [46]. The fluorinating agent not only provided hydrophobic properties, but also offered high resistance to chemicals, and was fire and radiation stable, which will be useful for various industrial applications.

Chen et al. studied fluorinated silane with poly(vinylidene fluoride) (PVDF) and then the graft polymerization process was applied to 3-trimethoxylpropyl methylacrylate with PVDF in order to couple with the fluorinated silane for better hydrophobic properties [47]. In this case, the polymeric nanofibers had both the fluorinated and triethoxysilane groups in the structure. The fluorinated polymeric groups mainly saturated with CF_2 and CF_3 provide a highly water-repellent property, while the triethoxysilane group mostly exhibited the chemical reactivity function in the polymer structures. Electrospun PVDF nanofibers with fluorinated silane were able to form a water contact angle above 144 degrees; however, the flat membrane made from PVDF provided a water contact angle of only 140 degrees. Fig. 4.7 shows several nanofiber polymers utilized in those experimental

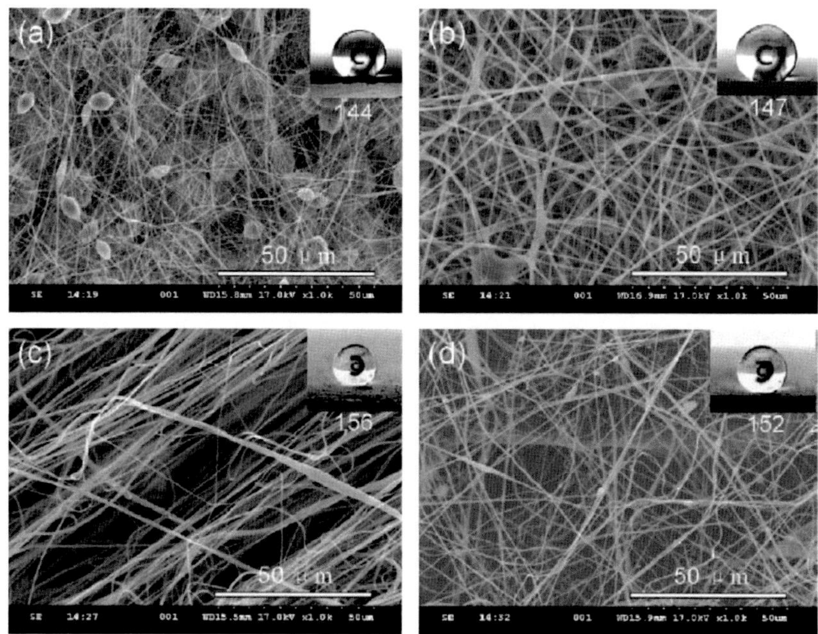

FIGURE 4.7

SEM images of membranes prepared from the following: (A) PVDF, (B) silk
fibron-SFPVDF, (C) PVDF/1H,1H,2H,2H-perfluorooctyltriethoxysilane (PFOTES),
and (D) SFPVDF/PFOTES using electrospinning method.

studies with various contact angle values [47]. The nanofibers with bead formations showed a water contact angle of 144 degrees (Fig. 4.7A), while the beadless
nanofibers showed higher water contact angle values of 147, 156, and 152 degrees,
respectively. This experimental study concluded that the crosslinked PVDF with
PFOTES lowered the overall surface energy of the nanofiber surfaces, and considerably improved the water contact angle values in superhydrophobic ranges.

4.2 ELECTROSPUN NANOFIBERS FOR METAMATERIALS AND LIGHT AND NOISE SENSITIVITY

Metamaterials are engineered materials that are made of multiple elements and
composites in the forms of metals and plastics. These materials are generally
organized in repeating patterns and decorations at known scales that are smaller
than the wavelengths of the light or sound exposed to those objects. The properties of metamaterials come from their design, structure, and precise tailoring abilities, which are mainly related to their geometry, size, shape, orientation, and

arrangement. These precisely controlled features can manipulate electromagnetic waves through absorbing, blocking, enhancing, or bending to make metamaterials or invisible materials (also known as negative-index metamaterials) [48]. There are a number of different applications of metamaterials, including medical devices, aircraft, optical filters, sensor detection, remote air vehicles, infrastructure monitoring, crowd control, high-frequency battlefield communication, smart solar power management, lenses for high-gain antennas, acoustics and seismic devices, shielding structures from earthquakes, and ultrasonic sensors. Research and development in metamaterials are primarily related to the interdisciplinary studies, such as electromagnetics, electrical engineering, solid-state physics, optics, optoelectronics, and fiberoptics, antennae engineering, microwave, material sciences, nanoscience, chemistry, and semiconductors [48].

Revathin et al. reported the barium-substituted magnesium ferrite $(Ba_{0.2}Mg_{0.8}Fe_2O_4)$–PVDF electrospun fiber-based tunable metamaterial structures for electromagnetic interference shielding in the microwave frequency region [49]. These nanocomposite materials are considered to be useful in controlling the flow of electromagnetic radiation for interference shielding purposes. Although the electrospinning method can increase the ferroelectric properties of PVDF nanofibers, adding barium magnesium ferrite into the PVDF can substantially change the magnetic property of the composite fibers and their mats. The authors stated that the magnetic and dielectric composite fibers at the microwave range can be designed for absorption and reflection by tuning these metamaterial structures during the electrospinning process for electromagnetic interference shielding purposes. The simulation studies indicated that the single negative metamaterial structure could become a double-negative metamaterial under the external magnetic field, which is useful for invisibility applications [49]. Micro- and nanoparticles, along with other inclusions, can be used to further change the absorbing, blocking, enhancing, or bending capabilities of electrospun nanofibers.

A recent study indicated that electrospun nanofibers can be utilized in aircraft noise reduction [50]. In this study, polyvinylpyrrolidone (PVP) was dissolved in ethanol while polystyrene (PS) was dissolved in dimethylformamide (DMF), and polyvinylchloride (PVC) was dissolved in dimethylacetamide (DMAc). Fig. 4.8 shows SEM images of PVP, PVC, and PS electrospun fiber. Table 4.1 gives the processing parameters of electrospinning of PVP, PS, and PVC, whereas Table 4.2 shows the weight, thickness, and sound absorption coefficient values of electrospun PVP, PVC, and PS fibers.

Fig. 4.9 shows the sound absorption of PS/PVP and PVC/PVP fibers as a function of frequencies. In this study, first all fiber samples were tested one by one for sound absorption between 200 and 6200 Hz frequency, but no sample showed promising results both at low frequency (200–500 Hz) and at high frequency (6200 Hz). Therefore, combinations of these fibers were used for the sound absorption test. In the PS/PVP sample, PS was placed first in a B&K type 4206 impedance tube for measuring sound absorption followed by the PVP fibers sample. Similarly, in the PVC/PVP sample, PVC was placed first followed by the

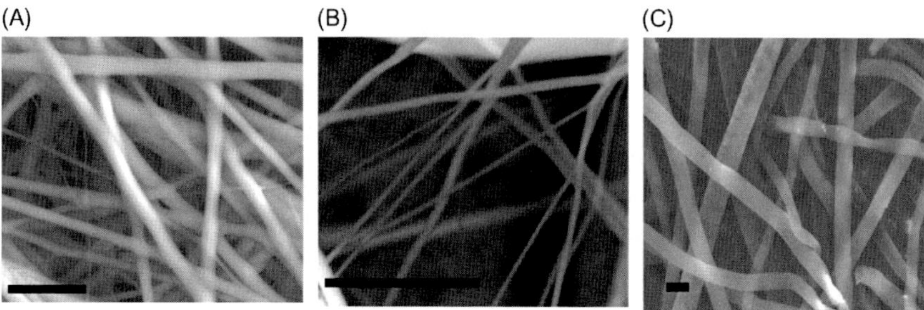

FIGURE 4.8

SEM images of (A) PVP, (B) PVC, and (C) PS electrospun fibers at various electrospinning conditions (scale bars are 10 μm).

Table 4.1 Electrospinning Process Parameters of PVC, PVP, and PS Fibers used for Sound Absorption Tests

Polymers w/ Solvent Ratio	Spinning Voltage (kV)	Screen Distance (cm)	Pump Speed (ml/h)	Fiber Diameter (nm)
PVP/ethanol (80:20)	15	20	3.0	~750
PVC/DMAc (80:20)	18	25	3.0	~250
PS/DMF (75:25)	25	25	1.5	~7000

Table 4.2 Weight, Thickness, and Sound Absorption Coefficient Values of Electrospun Fiber Samples

Sample	Sample Weight (g)	Sample Thickness (cm)	Sound Absorption Coefficient at 2000/6000 Hz
PS + PVP	1.0 + 1.5	3.29	0.97/0.96
PVP + PVC	1.5 + 0.5	2.52	0.98/0.99

PVP fiber sample. As can be seen in Table 4.2 and Fig. 4.9, both samples showed remarkable results at low frequency and at high frequency with the sound absorption coefficient reaching above 95% at 6200 Hz. These results are encouraging in view of the fact that the material generally used in sound absorption is melamine (polyimide), which is much heavier than nanofibers. The total weight of the PS/PVP sample is 2.5 g and thickness is 3.29 cm, while the total weight of the PVC/PVP sample is 2 g and thickness is 2.52 cm. The conventional material used in sound absorption is melamine (polyimide). A 1-inch (2.54 cm) melamine sample is approximately four times heavier than nanofiber samples. This clearly indicates that the high surface area of fibers and porous structures have substantial benefits

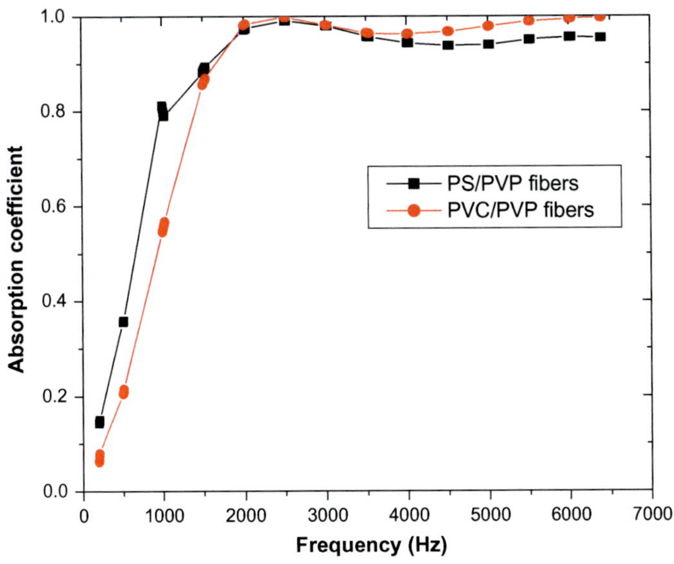

FIGURE 4.9

Sound absorption coefficient values of PVC/PVP fibers and PS/PVP fibers as a function of
frequencies between 0 and 6400 Hz.

for sound absorption and weight reduction, which is very critical for the fuel efficiency of aircraft [50]. These electrospun nanofibers can be reformulated using
the nanoscale inclusions, such as carbon nanotubes (CNTs), graphene, metallic
and ceramic nanofibers, and other nanoscale inclusions into the polymeric materials to improve the sound absorption properties.

Chen et al. reported about the full-color light-emitting electrospun nanofibers
manufactured using PFO/MEH-PPV/PMMA ternary blends of poly(9,9-dioctyl-
fluoreny-2,7-diyl) (PFO)/poly[2-methoxy-5-(2-ethylhexyloxy)-1,4-phenylenevinyl-
lene] (MEH-PPV)/poly(methyl methacrylate) (PMMA) [51]. The luminescence
electrospun (ES) nanofibers in the range of 30−35 nm were prepared from
chloroform solutions using a single capillary spinneret process. The test studies
indicated that the emission color of the ES nanofibers varied from white, blue,
greenish-yellow, yellowish-green, yellow, and orange when the MEH-PPV composition was increased in the fiber structures. The author also mentioned that the
nanofibers had higher photoluminescence efficiencies when compared to those of
the spin-coated films. It was concluded that different color light-emitting nanofibers could be fabricated by optimizing the composition of ingredients into the
nanofiber systems, which may be used for new light sources or sensory materials
for smart textiles [51].

Tang et al. investigated the highly reflective nanofiber films made by the electrospinning process and their applications on color uniformity and luminous efficacy improvement of white light-emitting diodes [52]. During the electrospinning

process, the authors fabricated poly(lactic-co-glycolic acid) (PLGA) nanofiber films with high reflectivity and scattering properties using different thickness nanofiber films and fiber diameters through changing the electrospinning time and solution concentration. The experimental studies indicated that the nanofiber mat reflectance and scattering ability were improved with the film thickness, while the fiber diameter had little effect on the reflectance and scattering properties at those scales. After optimization of the nanofiber film thickness and fiber diameters, the nanofiber films provided about 98.8% reflectance when compared to the $BaSO_4$ white plate. This study also indicated that the luminous efficiency was improved (about 12% at 350 mA), when the temperature was reduced. Thus, these novel highly reflective coatings of nanofiber films could be integrated into the lighting systems to improve their light efficacy and quality [52].

4.3 ELECTROSPUN NANOFIBERS FOR FIRE-RETARDANT FABRICS

Fire retardancy is a process of preventing fire from spreading to adjacent locations, while fire retardant is a substance that is employed to stop or slow down the fire spreading by reducing its intensity [53,54]. It is usually accomplished by reducing chemical reactions of the flames or delaying combustion. Organic substances, flammability, porosity, and oxygen level in the environment can substantially accelerate the fire and dense smoke and suspended particles. Generally, fire-retardant materials reduce the flammability through two actions: (1) physical action where fire is physically blocked and (2) chemical action where a chemical reaction is initiated to stop the fire [53]. Also, there are two simple steps to fabricating flame-retardant materials. The first is to prevent ignition by improving the heat capacity of the materials, and the second is to prevent spreading of the fire to other locations. Some ways of preventing fire spreading on the objects are as follows [53,54]:

- Formation of the char to eliminate fuel release and function of the char as a thermal insulation layer;
- Addition of the inclusions that scavenge the free radicals throughout fire, and disruption of the combustion process;
- Water contacting filler materials at elevated temperature;
- Noncombustible fillers that increase the volume and act as a thermal sink.

Recent studies have indicated a number of different ways of making electrospun nanofibers used in textile industries fire-retardant. Using electrospinning methods, different forms (e.g., polymeric, metallic, ceramic, and composite) of nanofibers can be produced and used as flame-retardant materials [54,55]. They can be nanofibrous membranes or coatings to advance the flame resistance, thermal stability, and other mechanical and electrical properties through mixing additives and fillers to the polymeric structures. The prevention of ignition and flame through noncombustible fillers and enhancing the heat capacity of the nanofibers are frequently employed approaches. Some of the most frequently used filler materials include MgO, TiO_2, ZnO, SiO_2, CaO, nanoclay, CNTs, graphene, zeolite, ammonium phosphate, melamine, pentaerythritol, and functionalized polyhedral oligomeric silsesquioxane (POSS) nanoparticles [54,55].

Lee et al. improved the thermal stability of electrospun PAN fibers using different additives (AP-PER-MEL and TiO_2) as flame-retardant materials [56]. The prepared nanocomposite fibers were subjected to flame/burning tests and the test results indicated that samples without any inclusions were completely burned with a high burning rate within 620°C. Nevertheless, the same nanofibers incorporated with nonflammable inclusions had much higher thermal stability (20 and 30%) results because of the synergy effects of inclusions/agents (AP-PER-MEL and TiO_2) into electrospun PAN fibers. It is estimated that for a pure polymeric material, the heat release capacity can be 300 J/g-K or less for the self-extinguishing of that material. According to the SEM images, it is concluded that the shape of the nanocomposite fibers were almost the same as the original samples before the flammability tests [56].

Moon et al. reported about the high flame resistance and strength of electrospun PAN nanofiber associated with CNTs and ochre (a natural clay earth pigment) as filler materials for different applications [57]. The heat release capacity of the PAN nanofibers was about 253 J/g-K, and in the presence of CNTs or ochre, the heat release capacity was reduced because of the nonflammable feature of the inclusions. The PAN nanofibers incorporated with 0.1% CNTs and 1% ochre and stabilized at 240°C reduced the heat release capacity to 24 J/g-K. The thermal stabilization process with the inclusions in nanofibers significantly increased the char formation and reduced the fire spreading during the burning tests. In addition to CNTs and ochre, boric acid, a strong flame-retardant inclusions in polymeric materials, were also added into the polyamide as nanocomposite prior to the electrospinning process [57].

Selvakumar et al. investigated the flame-retardant fabric systems of the polyamide and boric acid nanocomposite fibers produced through the electrospinning method [58]. In order to examine the efficiency of the electrospun polyamide/boric acid nanofibers, the fibers were placed on the cotton fabric. The fire-retardant properties of coated fabric systems with different nanoparticle inclusions were assessed for the ignition time and char formation. In the presence of 0.05 wt.% boric acid in nanofibers, a flame-spreading time of greater than 80% was achieved, while char formation of more than 75% was obtained, which indicates excellent fire protection capabilities of fabrics [58].

Wu et al. reported the flame retardancy of polyamide 6 (nylon 6), nanoclay, and intumescent nanocomposite fibers incorporated with montmorillonite clay platelets and intumescent as nonhalogenated flame-retardant additives [59]. Note that intumescent is a special ingredient that swells under the exposure of heat, while increasing its volume and decreasing its density. The authors specified that mixing and dispersion methods, as well as loading rate and exfoliation of nano-clay platelets in polymer, significantly affected the electrospinnability and thermal stability of the nanocomposite fibers. TGA studies indicated that the degradation temperatures and time of electrospun nanofibers were less than the pure nylon 6 samples, while the difference in residual char weight was significantly higher after the decomposition. The microscale calorimetry studies confirmed that the inclusions into the nanocomposite fibers played a major role in reducing the flammability of the nanofibers [59].

Some of the additives can scavenge free radicals in the polymers during burning and can disrupt the combustion process and lower the ignition temperature and time. Cai et al. studied the effects of ferric chloride ($FeCl_3$) on structure, surface morphology, and combustion behaviors of electrospun PAN composite nanofibers [60]. The purpose of the studies was to improve the fire resistance of electrospun PAN in the presence and absence of $FeCl_3$. The SEM analysis indicated that the diameters of electrospun nanocomposite nanofibers were reduced with the addition of $FeCl_3$ into the polymeric solution. This may be attributed to the electrical conductivity improvement of the polymer solutions when compared to the viscosity and surface tension changes. Atomic force microscopy (AFM) studies revealed that the surface morphology of the nanofibers changed from a smoother (without $FeCl_3$) to a rougher structure (with $FeCl_3$). The test results also indicated that loading of $FeCl_3$ into polymeric solution prior to the electrospinning process reduced the heat release rate and enhanced the combustion property of the composite nanofibers [60].

Highly porous and flame-retardant electrospun nanofibers as thermal insulation materials can delay the ignition time of the underlying materials. It was stated that a polymeric membrane of polyethylene oxide could be better thermal insulation materials when compared to the cast film of the same material with the same thickness [61]. Adding fire-resistant additives into the polymeric solution prior to the electrospinning process had a major impact on the fire-resistant capacity of the fibers and underlying materials. Recent studies indicated that electrospun cellulose fibers incorporated with magnesium hydroxide nanoparticles as blend or coating on the surface could significantly change the fire resistance and ignition times [62].

In addition to the fire-retardant applications of electrospun nanofibers on cloths, fabrics, and other underlying materials, nanofibers can also be used in rechargeable batteries and their separations, which requires a lot of fire-retardant properties. Although it rarely happens, battery-separating membranes sometimes cause fire due to failure at elevated temperatures and leak combustible electrolytes into the battery, where some exothermic reactions can take place, when highly reactive electrolytes interact with each other. Liu et al. investigated the electrospun core−shell microfiber separator with thermal-triggered flame-retardant properties for lithium-ion batteries using triphenyl phosphate (TPP) [63]. TPP is a highly popular organophosphorus-based flame-retardant material and is used as a core structure against heat formation. The outer shell of the electrospun nanofibers was made of poly(vinylidene fluoride-hexafluoropropylene)

(PVDF-HFP) because of its chemical inertness, resistance to electrolytes, and fairly low melting point (160°C) [63].

4.4 ELECTROSPUN NANOFIBERS FOR PROTECTIVE CLOTHING

Protective clothing is becoming more important for some specific work environments, including hospitals, battlefields, police stations, fire departments, and other occupational environments in which the risk of exposure to fires, chemicals, pathogens, and nanoparticles exists. Nanofibers with high flexibility, surface area, functionalization, light-weight, porosity, and pore size make them ideal candidates for protective clothing. It is easy to prevent the inhalation of micron-sized particles above 2.5 µm; nevertheless, nanoparticles between 1 and 100 nm are potentially dangerous as they can get directly absorbed through the skin and digestive systems. Electrospun nanofibers have been shown to effectively filter micro- and nanofibers into different contaminated water sources, such as water jet cutting, suspended nanoparticle dispersion, and lake water [25].

Faccini et al. developed a new set of protective clothing against nanoparticles based on electrospun nanofibers using electrospun polyamide 6 (PA6) [64]. The authors mentioned that nanoparticles of 20 and 80 nm were blocked by the membranes with up to 80% and 99% efficiencies, respectively. Prior to the nanoparticle penetration tests, electrospun nanofiber membranes were applied to the surfaces of viscose nonwoven fabrics through a hot-melt lamination process using a thermoplastic adhesive powder [64].

Gorji et al. reported electrospun nanofibers in protective clothing for different sectors [65]. In this study, the properties of electrospun nanofibers were investigated in detail, and their structures and potentials were analyzed for fire protection. It was stated that electrospun membranes with high elasticity, breathability, exceptionally lightweight structures, multifunctional properties, and filtration capabilities are major parameters for use in protective clothing. The major research efforts in this field include new materials development, characterization, and testing protective clothing [65].

Global transportation by airplanes and ships has been increasing the rate of spread of some life-threatening diseases in many countries. Finely textured fabrics associated with antibacterial agents can protect the wearers against those diseases and stop the spread of bacteria, fungi, viruses, and other pathogens (e.g., malaria, dengue fever, and Zika virus). Nanofibers are widely used materials to produce nanoscale fabrics; hence, fabrication of nanofibers with antibacterial substances will have many potential applications in the near future. The antibacterial substances can be dissolved into a polymeric solution (about 2 wt.%) before electrospinning, or sprayed onto the surface of electrospun nanofibers to impart antimicrobial properties to the nanofiber mats. The mechanical properties of the nanofibers must be sufficient not to break down/tear when wearing antibacterial mats [66]. A similar study was also performed by other investigators using Lanasol (brominated cyclic compound) as the antibacterial agent in poly(methyl methacrylate) (PMMA) and polyethylene oxide (PEO) nanofibers. The test results indicated that adding 4 wt.% in the fibers reduced the bacterial viability (*Staphylococcus aureus* subsp. *aureus*) by up to 99.99% [67].

Dhineshbabu et al. reported about the electrospun magnesia/nylon 6 hybrid nanofibers for protective clothing [68]. Magnesia (MgO) nanoparticles were produced via the ball mill grinding process and characterized through X-ray diffraction analysis, Brunauer–Emmett–Teller, transmission electron microscopy (TEM) methods prior to the electrospinning process. The produced MgO nanoparticles were dispersed in nylon 6 using a solvent, and electrospun to produce nanofiber film mats. Internal structures, dispersion of inclusions, and surface morphologies of nanocomposite mats were investigated through SEM and high-resolution TEM. The experimental studies indicated that the antibacterial properties, using *Staphylococcus aureus* and *Escherichia coli* and fire retardancy tests, of the produced nanofiber fabrics were relatively better when compared to those nanofibers made of nylon 6 nanofibers alone [68].

Agarwal et al. reported about the detoxification performance of electrospun fibers embedded with zeolite particles against a nerve agent stimulant (paraoxon). Zeolite particles in dispersion form were evenly sprayed onto the surface of electrospun fibers made of cellulose/polyethylene terephthalate (PET) blend [69]. Agarwal et al. also investigated the one-step synthesis of hollow strontium titanate and barium titanate structures for detoxification of nerve agents using a coaxial electrospinning process. A sintering process was applied to convert the nanofibers into hollow structures. A comparison study indicated that hollow structured nanofibers have better detoxification rate of nerve agent when compared to solid nanofibers, which may be attributed to the higher surface area of the hollow structured nanofibers [70].

Teli et al. recently developed the multifunctional nonwoven fabrics by an electrospinning process for medical protection [71]. In this study, a PVA and starch solution was prepared with the addition of citronella essential oil for emulsion formation in a small quantity prior to the electrospinning process. The prepared nonwoven fabrics were proposed to use as face masks to protect bacterial and viral entry into the nose and mouth during medical treatments. The test studies indicated that electrospun nanofibers of PVA, starch, and essential citronella oil provided antibacterial and mosquito-repellent properties. In addition to the face masks, these nanofibers could also be used for nose masks, napkins, head masks, wound covers, and so on [71].

4.5 CONCLUSIONS

In this chapter, nanofibers, their properties, and industrial applications in textile and medical fields were investigated in detail. In the first part of the chapter, superhydrophobic electrospun nanofibers for nonwettable surfaces were investigated, and other related subjects, such as jet formation in electrospinning, electrospinning of nanofibers, and the theory of superhydrophobicity and super-hydrophobic nanofibers were studied. In the second part of the chapter, electrospun nanofibers used for metamaterials, as well as light and noise sensitivity properties of those fibers were explored. In the final part of the chapter, electrospun nanofibers for fire-retardant fabrics and protective clothing were analyzed and recent developments in the field were reported. These studies indicated that electrospun nanofibers have a huge potential in the textile industry and research and development in these fields have been growing substantially.

REFERENCES

[1] S. Cavaliere, Electrospinning for Advanced Energy and Environmental Applications, CRC Press, Boca Raton, FL, 2015.

[2] W.S. Khan, R. Asmatulu, "Magnetic Properties of Electrospun PVP and PAN Nanocomposite Fibers Associated with NiZn-Ferrite Nanoparticles," SAMPE Fall Technical Conference, Fort Worth, TX, October 17–20, 2011, 9 pages.

[3] S. Ramakrishna, K. Fujihara, W.E. Teo, T.C. Lim, Z. Ma, An Introduction to Electrospinning and Nanofibers, World Scientific, London, UK, 2005.

[4] Y. Gogotsi, Nanomaterials Handbook, CRC Press, Boston, 2006.

[5] R. Andrews, D. Jacques, A.M. Rao, T. Rantell, F.Y. Chen, J. Chen, et al., Nanotube composite carbon fibers, Appl. Phys. Lett. 75 (1999) 1329–1331.

[6] X.L. Xie, Y.W. Mai, X.P. Zhou, Dispersion and alignment of carbon nanotubes in polymer matrix: a review, Mater. Sci. Eng. R Reports 49 (2005) 89–112.

[7] A. Theron, E. Zussman, A.L. Yarin, Electrostatic field-assisted alignment of electrospun nanofibers, Nanotechnology 12 (2001) 384–390.

[8] L. Yeo, J. Friend, Electrospinning carbon nanotube polymer composite nanofibers, J. Exp. Nanosci. 1 (2006) 177–209.

[9] E. Lenk, "Melt spinning apparatus," US Patent: 4704077, November 3, 1987.

[10] K. Cantor, Blown Film Extrusion, An Introduction - Book Review, Hanser, 2006.

[11] G. Taylor, Electrically driven jets, Proc. Roy. Soc. Lond. A 313 (1969) 453–475.

[12] D. Reneker, A. Yarin, H. Fong, S. Koombhongse, Bending instability of electrically charged; liquid jets of polymer solutions in electrospinning, J. Appl. Phys. 87 (2000) 4531–4557.

[13] A. Yarin, S. Koombhongse, D. Reneker, Bending instability in electrospinning of nanofibers, J. Appl. Phys. 89 (2001) 3018–3026.

[14] W.S. Khan, R. Asmatulu, M. Ceylan, A. Jabbarnia, Recent progress on conventional and non-conventional electrospinning processes, Fibers Polym. 14 (2013) 1235–1247.

[15] L. Saeednia, L. Yao, M. Berrendt, K. Cluff, R. Asmatulu, Structural and biological properties of thermosensitive chitosan-graphene hybrid hydrogels for sustained drug delivery applications, J. Biomed. Mater. Res. Part A 105 (2017) 2381–2390.

[16] N. Nuraje, W.S. Khan, M. Ceylan, Y. Lie, R. Asmatulu, Superhydrophobic electrospun nanofibers, J. Mater. Chem. A 1 (2013) 1929–1946.

[17] R. Asmatulu, M. Ceylan, N. Nuraje, Study of superhydrophobic electrospun nano-composite fibers for energy systems, Langmuir 27 (2) (2011) 504–507.

[18] M. Ceylan, R. Asmatulu, "Enhancing the Superhydrophobic Behavior of Electrospun Fibers via Graphene Addition and Heat Treatment," ASME International Mechanical Engineering Congress and Exposition, Vancouver, Canada, November 12–18, 2010, 7 pages.

[19] I. Alarifi, Fabrication and Characterization of Electrospun Carbonized Polyacrylonitrile Fibers as Strain Gauges in Composite Aircraft for Structural Health Monitoring Applications, Ph.D. Dissertation, Wichita State University, April 4, 2017.

[20] A. Alharbi, I.M. Alarifi, W.S. Khan, R. Asmatulu, Highly Hydrophilic Electrospun Polyacrylonitrile / Polyvinypyrrolidone Nanofibers Incorporated with Gentamicin as Filter Medium for Dam Water and Wastewater Treatment, Journal of Membrane and Separation Technology 5 (2016) 38–56.

[21] I.M. Alarifi, W.S. Khan, A.K.M.S. Rahman, Y. Kostogorova-Beller, R. Asmatulu, Synthesis, analysis and simulation of carbonized electrospun nanofibers infused carbon prepreg composites for improved mechanical and thermal properties, Fibers Polym. 17 (2016) 1449–1455.

[22] I.M. Alarifi, A. Alharbi, W.S. Khan, A.K.M.S. Rahman, R. Asmatulu, Mechanical and thermal properties of carbonized PAN nanofibers cohesively attached to surface of carbon fiber reinforced composites, Macromol. Symp. 365 (2016) 140–150.

[23] I.M. Alarifi, A. Alharbi, O. Alsaiari, R. Asmatulu, Training the engineering students on nanofiber-based SHM systems, Trans. Tech. STEM Educ. 1 (2016) 59–67.

[24] I.M. Alarifi, A. Alharbi, W.S. Khan, A. Swindle, R. Asmatulu, Thermal, electrical and surface properties of electrospun polyacrylonitrile nanofibers for structural health monitoring, Materials 8 (2015) 7017–7031.

[25] R. Asmatulu, H. Muppalla, Z. Veisi, W.S. Khan, A. Asaduzzaman, N. Nuraje, Study of hydrophilic electrospun nanofiber membranes for filtration of micro and nanosize suspended particles, Membranes 3 (2013) 375–388.

[26] A. Jabbarnia, W.S. Khan, A. Ghazinezami, R. Asmatulu, Investigating the thermal, mechanical and electrochemical properties of PVdF/PVP nanofibrous membranes for supercapacitor applications, J. Appl. Polym. Sci. (2016). Available from: https://doi.org/10.1002/app.43707.

[27] A. Jabbarnia, W.S. Khan, A. Ghazinezami, R. Asmatulu, Tuning the ionic and dielectric properties of electrospun PVdF/PVP nanofibers with carbon black nanoparticles for supercapacitor applications, Int. J. Eng. Res. Appl. 6 (2016) 65–73.

[28] A. Jabbarnia, R. Asmatulu, Synthesis and characterization of PVdF/PVP-based electrospun membranes as separators for supercapacitor applications, J. Mater. Sci. Technol. Res. 2 (2015) 43–51.

[29] T. Darmanini, F. Guittard, Superhydrophobic and superoleophobic properties in nature, Mater. Today 18 (2015) 13.

[30] R.J. Crawford, E.P. Ivanova, Superhydrophobic Surfaces, Elsevier, Amsterdam, 2015.

[31] K.Y. Law, H. Zhao, Surface wetting: characterization, Contact Angle, and Fundamentals, Springer, London, UK, 2016.

[32] J. Zimmermann, F.A. Reifler, G. Fortunato, L.C. Gerhardt, S. Seeger, A simple, one-step approach to durable and robust superhydrophobic textiles, Adv. Funct. Mater. 18 (2016) 3662–3669.

[33] S. Jurak, E. Jurak, M.M. Rahman, R. Asmatulu "Nanoscale Superhydrophobic Coating for Corrosion Mitigations," in Nanocomposites, Elsevier, Editor M. Ram (in press).

[34] S. Pelfrey, T. Cantu, M.R. Papantonakis, D.L. Simonson, R.A. McGill, J. Macossay, Microscopic and spectroscopic studies of thermally enhanced electrospun PMMA micro- and nanofibers, Polym. Chem. 1 (2010) 866−869.

[35] P.C. Hiemenz, T.P. Lodge, Polymer Chemistry, Second ed., CRC Press, Boca Raton, FL, 2007.

[36] G. Cao, Y. Wang, Nanostructures and Nanomaterials: Synthesis, Properties, and Applications, Second ed., World Scientific, Boston, 2011.

[37] B. Rogers, J. Adams, S. Pennathur, Nanotechnology Understanding Small Systems, Second ed., CRC Press, New Yok, 2011.

[38] S. Mohammad, M.N. Uddin, G. Hwang, R. Asmatulu, Superhydrophobic PAN nanofibers for gas diffusion layers of proton exchange membrane fuel cells for cathodic water management, Int. J. Hydrog. Energy (2017). Available from: https://doi.org/10.1016/j.ijhydene.2017.07.229).

[39] J. Nookala, R. Asmatulu, "Effects of Tilt Angles on Self-cleaning of Superhydrophobic Composite Surfaces," CAMX Conference, Anaheim, CA, September 26−29, 2016, 10 pages.

[40] S. Kasaragadda, R. Asmatulu, "Effects of Surface Hydrophobicity on Moisture Ingression of Fiber Reinforced Laminate Composites," CAMX Conference, Dallas, TX October 27−29, 2015, 9 pages.

[41] I.M. Alarifi, A. Alharbi, W.S. Khan, R. Asmatulu, Carbonized electrospun PAN nanofibers as highly sensitive sensors in SHM of composite structures, J. Appl. Polym. Sci. (2015). Available from: https://doi.org/10.1002/app.43235.

[42] J. Zheng, A. He, J. Li, J. Xu, C.C. Han, Studies on the controlled morphology and wettability of polystyrene surfaces by electrospinning or electrospraying, Polymer 47 (2006) 7095−7102.

[43] M. Kang, R. Jung, H.-S. Kim, H.-J. Jin, Preparation of superhydrophobic polystyrene membranes by electrospinning, Colloids Surf. A Physicochem. Eng. Aspects 314 (2008) 411−414.

[44] K. Acatay, E. Simsek, C. Ow-Yang, Y.Z. Menceloglu, Tunable, superhydrophobically stable polymeric surfaces by electrospinning, Angew. Chem. Int. Ed. 43 (2004) 5210−5213.

[45] S. Agarwal, S. Horst, M. Bognitzki, Electrospinning of fluorinated polymers: formation of superhydrophobic surfaces, Macromol. Mater. Eng. 291 (2006) 592−601.

[46] A. Singh, L. Steely, H.R. Allcock, Poly[bis(2,2,2-trifluoroethoxy) phosphazene] superhydrophobic nanofibers, Langmuir 21 (2005) 11604−11607.

[47] Y. Chen, H. Kim, Preparation of superhydrophobic membranes by electrospinning of fluorinated silane functionalized poly (vinylidene fluoride), Appl. Surf. Sci. 255 (2009) 7073−7077.

[48] F. Capolino, Theory and Phenomena of Metamaterials − Metamaterials Handbook, CRC Press, New York, 2017.

[49] V. Revathin, S.D. Kumar, V. Subramanian, M. Chellamuthu, BMFO-PVDF electrospun fiber based tunable metamaterial structures for electromagnetic interference shielding in microwave frequency region, Eur. Phys. J. Appl. Phys. 72 (2005) 7.

[50] W.S. Khan, R. Asmatulu, M.B. Yildirim, Acoustical properties of electrospun fibers for aircraft interior noise reduction, J. Aerospace Eng. 25 (3) (2012) 376−382.

[51] H.C. Chen, C.T. Wang, C.L. Liu, W.C. Chen, Full color light-emitting electrospun nanofibers prepared from PFO/MEH-PPV/PMMA ternary blends, J. Polym. Sci. Part B Polym. Phys. 47 (2017) 463−470.

[52] Y. Tang, G. Liang, J. Chen, S. Yu, Z. Li, L. Rao, et al., Highly reflective nanofiber films based on electrospinning and their application on color uniformity and luminous efficacy improvement of white light-emitting diodes, Optics Express 25 (2017) 20598−20611.

[53] A. Ghazinezami, W.S. Khan, A. Jabbarnia, R. Asmatulu, Impacts of nanoscale inclusions on fire retardancy, thermal stability, and mechanical properties of polymeric PVC nanocomposites, J. Thermal Eng. 3 (2017) 1308−1318.

[54] D.Y. Wang, Novel Fire Retardant Polymers and Composite Materials, Woodhead Publishing, London, UK, 2017.

[55] W.E. Teo, S. Ramakrishna, A review on electrospinning design and nanofiber assemblies, Nanotechnology 17 (2006) 89−106.

[56] Y.S. Lee, M. Kim, J. Sun, S. Jin, Improvement of thermal stability of electrospun PAN fibers by various additives, Carbon Lett. 9 (2008) 200−202.

[57] S.C. Moon, T. Emrick, High flame resistant and strong electrospun polyacrylonitrile-carbon nanotubes-ochre nanofibers, Polymer 54 (2013) 1813−1819.

[58] N. Selvakumar, A. Azagurajan, T.S. Natarajan, M.M.A. Khadir, Flame-retardant fabric systems based on electrospun polyamide/boric acid nanocomposite fibers, J. Appl. Polym. Sci. 126 (2012) 614−619.

[59] H. Wu, M. Krifa, J.H. Koo, Flame retardant polyamide 6/nanoclay/intumescent nanocomposite fibers through electrospinning, Textile Res. J. 84 (2014) 1106−1118.

[60] Y. Cai, D. Gao, Q. Wei, H. Gu, S. Zhou, F. Huang, et al., Effects of ferric chloride on structure, surface morphology and combustion property of electrospun polyacrylonitrile composite nanofibers, Fibers Polym. 12 (2011) 141−150.

[61] H. Jiang, Y. Jiao, M.R. Aluru, L. Dong, Electrospun nanofibrous membranes for temperature regulation of microfluidic seed growth chips, J. Nanosci. Nanotechnol 12 (2012) 6333−6339.

[62] Y.Y. Zheng, J.J. Miao, N. Maeda, D. Frey, R.J. Linhardt, T.J. Simmons, Uniform nanoparticle coating of cellulose fibers during wet electrospinning, J. Mater. Chem. 2 (2014) 15029−15034.

[63] K. Liu, W. Liu, Y. Qiu, B. Kong, Y. Sun, Z. Chen, et al., Electrospun core-shell microfiber separator with thermal-triggered flame-retardant properties for lithium-ion batteries, Sci. Adv. 3 (2017) e1601978.

[64] M. Faccini, C. Vaquero, D. Amantia, Development of protective clothing against nanoparticle based on electrospun nanofibers, J. Nanomater. 2012 (2012) 892−894.

[65] M. Gorji, R. Baghersadeh, H. Fashandi, Electrospun nanofibers in protective clothing, Electrospun Nanofibers, A volume in Woodhead Publishing Series in Textiles, 2017, pp. 571−598.

[66] P. Kampeerapappun, Preparation characterization and antimicrobial activity of electrospun nanofibers from cotton waste fibers, Chiang Mai J. Sci. 39 (2012) 712−722.

[67] R.L. Andersson, A. Martinez-Abad, J.M. Lagaron, U.M. Gedde, P.E. Mallon, R.T. Olsson, et al., Antibacterial properties of tough and strong electrospun PMMA/PEO fiber mats filled with lanasol-A naturally occurring brominated substance, Int. J. Mol. Sci. 15 (2014) 15912−15923.

[68] N.R. Dhineshbabu, G. Karunakaran, R. Suriyaprabha, P. Manivasakan, V. Rajendran, Electrospun MgO/Nylon 6 hybrid nanofibers for protective clothing, Nano-Micro Lett. 6 (2014) 46−54.

[69] S.R. Agarwal, S. Sundarrajan, R. Ramakrishna, Functionalized cellulose: PET polymer fibers with zeolites for detoxification against nerve agents, J. Inorganic Mater. 27 (2012) 332–336.

[70] S.R. Agarwal, S. Sundarrajan, A. Venkatesan, S. Ramakrishna, One-step synthesis of hollow titanate (Sr/Ba) ceramic fibers for detoxification of nerve agents, J. Nanotechnol. 2012 (2012) 7.

[71] M.D. Teli, P. Mathur, P. Chavan, Development of multifunctional non-woven fabrics by electro spinning for medical protection, Int. Res. J. Eng. Technol. 4 (2017) 63–68.

Electrospun nanofibers for agriculture and food industries

5

CHAPTER OUTLINE

5.1 NANOFIBERS FOR AGRICULTURAL APPLICATIONS

5.1.1 ELECTROSPINNING OF NANOFIBERS

Electrospinning is a process of producing micro- and nanoscale fibers in organic and inorganic forms for various industrial applications. Polymeric solutions are usually prepared using appropriate polymers [e.g., polyacrylonitrile (PAN), polyvinyl chloride, polyethylene glycol, plycaprolactone (PCL), poly(vinyl alcohol), polyvinylidene difluoride (PVDF), poly(methyl methacrylate), poly(lactic-co-glycolic acid) (PLGA), carboxymethyl cellulose (CMC), polyvinylpyrrolidone (PVP), chitosan, collagen, gelatin, poly(ethylene) oxide (PEO), polysaccharides, enzymes, proteins, polyesters, etc.], and solvents (dimethylformamide, dimethylacetamide, ethanol, acetone, toluene, tetrahydrofuran, acetonitrile, chloroform, dimethylsulfoxide, etc.) prior to the electrospinning process. In the electrospinning process, a DC voltage usually between 10 and 25 kV DC is used to overcome the surface tension of the prepared solutions and capillary forces. In addition to those polymers, some other polymeric materials, such as sol-gel, hydrogel, and silk can also be used in the electrospinning process to produce organic and inorganic micro- and nanofibers [1−7].

Synthesis and Applications of Electrospun Nanofibers. DOI: https://doi.org/10.1016/B978-0-12-813914-1.00005-5

When the applied electric voltage is amplified above the critical voltage limit, also known as the threshold intensity, the semicircular profile of the polymeric solution at the capillary tip starts elongating in a form called a Taylor cone [1]. The polymeric jet travels in a straight path, known as the jet length, where instability of the fiber takes place. During this time, fiber stretching occurs to form micro- and nanoscale fibers based on the selected system and process parameters. Throughout the process, a large portion of the solvent in the polymeric solution evaporates to form solid fibers. The morphology and diameter of electrospun fibers rely on various parameters, such as applied voltage, concentration, viscosity, pump speed, and syringe tip to collector screen distance [8−14]. Fig. 5.1 shows a schematic view of the electrospinning method for nanofiber fabrication, and the electrospun PAN nanofiber produced at 25 kV DC voltage, 2 ml/h pump speed, and 25 cm tip-to-collector distance [1−5].

After the fabrication of the nanofibers, a number of different characterization techniques are applied to determine their physical, chemical, physicochemical, and biological properties. Some of the major characterization techniques include scanning electrospun microscopy (SEM), transition electron microscopy (TEM), X-ray diffraction (XRD), differential scanning calorimeter (DSC), water contact angle (goniometer), UV-Vis spectroscopy, Raman spectroscopy, Fourier-transform infrared spectroscopy (FTIR), nuclear magnetic resonance (NMR), and gas chromatography. From these tests, the degree of crystallinity, chemical bonds, interfacial interactions, elemental distributions, sizes, and shapes of fibers and cavities, and contaminations will be determined.

In order to change the physical, chemical, and other properties of the electrospun nanofibers, some micro- and nanosize inclusions can be added into the

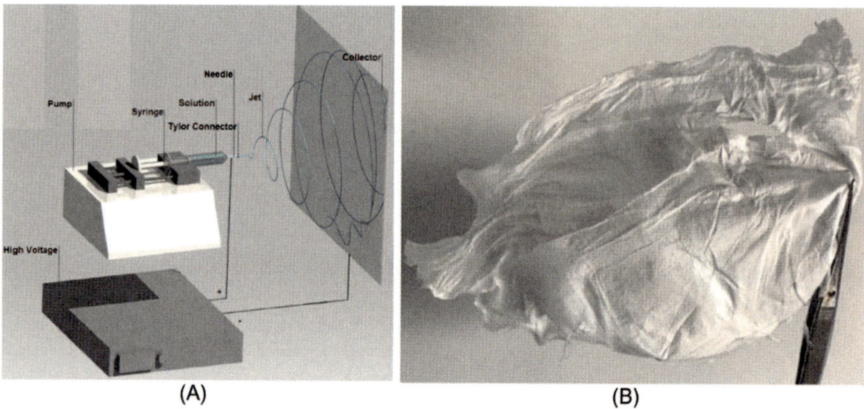

FIGURE 5.1

(A) Schematic view of the electrospinning method for nanofiber fabrications, and (B) electrospun PVP nanofiber produced at 25 kV DC voltage, 2 ml/h pump speed, and 25 cm screen tip-to-collector distance.

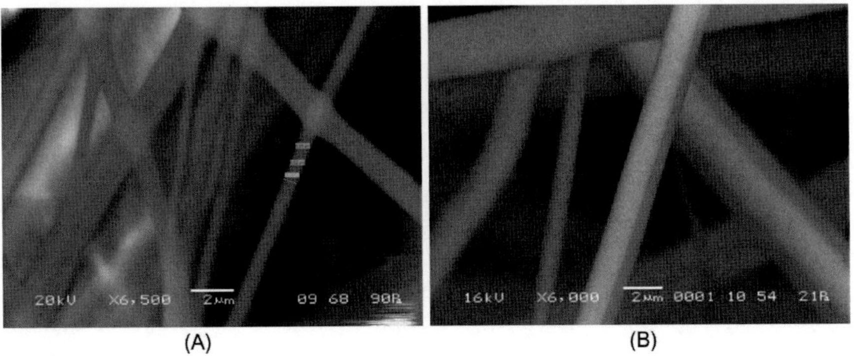

(A)   (B)

FIGURE 5.2

SEM images showing (A) PVP only and (B) 2 wt.% MWCNT-reinforced nanocomposite fibers.

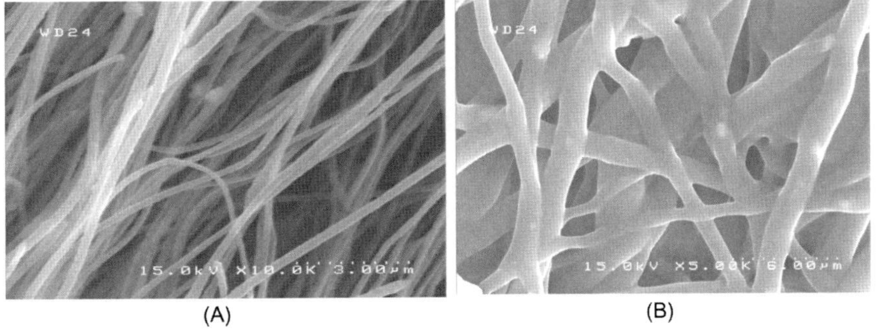

(A)   (B)

FIGURE 5.3

SEM images showing (A) PAN and (B) PVP fibers with 4 wt.% $NiZnFe_2O_4$ magnetic nanoparticles.

polymeric solutions before the electrospinning process [15,16]. Multiwalled carbon nanotubes (MWCNTs) and NiZn ferrite nanoparticles in the range of 0.5, 1, and 2% were dispersed in PVP and PAN using ethanol and DMF, and then the dispersions were electrospun at a constant 3 ml/h mass flow rate, 15 cm distance, and 18 kV applied DC voltage. Fig. 5.2 shows SEM images of the PVP only and 2% nanocomposite fibers and their morphologies. As can be seen, the average fiber diameters of 0% and 2 wt.% MWCNTs added nanocomposite fibers are 500–700 nm and 1.2–1.6 μm, respectively [15]. An increase in the diameters as a function of nanoscale inclusions in polymeric solutions is primarily because of the increase in the solution's viscosity, which may limit the starching of fibers during traveling between the tip and collector screen. Fig. 5.3 shows SEM images of PAN and PVP fibers with 4 wt.% nickel zinc ferrite nanoparticles. As can be seen, the average diameters of PAN and PVP fibers are around 250 and 750 nm, respectively.

5.1.2 AGRICULTURAL APPLICATIONS OF NANOFIBERS

Because of the tailoring properties, high porosity and surface areas, ease of active ingredient additions, and flexibility of electrospun nanofibers, many researchers and scientists have been investigating the major applications of nanofibers in food industry and agricultural applications. The other advantages of using electrospun nanofibers over the other fiber-spinning applications are that electrospun nanofiber spinnerets can be pointed out in any directions with extended distance to cover a greater surface area. This will have a lot of flexibility for food and plant protection when the electrospinning process is cost-effective. The active ingredients into the electrospun nanofibers, including pheromones, fungicides, herbicides, insecticides, pesticides, and hormones can be adjusted based on the need [17,18].

5.1.2.1 Pest control and seed germination

This is a method of controlling insect pests by using pheromones to lead the pests to traps for plant protection and high agricultural yields. Electrospun nanofibers with extremely high surface areas and flexibility can successfully release the active ingredients for plant protection against *Lobesia botrana* (grape vine moth) [19]. During the electrospinning process, a complex emulsion mixture of 1 wt.% oligolactide/pheromone/Brij S20 dispersion/16 wt.% polyhexyleneadipate-block-methoxypolyethyleneglycol dispersion/polyethylene oxide was used to produce these nanofibers.

Bisotto-de-Oliveira et al. investigated the electrospun nanofiber as a delivery system for the synthetic attactant trimedlure incorporated with *Ceratitis capitata* Wied (Diptera, Tethritidae) for controlling medfly population [20]. Prior to the electrospinning process, polyethylene glycol (PEG-PCL), each of the polymeric materials, such as ethylcelluose, polycaprolactone, and polyvinyl acetate-vinyl pyrrolidone were mixed with the trimedlure in appropriate organic solvents. Approximately, 10 wt.% of active ingredient was dissolved in the mixture, and tested in the farm field to observe the effectiveness of the produced membrane. The test results indicated that the treated membrane was highly effective in attracting the flies [20].

It has been reported that a number of different polymers could be electrospun after selecting appropriate polymers/blends and solvents for encapsulation of pheromones to determine the best possible combinations for the attraction of targeted pests. Various parameters, including concentrations of active agents, as well as polymers and solvents, active ingredient release rates and pump speed, applied voltage, and tip-to-collector distance could be investigated to produce suitable and functional nanofibers. Bisotto-de-Oliveira et al. stated that the electrospun nanofibers incorporated with synthetic sex pheromone of *Grapholita molesta* (Lepidoptera, Tortricidae) could be used for agricultural purposes [20].

Recent studies have shown that electrospun nanofibers can be used for the fungicide treatment of seeds. Castaneda et al. studied an innovative rice seed coating (*Oryza sativa*) with electrospun polymers, nanofibers, and microparticles [21]. In this study, ethyl cellulose nanofibers incorporated with commercially available fungicide (e.g., Vitavax Thiram SC-200 and Carbex 500) were sprayed over rice seeds for improved seed germination purposes. The test results indicated that the germination rate was improved to 95% in the presence of an electrospun nanofiber coating, while the germination rate of the same seeds without any nanofibers (rice seeds without any coating) was only 87%. The authors stated that in addition to fungicide in the nanofibers, other chemical ingredients could be added in the same nanofibers for multiple purposes [21].

5.1.2.2 Fertilizer application

Fertilizing is a new application of electrospun nanofibers and has been tested recently. Some of the fertilizers were dissolved in appropriate solvents and added into the electrospinning solution to produce nanofibers with the fertilizers. The new studies indicated that based on the polymer types and structures of the nanofibers, the fertilizer release rates might be tailored for plant growth [22,23]. It was stated that the electrospun nanofiber network was more robust and less likely to be washed away when compared with the loose fine particles. An electrospinning process can also lower the amount of losses of fertilizers during the farming seasons, and prevent fertilizer pollution in surface and groundwater sources by minimizing fertilizer releases. When the dry seasons come, nanofibers loaded with layers of fertilizers can be electrospun on top of the soil to hold it with the seed until the rain or irrigation water arrives to the region. Krishnamoorthy et al. reported that seeds coated with electrospun nanofiber and various concentrations of fertilizers improved the germination level of seeds [24].

5.1.2.3 Pollution and contamination controls

Even though some of the active ingredients used in agricultural applications, including fertilizers, hormones, herbicides, pesticides, and fungicides can improve the yield of crops, they can also be major sources of air, soil, and water pollution, and can be detrimental to the farming fields and crops itself. Fungicides, pesticides, and their residues are primary concerns for consumers in the market, while excess use of fertilizers can be washed into lakes, creeks, and rivers to disrupt the ecosystem and other living microorganisms, insects, animals, and plants. Herbicides are often used to decrease the growth rates of weeds in crop lands, but they can act against seed germination or seedlings if the water source is contaminated with them.

Palvannan et al. reported the efficient transformation of phenyl urea herbicide chloroxuron immobilized on polyurethane nanofibers for the removal of herbicide contaminants [25]. In the first step, zein/polyurethane nanofibers were produced using an electrospinning process, and then laccase (copper-containing oxidase enzymes of fungi, plants, and microorganisms) was accumulated on the surfaces of those electrospun nanofibers using glutaraldehyde to crosslink laccase in the solution. The retention tests conducted on the nanofiber surface confirmed that the degradation of chloroxuron could go up to 25 reuse cycles [25].

Other major sources of water contamination are related to affluent, mainly coming from animal farms (e.g., cow, buffalo, pig, goat, sheep, chicken, turkey, etc.) [26]. Thus, some of the contaminations can be naturally occurring from wild animals, which can be easily eliminated by nature; however, because of the larger number of animal farms in specific locations, this can drastically increase the concentration of the contaminants, which can directly affect other animals and ecosystems near the water sources. Pule et al. studied electrospun fibers incorporated with gold nanoparticles for on-site detection of 17β-estradiol (hormone) for dairy farm run-offs [26]. This colorimetric probe with a detection limit of 100 ng/ml showed that the natural steroidal estrogens greatly affected the fish reproduction system and gender at higher concentrations. The system works based on the color changes through the contact points. For example, upon contact with 17β-estradiol, the nanofiber probe changes color into pink, while for other substances, including 4-tert-octylphenol, deltamethrin, p,p'-DDE, cholesterol, and nonylphenol the same probe changes color into a brownish appearance. It was concluded that the electrospun nanofiber probe had a very high sensitivity in detecting 17β-estradiol [26].

5.1.2.4 Seed development research

Understanding seed development and root growth has inspired many scientists and engineers to study more, for better plant development through the environmental stimuli. One of the primary challenges for root growth is to regulate the temperature during seed development, which needs to be managed using the new method. In order to control the temperature of *Arabidopsis* seeds during the growth process, Jiang et al. developed novel electrospun polyethylene oxide fiber membrane and tested it on seeds [27]. In this study, the electrospun nanofibers were chosen because of their better insulation capabilities over the other cast membranes. It was reported that the electrospun nanofibers provided much better insulation properties for the same thickness. The insulation could be easily controlled by changing the thickness of the nanofiber membranes; thus, these nanofiber membranes could reveal various growth rate variations at different temperature changes [27].

5.1.2.5 Microbiome introduction

Microbiomes (e.g., bacteria, microbes, and other microorganisms) play curtailing roles in the soil ecosystem and plant growth. A number of different factors affect plant microbiome inoculations in crop soil, including soil temperature, moisture, colonization in the root system, as well as long-term excessive heat, dryness, and longer flooding seasons [28]. Effective microbiome inoculation can be another solution for sustainable crop production, as well as solving some of the environmental concerns through reducing the chemical fertilizers utilized for croplands [29]. The presence of certain microbiomes in the plant root system is also critically important for some economically sound crops. Electrospun nanofibers can be a useful encapsulation method for those microbiomes to enhance the viability of the microorganisms in harsh environmental conditions [18].

Damasceno et al. studied the rhizobia survival rates in seeds coated with PVA electrospun nanofibers. The authors encapsulated rhizobia, an interesting bacterium found in legumes, to protect them from harsh environmental conditions, such as excessive heat and dehydration [23]. The test results indicated that PVA nanofibers, combined with rhizobia, showed significantly higher rhizobia survival rates when compared to the base tests (unprotected rhizobia) after 48 h storage time. Other studies were also conducted to find out if the encapsulated bacteria in the nanofibers had any impact on the formation of nodules in soybean. The number of nodules between rhizobia-encapsulated nanofiber and the positive control (unprotected rhizobia) over a 30-days period were very similar. Thus, this study indicated that rhizobia encapsulation of the nanofibers could be a viable option for storage, transportation, and delivery of the bacteria to the root systems of the plants [23].

De Gregorio et al. reported about the beneficial rhizobacteria (PGPR) of electrospun nanofibers for soybean seed bioinoculants [30]. The plant growth rhizobacteria, *Pantoea agglomerans* ISIB55 and *Burkholderia caribensis* ISIB40, were electrospun with the PVA fibers and tested after coating of soybean. The germinated soybean plants were effectively colonized by the encapsulated bacteria in PVA nanofibers after 30 days storage. Test results confirmed that the germination rate, root length, and dry weight of the root, as well as leaf number and dry weight of the shoot, were drastically improved based on the selected PGPRs in the electrospun nanofibers [18,30].

Natural occurring biocontrol agents and some microorganisms (e.g., *Trichoderma* and *Bacillus subtilis*) can be used to protect crops against pathogens in the air, water, and soil [18]. The microorganisms can be encapsulated to enhance their efficiency and viability during storage and processing, so that they can be effectively utilized for a longer time against harsh environmental conditions. In addition to that, the encapsulation process will be beneficial for slow-releasing active agents and handling of microorganisms. Note that burst release will reduce the long-term potency of the agents during the farming and dry seasons [18,31].

Spasova et al. studied the encapsulation efficiency of electrospun nanofibers incorporated with the fungal spores of *Trichoderma viride* for plant biocontrol purposes [31]. During the electrospinning process, *Trichoderma* were dissolved in the electrospinning solutions with different ratios of chitosan−PEO and chitosan−polyacrylamide (PAAm) to facilitate nanofiber formation with all the specified ingredients. The *Trichoderma* viability test results showed that the spores had high viability rates at high DC electric fields during electrospinning. The produced hybrid nanofibers exhibited a very effective antifungal property against some of the microorganisims (e.g., *Fusarium* and *Alternaria* strains) [18]. In addition, the prepared mixture of solutions was directly applied to the root system and leaves of plants using the electrospinning method. It is hypothesized that the direct allocation of electrospun nanofibers on plants could protect them against harmful microorganisms without damaging their root systems or affecting their nutrition. The proposed system can be easily applied on larger-scale farming fields, especially when industrial electrospinning instruments are available [31].

5.1.2.6 Drug delivery in animals

Recent studies have exhibited that applications of electrospun fibers in drug delivery and biomedical engineering have been increasing drastically because of the outstanding properties of nanofibers and the ease of applications in many engineering fields [3,5,32−35]. The same approaches could be considered for agricultural applications and farm animals and pets. Karuppannan et al. reported about the fabrication and characterization of progesterone-loaded electrospun nanofibers for drug delivery and other related applications [36]. The authors mentioned that the higher concentrations of progesterone in the electrospinning solution were found to disrupt nanofiber production because of the increased viscosity prior to the electrospinning process. Nevertheless, the progesterone hormone release could be sustained over a week with 87% of the release rate even at the lowest concentration of hormone. Increasing the progesterone concentrations increased the shelf-life of the hormone release [36]. The hormone release approach can be applied for various domestic farm animals, pests, trees, and other plants [18].

5.1.2.7 In vitro pollen germination

Pollen germination is an important process for sustainable plant growth and crop yields. During pollen germination, the supporting medium is usually in the form of liquid or gel, so the nanofiber membranes can be employed as supporting materials for pollen germination of the seeds on liquid medium. Bodhipadma et al. reported the use of electrospun polyvinylidene fluoride (PVDF) and polylactic acid (PLA) nanofiber membranes for the enhancement of in vitro germination rates of *Artabotrys hexapetalus* pollen [37]. When the thickness of electrospun membranes is less than 8 μm, more than 65% of germination was recorded when compared to the liquid-only medium (less than 60%) and agar gel medium (less than 50%). Nevertheless, higher membrane thickness (18 μm) provided a substantial reduction (less than 3%) in the germination rates, which may be attributed to the thick membrane blockage for the pollen to the liquid medium. The pollen tube length with the electrospun membrane and liquid medium was more than 500 μm; however, the same lengths for the agar gel and liquid-only media tests were 390 and 140 μm, respectively [37].

5.1.2.8 Bionanosensors for pesticide detection

Recent studies have indicated that one of the most widely used detection methods for sensing of pesticides is the enzyme-based acetylcholinesterase (AChE). This molecule converts the acetylcholine neurotransmitter into choline through hydrolysis reactions [38]. The detection mechanism is related to the inhibition of AChE activity in the presence of carbamate insecticides and organophosphorus, which is the same mechanism as for killing the insects. It is stated that the amount of enzyme inhibition is comparative to the concentration of pesticides in the environment [38–40]. When the insects are exposed to insecticides, inhibition of enzyme activity will improve the acetylcholine concentration, resulting in insect death [18].

Immobilization of enzymes in a solution is a major step for preparing enzyme-based biosensors. Electrospun nanofibers can be used as supporting materials to immobilize the enzyme since they increase the loading capacity of enzyme molecules, and the number of recognition sites in biosensing devices. The high surface area nanofibers can drastically enhance the sensitivity and accuracy of the sensors. Moradzadegan et al. studied the immobilization of acetylcholinesterase using electrospun PVA/bovine serum albumin (BSA) nanofiber membranes after the crosslinking process with glutaraldehyde solution [41]. The test results demonstrated that the immobilization yield was approximately 100%, of which 40% of immobilized enzyme maintained their original activity throughout the tests when compared to the free enzyme without any immobilization. The test results conducted over a period of 100 days also revealed that the preservation rates of the immobilized enzyme were 90 and 34% at 4 and 30°C, respectively [18]. However, the reusability tests confirmed that the enzyme activity was reduced about 30% after 10 reuse tests. Stoilova et al. functionalized the electrospun mats from the copolymers of styrene–maleic anhydride for immobilization of AchE [38]. AChE molecules were covalently bonded to the surfaces of nanofibers using glutaraldehyde prior to the testing and detection.

It was found that the storage and thermal stability of the immobilized enzyme could be considerably enhanced when compared to that of the free enzyme in the solution. The reusability of the immobilized enzyme on nanofibers demonstrated that 65% of the enzyme activity was lost after 10 cycle tests. Stoilova et al. also investigated the electrospun PAN nanofiber membranes employed for the immobilization of acetylcholinesterase after the chemical modification of the surfaces of nanofibers with chitosan [42]. The immobilized enzyme on the surface of nanofiber mats provided excellent storability, with 60% of the initial activity after 40 days of storage time. Also, the immobilized enzyme provided 45% of its original activity after 10 reuses, which is still acceptable for detection and analysis purposes of AChE enzyme.

5.1.2.9 DNA extraction in agricultural research

Proper extraction of DNA molecules from plants is critical for the genetic investigation and classification of those species. It was indicated that classical DNA extraction methods currently applied in agricultural are time-consuming and tedious [18]. Nam et al. prepared the highly magnetic NiZn—ferrite nanofibers through an electrospinning method for the DNA studies [43]. The produced nanofibers are used to separate DNA molecules from other molecules under the external magnetic fields. In order to prepare magnetic nanofibers, some compounds of iron chloride, nickel acetate, and zinc acetate were stoichiometrically mixed with PVP solution in a solvent, and electrospun at known voltage, pump speed, and tip-to-collector distance. The produced nanofibers of PVP with the metal compounds were calcined to decompose the organic portion of the nanofibers. The separation study indicated that approximately 50% of DNA attached to magnetic nanofibers was properly separated [43].

Demirci et al. modified the surfaces of electrospun cellulose acetate nanofibers via reversible addition-fragmentation chain transfer (RAFT) polymerization process for the DNA extraction [44]. In this process, nanofibers were produced via the electrospinning method, and then the membranes with positive charges were fabricated using grafting of poly [(ar-vinylbenzyl) trimethylammonium chloride] on the surface of the nanofibers. The positively charged membrane surface will attract the negatively charged DNA molecules because of the phosphate groups in the DNA structures. It was stated that maximum DNA attraction/absorption on the nanofiber surfaces was observed within 90 min of the duration. After the five cycles of experiments, the absorption capacity (or reusability) was reduced to 46%. The prepared nanofibers can be used for DNA extractions of the plants, insects, and mammals for further studies to identify the genetic structures and classifications of the species.

5.2 PREPARATION OF PROTECTIVE CLOTHES FOR FARM WORKERS

Public concerns have been rising for those people (e.g., farmers, military personnel, and agriculture and food science scientists, engineers, and workers) who are exposed to harmful fungicides, herbicides, insecticides, pesticides, other chemicals and biological agents, and nerve gases. Therefore, it has been forcing many government agencies and private companies to develop new and efficient protective clothes against those toxic materials [45]. Along with their good barrier properties for harmful agents, those protective cloths should possess suitable air/vapor permeability for satisfactory heat transfer and thermal security for the workers; however, this is a challenging job to combine all the properties in one single protective suite [18,45—47]. It was reported that a large portion of traditional protective clothes display good barrier and protection properties, but relatively weak moisture and air permeability, which increases the chance of hyperthermia for workers, especially during the hot farming seasons.

Lee and Obendorf developed protective textile materials as barriers to liquid penetration using a melt-electrospinning process [46], while Raza et al. investigated protective clothing based on membranes of electrospun nanofibrous mats [47]. These studies stated that electrospun nanofiber membranes could concurrently provide high protection and good breathability properties because of the extraordinary surface texture, flexibility, light weight, high resistance to particle penetration, low toxicity, and small porosity (or small pore sizes) of the electrospun nanofibers [46,47]. In addition to these advantages, electrospun nanofibers can be directly applied on the surfaces of previously made protective clothes and sandwiched between the clothes. The electrospun nanofibers can be modified or functionalized by incorporating appropriate materials/chemicals to absorb toxins and other harmful materials before they reach the skin [47]. For instance, some of the protective clothes made of electrospun PAN nanofibers were effectively modified through layer-by-layer detoxifying coating agents to eliminate nerve gases. Because of the lower rate of production and weaker mechanical strength, nanofiber-based protective clothing has limited applications in the agriculture industry. However, developing better instrumentations and new production methods will overcome these obstacles [18].

Highly inert and durable electrospun polypropylene (PP) nanofibers were sprayed on the surface of nonwoven substrate to prepare protective clothes, owing to the acceptable breathability and high barrier properties of the nanofibers. The melt electrospinning method was selected to prepare nanofibers because of the low solubility of PP in most organic solvents. This spinning technique does not require solvent during the spinning process, so it is environmentally friendlier and its products can be used in many biomedical, agricultural, and food science applications [48]. Microscopy analysis conducted on the fiber surface showed that the surfaces of the nanofibers are relatively smooth [18].

The electrospun nanofiber film thickness and lamination are important parameters for barrier performance and air/moisture permeability of nanofiber-based clothes. Air and vapor permeability were decreased because of the laminations and layer thickness; however, it is stated that this could be acceptable when compared to the currently used protective clothes, which have considerably lower permeability [18,46]. In order to determine the protective performances of electrospun nanofibers, mixtures of two different pesticides with various viscosity and surface tensions were applied to the surfaces of the clothes. The pesticides penetrated on the clothes were extracted for chromatography/mass spectrometry (GC/MS) analyses. The test results confirmed that the surface tension of the pesticides was an important parameter for the penetration of toxic chemicals through the clothes: no penetration was observed with the high surface tension liquid mixture, while considerable penetration was noted with the low surface tension mixture. The authors also stated that in order to obtain both higher protection and higher breathability, nanofibers with low thickness and laminated electrospun nanofibers with high thickness might be appropriate for tailoring options for both high and low surface tension pesticides in liquid [18,46].

Lee and Obendorf produced the electrospun polyurethane (PU) nanofiber webs to fabricate protective textile materials as barriers against the liquid penetration [49]. In this study, PU nanofibers were directly electrospun on the surface of PP nonwoven fabrics for improved barrier properties and mechanical strengths of the cloths. It was determined that the diameters of the solvent-based nanofibers were smaller than that of melt-electrospinning fibers [46]. The web density of the nanofibers was also investigated and the test results indicated that there were no major changes in the transport of water/moisture; however, the air permeability was decreased as a result of improved web density. When compared to the traditional protective materials, nanofiber-based cloths were still better in terms of air permeability and water vapor transportation. The penetration tests of pesticide liquid/vapor (mist) indicated that the penetration value with low surface tension was 80–100% in the control sample (nonwoven substrate). This value was reduced to nearly zero for the layered nanofiber substrates. Even with the high surface tension mixture of the control sample, the penetration was less than 10%, which is substantially better when compared to the currently used methods [46].

It was indicated that the pore size of the layered nanofiber system is considerably smaller than the other nanoscale substrates. The pore size and density can be tailored based on the selected materials and process and system parameters of the electrospinning process. Lee and Obendorf investigated the transport properties of layered fabric systems (PU nanofibers) obtained using electrospun nanofibers and compared the test results with other traditional materials in terms of barrier properties and air/moisture permeability [50]. The conventional nonwoven fabrics and microporous materials have been used as traditional materials for a while, with various permeability properties. The protection ability was based on the pore size measurements where microporous materials had the lowest, while conventional nonwoven materials had the highest pore size [50].

5.3 NANOFIBERS FOR THE FOOD INDUSTRY AND FOOD PACKAGING

5.3.1 ELECTROSPINNING FOR THE FOOD INDUSTRY

Noruzi reported the recent development on the applications of electrospun nanofibers in agriculture and the food industry [18]. It was stated that many exciting characteristics of nanofibers produced through electrospinning process, including flexibility, porosity, high surface-to-volume ratio, biocompatibility, biodegradability, and safety make these nanofibers appropriate candidates for utilization in a number of applications, such as filtration, drug, DNA and gene delivery, wound dressing, tissue engineering, scaffolding, energy conversion, and bio- and nanosensors. However, direct agricultural and food applications of electrospun nanofibers are relatively new, and have been drastically growing [18]. The author specified that electrospun nanofibers could be used in many agriculture and food science applications, such as protective clothes for farm workers, plant protections using hormone-loaded nanofibers, and encapsulation of biocontrol agents, deoxyribonucleic acid extraction, encapsulation of agrochemical materials, developing bio- and nanosensors for pesticide detection, measurement and preconcentration of pesticides in farm products, environmental sampling for soil and fertilization, treatment of irrigation water, manufacturing and applications of food packaging materials, as well as filtration of beverage products (e.g., water, juice, alcoholic drinks, etc.), and other related separations [18]. These studies indicated that the use of agriculture and food science applications of electrospun nanofibers could address some of the major problems and concerns in agriculture and food industries.

Rezaei et al. reported the major applications of cellulosic nanofibers produced through the electrospinning method in food science and their potential benefits and risks in the industry [51]. It was noted that cellulosic materials are available anywhere on Earth, plentiful, biodegradable, and low-cost compared to other options. Electrospinning of cellulosic materials will provide extraordinary physical, chemical, physicochemical, and biological properties of nanofibers for many food and other agricultural industries. The authors had a major review on the electrospun nanofibers and their use in food science, some of which are summarized as: (1) immobilization of vitamins, DNA, enzymes, and antimicrobials; (2) drug delivery and controlled drug release; (3) bio- and nanosensors for food and soil contamination; (4) filtration and separation of suspended particles in drinking water and juice; and (5) thin-film and composite reinforcements. The authors also specified that there were some potential risks (e.g., trapped solvents and moistures in fibers, contaminations, pH-related acidity and alkalinity, allergic reactions on skin, and degradation) of using nanofibers in food science and industries [51].

One of the major applications of electrospun nanofibers is to encapsulate food to protect it against contamination, oxidation, UV light, odor control, and moisture [52], because electrospinning is a cost-effective and simple method, and does not require external heating or equipment. The other major applications include improved probiotics viability and handleability [53], preparing factional food products [18], immobilization of cells and enzymes [54], and controlled release of food ingredients to improve the shelf-life [54,55]. Some of the traditionally used encapsulation techniques (e.g., spray-drying) require high temperatures; however, high temperature can decompose food and change the taste. Antioxidants (e.g., carotenes and gallic acid) could be added into zein nanofibers and applied on food products to increase the food product quality, stability, and lifetime [52]. Note that zein is a biodegradable protein and edible polymer mainly originated in corn, which can be easily dissolved in ethanol and blended with antioxidants and active agents prior to the electrospinning process [18]. Melt electrospinning can be applied to produce zein nanofibers for food protection.

UV light exposure tests were also conducted on encapsulated carotene samples to determine the photo-oxidation/degradation behaviors. The test results indicated that after 60 min of UV exposure, the carotene content in the nanofibers was not changed much; however, the carotene content of the control samples was drastically reduced. The carotene release rate in a solution was mainly attributed to the aging time of electrospun nanofibers and relative humidity in the test chamber [56]. The crosslinking, oxidation, and natural light may also accelerate the aging process and change the release rate of the agents in nanofibers after longer exposures.

Lopez-Rubio et al. produced electrospun PVA nanofibers incorporated with probiotic bacteria (*Bifidobacterium* strains) because of the major health benefits, including reduction of serum cholesterol and gastrointestinal infections, and improvements in lactose metabolism and the immune system [53]. Instead of using a conventional electrospinning process, the authors chose to use coaxial electrospinning for better encapsulation and controlled release profiles. The produced nanofibers are considered for utilization in the food industry for improved viability (in the temperature ranges of 20, 4, and $-20°C$) of these good bacteria during processing, transportation, and storage. At lower temperature ranges, the survival rate of *Bifidobacterium* in nanofibers is significantly higher.

Uyar et al. produced electrospun functional PMMA nanofibers associated with cyclodextrin (CD)−menthol inclusions for improved thermal stability of nanofiber systems [57]. The authors reported that the electrospun nanofibers were highly uniform and bead-free. The characterization studies indicated that a strong interaction between β- and γ-CD with menthol was observed; however, the same interaction was not noticed between α-CD and menthol. The blend of PMMA/menthol/β-CD and PMMA/menthol/γ-CD showed better thermal stability, while there were some thermal decompositions on PMMA/menthol/α-CD samples [18,57].

Kayaci and Uyar recently reported about electrospinning of PVA polymer encapsulated with vanillin/cyclodextrin inclusions for the purposes of high-temperature stability and prolonged shelf-life of vanillin (a widely used volatile flavor) [58]. The diameter of the PVA nanofibers was mainly between 100 and 250 nm. In these studies, three different CDs, including α, β, and γ, were investigated using the thermogravimetric (TGA) method, and test results revealed that there was significantly higher thermal stability for the PVA/vanillin/CD nanofibers when compared to the other prepared samples [58,59].

Lemma et al. investigated the controlled release rate of retinyl acetate (RA) from β-cyclodextrin (β-CD) functionalized PVA nanofibers to enhance the thermal stability [60]. The experimental studies conducted on the PVA/β-CD complex encapsulated with RA showed that the thermal stability of RA was significantly increased when compared to the base case (nonencapsulated). Furthermore, it was also observed that RA molecules were more stable in PVA/RA/β-CD nanofibers than in PVA/RA nanofibers, indicating that β-CD has a better impact on the stability of RA molecules. The authors also studied the RA release rate of electrospun nanofiber complex for 100 days and reported that the release rate was considerably slower in PVA/RA/β-CD nanofibers compared to the other nanofibers (PVA/RA). Approximately 80% of the RA was released in the 100 days of experimental duration [60].

5.3.2 ELECTROSPINNING FOR THE PACKAGING INDUSTRY

Because of industrialization and globalization, packaging of food and food products necessitates longer shelf-life, quality, and food safety/monitoring during manufacturing, transportation, storage, and consumption. In order to address some of the major concerns, electrospun nanofibers are considered for food packaging and beverage industries. The nanofibers can keep many properties of foods and beverages, including taste, flavor, color/appearance, texture, consistency, as well as barrier, mechanical, and antimicrobial properties during transport and storage. New studies were mainly focused on the moisture absorption, oxygen scavenging, antibacterial property, and barrier packaging, which represents most of the active packaging market [61]. In the active packaging process, antimicrobial agents (e.g., silver nanoparticles, gentamicin, allyl isothiocyanate, eugenol, benzoic acid, sorbic acid, acetic acid, hydrogen peroxide, and phosphates) combined with electrospun nanofibers can diffuse into the food product to prevent the proliferation of microorganisms during storage and transportation. This approach will eventually extend the food product shelf-life and reduce food-related risks for consumers [61].

Vega-Lugo and Lim studied the controlled release mechanism of allyl isothiocyanate (AITC) incorporated with soy protein and PLA electrospun fibers [62]. It was stated that AITC is a natural and safe antimicrobial agent and is used to decrease its volatility during the active food packaging. Two methods were adapted to combine AITC with soy protein/PEO and PLA nanofibers: (1) direct addition of AITC to the electrospinning solution prior to electrospinning, and (2) complexation of AITC with β-CD before the electrospinning process. The release studies showed that AICT release from the fibers was nearly zero when 0% relative humidity was selected; however, the release was considerably higher when the humidity and CD content were increased, which indicates that the release mechanisms were primarily based on the humidity and CD agent of the fibers/environment. It was commented that the prepared materials could be used for food packaging [62].

It was reported that eugenol (EG), found in the clove plant, is a natural material with strong antifungal, antibacterial, and antioxidant properties, and can be a great candidate for packaging material applications. Nevertheless, this natural material cannot be utilized for the same purposes because of its instability and volatile nature [18]. Kayaci et al. performed a number of tests to improve the thermal stability of EG in electrospun PVA nanofibers [59]. The test results showed that there was no substantial increase in its thermal stability, although there was a strong interaction between EG and β-CDs based on the FTIR spectrum studies.

Neo et al. studied the encapsulation efficiency of food-grade antioxidant nanofibers using a zein−gallic acid system for packaging material applications [63,64]. Strong antimicrobial properties of these nanofibers were proposed to use for food packaging against Gram-positive and Gram-negative bacteria. XPS studies showed that some of gallic acid was present on the surfaces of the zein nanofibers, which may not be desirable for the complete encapsulation of the gallic acid into the nanofiber structures. Cytotoxicity studies also indicated that the prepared zein-gallic acid nanofibers are highly biocompatible, and could be used for the packaging industry.

In addition to those studies, many other studies have also been conducted on various biodegradable nanofibers and materials for food packaging applications. Some of them include triclosan (cyclodextrin inclusion complexation) in PLA nanofibers, red raspberry extract (rich in anthocyanins) in soy protein isolate nanofibers, geraniol−CD encapsulated in PVA nanofibers, and PVA/chitosan/tea extract nanofibers [18]. Since this is a developing field, more studies will likely be conducted on these subjects in the future.

5.3.3 ELECTROSPINNING FOR THE FILTRATION OF BEVERAGE PRODUCTS

Recent studies showed that electrospun nanofibers could be used for the filtration of beverages, drinks, and some other juice products because of the high surface area and porous structure of nanofiber membranes [18]. It was stated that highly permeable nanofibers could create high flow rates to filter some of the suspended particles [65]. Veleirinho and Lopes-da-Silva developed and applied electrospun poly(ethylene terephthalate) nanofiber membranes for the clarification of apple juice [66]. The comparison studies showed that the electrospun nanofiber membranes provided better performance on clarification than other currently utilized ones (clarification agents and ultrafiltration). The filtration rates of the nanofiber systems were substantially higher when compared to other options at the same applied pressures. The turbidity, appearance, and color of produced apple juice were identical to those of the apple juice produced through the traditional filtration techniques.

Fuenmayor et al. filtered apple juice using nylon nanofibrous membranes (average fiber diameter of 95 nm) [65]. During the tests, the effects of nanomembrane thickness on yield, clarification, and filtration time were explored for comparison purposes. The test results revealed that the fiber membrane thickness had a major impact on the clarification yield. At higher thickness, the clarification was better, but it also increased the filtration time due to the change in pressure difference. This nanomembrane is durable and robust, and also has suitable mechanical properties, which are necessary during the filtration of beverages. The test results indicated that electrospun nylon membrane provided similar or better efficiency in removing turbidity and color from the apple juice when compared to the commercially used PA membrane [65].

Lemma et al. investigated the removal rate of bacteria and yeast in water and beer using electrospun nylon nanofiber membranes [67]. In this study, the beer samples inoculated with two different bacteria (*Flavobacterium johnsoniae* and *Iodobacter fluviatilis*) and a yeast (*Saccharomyces cerevisiae*) were analyzed using nylon nanomembrane. The test results confirmed that more than 95% of turbidity (mainly microorganisms) in beer samples was removed after the nanofiltration process. Nevertheless, more applied pressure was required to remove bacteria from the liquids because of the smaller dimensions of the bacteria, resulting in higher capillary pressure on the surface of the nanomembranes [67].

5.4 CONCLUSIONS

In the first part of this chapter, electrospinning and various applications of electrospun nanofibers in agriculture and food industries were investigated. In the second part, preparation and applications of nanofibers for protective clothes for farm workers were studied, and recent development in the fields were provided. In the last part of the chapter, applications of electrospun nanofibers for the food industry, food packaging, and filtration of beverages and other juices were analyzed in detail. These studies also indicated that biodegradable and biocompatible electrospun nanofibers could be utilized as the delivery system in foods for nutrients to protect products during processing, transportation, and storage. There are also many advantages to using electrospun nanofibers and their membranes when compared with traditionally used products.

REFERENCES

[1] M. Ceylan, R. Asmatulu, "Enhancing the Superhydrophobic Behavior of Electrospun Fibers via Graphene Addition and Heat Treatment," ASME International Mechanical Engineering Congress and Exposition, Vancouver, Canada, November 12−18, 2010, 7 pages.

[2] M. Ceylan, "Superhydrophobic Behavior of Electrospun Nanofibers with Variable Additives," M.S. Thesis, Wichita State University, December, 2009.

[3] W.S. Khan, R. Asmatulu, M. Ceylan, A. Jabbarnia, Recent progress on conventional and non-conventional electrospinning processes, Fibers Polymers 14 (2013) 1235−1247.

[4] L. Saeednia, L. Yao, M. Berrendt, K. Cluff, R. Asmatulu, Structural and biological properties of thermosensitive chitosan-graphene hybrid hydrogels for sustained drug delivery applications, J. Biomed. Mater. Res. Part A 105 (2017) 2381−2390.

[5] N. Nuraje, W.S. Khan, M. Ceylan, Y. Lie, R. Asmatulu, Superhydrophobic Electrospun Nanofibers, Journal of Materials Chemistry A 1 (2013) 1929−1946.

[6] R. Asmatulu, M. Ceylan, N. Nuraje, Study of superhydrophobic electrospun nanocomposite fibers for energy systems, Langmuir 27 (2) (2011) 504−507.

[7] Z. Veisi, "Detection of Cox-2 Enzyme using Highly Sensitive Electrospun Polyaniline Nanofiber-based Biosensor," M.S. Thesis, Wichita State University, May, 2013.

[8] M.R. Anwar, "Electrospun Fiber-based Biosensors for Ultrasensitive Protein Detection," M.S. Thesis, Wichita State University, July, 2011.

[9] I. Alarifi, "Fabrication and Characterization of Electrospun Carbonized Polyacrylonitrile Fibers as Strain Gauges in Composite Aircraft for Structural Health Monitoring Applications," Ph.D. Dissertation, Wichita State University, April 4, 2017.

[10] I.M. Alarifi, A. Alharbi, W.S. Khan, R. Asmatulu, Carbonized electrospun PAN nanofibers as highly sensitive sensors in SHM of composite structures, J. Appl. Polymer Sci. (2015). Available from: https://doi.org/10.1002/app.43235.

[11] I.M. Alarifi, W.S. Khan, A.K.M.S. Rahman, Y. Kostogorova-Beller, R. Asmatulu, Synthesis, analysis and simulation of carbonized electrospun nanofibers infused carbon prepreg composites for improved mechanical and thermal properties, Fibers Polymers 17 (2016) 1449−1455.

[12] I.M. Alarifi, A. Alharbi, W.S. Khan, A.K.M.S. Rahman, R. Asmatulu, Mechanical and thermal properties of carbonized PAN nanofibers cohesively attached to surface of carbon fiber reinforced composites, Macromol. Symp. 365 (2016) 140−150.

[13] I.M. Alarifi, A. Alharbi, O. Alsaiari, R. Asmatulu, Training the engineering students on nanofiber-based SHM systems, Trans. Tech. STEM Educ. 1 (2016) 59−67.

[14] I.M. Alarifi, A. Alharbi, W.S. Khan, A. Swindle, R. Asmatulu, Thermal, electrical and surface properties of electrospun polyacrylonitrile nanofibers for structural health monitoring, Materials 8 (2015) 7017−7031.

[15] W.S. Khan, R. Asmatulu, M.M. El-Tabey, Dielectric properties of electrospun PVP and PAN nanocomposite fibers, J. Nanotechnol. Eng.d Med. 1 (2010) 6.

[16] W.S. Khan, R. Asmatulu, I. Ahmed, T.S. Ravigururajan, Thermal conductivities of electrospun PAN and PVP nanocomposite fibers incorporated with MWCNTs and NiZn ferrite nanoparticles, Int. J. Thermal Sci. 71 (2013) 74−79.

[17] http://electrospintech.com/agriintro.html#.WqnrAU2G_IV (accessed 14.03.2018).

[18] M. Noruzi, Electrospun nanofibres in agriculture and the food industry: a review, J. Sci. Food Agric. 96 (14) (2016) 4663−4678.

[19] P. Bansal, "Water-Based Polymeric Nanostructures for Agricultural Applications" PhD Thesis 2010 Philipps-Universitat Marburg.

[20] R. Bisotto-de-Oliveira, B.C. De Jorge, I. Roggia, J. Sant'Ana, C.N. Pereira, Nanofibers as a vehicle for the synthetic attactant TRIMEDLURE to be used for ceratitis capitata wied: (Diptera, Tethritidae) capture, J. Res. Updates Polym. Sci. 3 (2014) 40−47.

[21] L.M.F. Castaneda, C. Genro, I. Roggia, S.S. Bender, R.J. Bender, C.N. Pereira, Innovative rice seed coating (*Oryza sativa*) with polymer nanofibres and microparticles using the electrospinning method, J. Res. Updates Polym. Sci. 3 (2014) 33−39.

[22] D.D. Castro-Enriquez, F. Rodriguez-Felix, B. Ramirez-Wong, P.I. Torres-Chavez, M. M. Castillo-Ortega, D.E. Rodriguez-Felix, et al., Preparation, characterization and

release of urea from wheat gluten electrospun membranes, Materials 5 (2012) 2903–2916.

[23] R. Damasceno, I. Roggia, C. Pereira, E. Sa, Rhizobia survival in seeds coated with polyvinyl alcohol (PVA) electrospun nanofibers, Can. J. Microbiol 59 (11) (2013) 716–719.

[24] V. Krishnamoorthy, S. Rajiv, An eco-friendly top down approach to nutrient incorporated electrospun seed coating for superior germination potential, J. Adv. Appl. Sci. Res. 1 (2017) 16.

[25] T. Palvannan, T. Saravanakumar, A.R. Unnithan, N.J. Chung, D.H. Kim, S.M. Park, Efficient transformation of phenyl urea herbicide chloroxuron by laccase immobilized on zein polyurethane nanofiber, J. Mol. Catal. B Enzym. 99 (2014) 156–162.

[26] B.O. Pule, S. Degni, N. Torto, Electrospun fibre colorimetric probe based on gold nanoparticles for on-site detection of 17β-estradiol associated with dairy farming effluents, Water SA 41 (2015) 27–32.

[27] H. Jiang, Y. Jiao, M.R. Aluru, L. Dong, Electrospun nanofibrous membranes for temperature regulation of microfluidic sccd growth chips, J. Nanosci. Nanotechnol. 12 (8) (2012) 6333–9339.

[28] R.D. Souza, A. Ambrosini, L.M. Passaglia, Plant growth-promoting bacteria as inoculants in agricultural soils, Genetics Mol. Biol. 38 (4) (2015) 401–419.

[29] B.J. Alves, R.M. Boddey, S. Urquiaga, The success of BNF in soybean in Brazil, Plant Soil 252 (1) (2003) 1–9.

[30] P.R. De Gregorio, G. Michavila, L. Ricciardi Muller, C. de Souza Borges, M.F. Pomares, L.E. Saccol de Sá, et al., Beneficial rhizobacteria immobilized in nanofibers for potential application as soybean seed bioinoculants, PLoS One 12 (5) (2017) e0176930.

[31] M. Spasova, N. Manolova, M. Naydenov, J. Kuzmanova, I. Rashkov, Electrospun biohybrid materials for plant biocontrol containing chitosan and *Trichoderma viride* spores, J. Bioact. Compat. Polym. 26 (2011) 48–55.

[32] S.M. Hughes, A. Pham, K.H. Nguyen, R. Asmatulu, Training undergraduate engineering students on biodegradable PCL nanofibers through electrospinning process, Trans. Tech. STEM Educ. 1 (2016) 19–25.

[33] J. Jiang, M. Ceylan, T. Jai, L. Yao, R. Asmatulu, S.Y. Yang, Poly-ε-caprolactone electrospun nanofiber mesh as a gene delivery tool, AIMS Bioeng. 3 (2016) 528–537.

[34] R. Asmatulu, S. Patrick, M. Ceylan, I. Ahmed, S.Y. Yang, N. Nuraje, Antibacterial polycaprolactone/natural hydroxyapatite nanocomposite fibers for bone scaffoldings, J. Bionanosci. 9 (2015) 1–7.

[35] Y. Li, M. Ceylan, B. Sherstha, H. Wang, Q.R. Lu, R. Asmatulu, et al., Nanofibers support oligodendrocyte precursor cell growth and function as a neuron-free model for myelination study, Biomacromolecules 15 (2014) 319–326.

[36] C. Karuppannan, M.J. Sivaraj, G. Kumar, R.S. Seerangan, S. Balasubramanian, D.R. Gopal, Fabrication of progesterone-loaded nanofibers for the drug delivery applications in bovine, Nanoscale Res. Lett. 12 (2017) 116–178.

[37] K. Bodhipadma, S. Noichinda, N. Chanunpanich, W. Sukthavornthum, D.W.M. Leung, The utility of electrospun nanofibre mats for in vitro germination of *Artabotrys hexapetalus* pollen, Science Asia 42 (2016) 178–183.

[38] S. Liu, L. Yuan, X. Yue, Z. Zheng, Z. Tang, Recent advances in nanosensors for organophosphate pesticide detection, Adv. Powder Technol. 19 (2008) 419–441.

[39] D. Liu, W. Chen, J. Wei, X. Li, Z. Wang, X. Jiang, A highly sensitive, dual-readout assay based on gold nanoparticles for organophosphorus and carbamate pesticides, Anal. Chem. 84 (2012) 4185–4191.

[40] X. Gao, G. Tang, X. Su, Optical detection of organophosphorus compounds based on Mn-doped ZnSe d-dot enzymatic catalytic sensor, Biosens. Bioelectron. 36 (2012) 75–80.

[41] A. Moradzadegan, S.O. Ranaei-Siadat, A. Ebrahim-Habibi, M. Barshan-Tashnizi, R. Jalili, S.F. Torabi, Immobilization of acetylcholinesterase in nanofibrous PVA/BSA membranes by electrospinning, Eng. Life Sci. 10 (2010) 57–64.

[42] O. Stoilova, N. Manolova, K. Gabrovska, I. Marinov, T. Godjevargova, D.G. Mita, Electrospun polyacrylonitrile nanofibrous membranes tailored for acetylcholinesterase immobilization, J. Bioact. Compat. Polym. 25 (2010) 40–57.

[43] J.H. Nam, Y.H. Joo, J.H. Lee, J.H. Chang, J.H. Cho, M.P. Chun, Preparation of NiZn-ferrite nanofibers by electrospinning for DNA separation, J. Magnet MagnetMate 321 (2009) 1389–1392.

[44] S. Demirci, A. Celebioglu, T. Uyar, Surface modification of electrospun cellulose acetate nanofibers via RAFT polymerization for DNA adsorption, Carbohydr. Polym. 113 (2014) 200–207.

[45] L. Chen, L. Bromberg, J.A. Lee, H. Zhang, H. Schreuder-Gibson, P. Gibson, Multifunctional electrospun fabrics via layer-by-layer electrostatic assembly for chemical and biological protection, Chem. Mater. 22 (2010) 1429–1436.

[46] S. Lee, S.K. Obendorf, Developing protective textile materials as barriers to liquid penetration using melt-electrospinning, J. Appl. Polym. Sci. 102 (2006) 3430–3437.

[47] A. Raza, Y. Li, J. Sheng, J. Yu, B. Ding, Protective clothing based on electrospun nanofibrous membranes, in: B. Ding, J. Yu (Eds.), ElectrospunNanofibers for Energy and Environmental Applications, Springer, Berlin Heidelberg, 2014, pp. 355–369.

[48] J. Lyons, C. Li, F. Ko, Melt-electrospinning part I: processing parameters and geometric properties, Polymer 45 (2004) 7597–7603.

[49] S. Lee, S.K. Obendorf, Use of electrospun nanofiber web for protective textile materials as barriers to liquid penetration, Textile Res. J. 77 (2007) 696–702.

[50] S. Lee, S.K. Obendorf, Transport properties of layered fabric systems based on electrospun nanofibers, Fibers Polym. 8 (2007) 501–506.

[51] A. Rezaei, A. Nasirpour, M. Fathi, Application of Cellulosic Nanofibers in Food Science Using Electrospinning and Its Potential Risk, Comprehensive Reviews in Food Science and Food Safety 14 (2015) 269–284.

[52] A. Fernandez, S. Torres-Giner, J.M. Lagaron, Novel route to stabilization of bioactive antioxidants by encapsulation in electrospun fibers of zein prolamine, Food Hydrocolloids 23 (2009) 1427–1432.

[53] A. López-Rubio, E. Sanchez, Y. Sanz, J.M. Lagaron, Encapsulation of living bifidobacteria in ultrathin PVOH electrospun fibers, Biomacromolecules 10 (2009) 2823–2829.

[54] V. Nedovic, A. Kalusevic, V. Manojlovic, S. Levic, B. Bugarski, An overview of encapsulation technologies for food applications, Proc. Food Sci. 1 (2011) 1806–1815.

[55] M.A. Augustin, Y. Hemar, Nano- and micro-structured assemblies for encapsulation of food ingredients, Chem. Soc. Rev. 38 (2009) 902−912.

[56] Y. Li, L.T. Lim, Y. Kakuda, Electrospun zein fibers as carriers to stabilize (−)-epigallocatechin gallate, J. Food Sci. 74 (2009) C233−C240.

[57] T. Uyar, Y. Nur, J. Hacaloglu, F. Besenbacher, Electrospinning of functional poly (methyl methacrylate) nanofibers containing cyclodextrin−menthol inclusion complexes, Nanotechnology 20 (2009) 125703.

[58] F. Kayaci, T. Uyar, Encapsulation of vanillin/cyclodextrin inclusion complex in electrospun polyvinyl alcohol (PVA) nanowebs: prolonged shelf-life and high temperature stability of vanillin, Food Chem. 133 (2012) 641−649.

[59] F. Kayaci, O.C. Umu, T. Tekinay, T. Uyar, Antibacterial electrospun poly(lactic acid) (PLA) nanofibrous webs incorporating triclosan/cyclodextrin inclusion complexes, J. Agric. Food Chem. 61 (2013) 3901−3908.

[60] S.M. Lemma, M. Scampicchio, P.J. Mahon, I. Sbarski, J. Wang, P. Kingshott, Controlled release of retinyl acetate from β-cyclodextrin functionalized poly(vinyl alcohol) electrospun nanofibers, J. Agric. Food Chem. 63 (2015) 3481−3488.

[61] C. Tekmen, "Nanofibers in Food Packaging," Technical Note, Japan (online accessed 17.03.2018).

[62] A.C. Vega-Lugo, L.T. Lim, Controlled release of allyl isothiocyanate using soy protein and poly(lactic acid) electrospun fibers, Food Res. Int. 42 (2009) 933−940.

[63] Y.P. Neo, S. Ray, J. Jin, M. Gizdavic-Nikolaidis, M.K. Nieuwoudt, D. Liu, Encapsulation of food grade antioxidant in natural biopolymer by electrospinning technique: a physicochemical study based on zein−gallic acid system, Food Chem. 136 (2013) 1013−1021.

[64] Y.P. Neo, S. Swift, S. Ray, M. Gizdavic-Nikolaidis, J. Jin, C.O. Perera, Evaluation of gallic acid loaded zein sub-micron electrospun fibre mats as novel active packaging materials, Food Chem. 141 (2013) 3192−3200.

[65] C.A. Fuenmayor, S.M. Lemma, S. Mannino, T. Mimmo, M. Scampicchio, Filtration of apple juice by nylon nanofibrous membranes, J. Food Eng. 122 (2014) 110−116.

[66] B. Veleirinho, J. Lopes-da-Silva, Application of electrospun poly(ethylene terephthalate) nanofiber mat to apple juice clarification, Process Biochem. 44 (2009) 353−356.

[67] S.M. Lemma, A. Esposito, M. Mason, L. Brusetti, S. Cesco, M. Scampicchio, Removal of bacteria and yeast in water and beer by nylon nanofibrous membranes, J. Food Eng. 157 (2015) 1−6.

FURTHER READING

O. Stoilova, M. Ignatova, N. Manolova, T. Godjevargova, D. Mita, I. Rashkov, Functionalized electrospun mats from styrene−maleic anhydride copolymers for immobilization of acetylcholinesterase, Eur. Polym. J. 46 (2010) 1966−1974.

Electrospun nanofibers for energy applications

6

CHAPTER OUTLINE

6.1 NANOTECHNOLOGY IN ENERGY GENERATION

Nanotechnology is an emerging field, with applications in almost all scientific and research areas, such as communication, information, electronics, energy, biology, medical science, and many others. It is an interdisciplinary science whose full potential is yet to be explored. However, it is expanding to all disciplines of science and technology at an alarming rate. Nanotechnology is the art and science of manipulating matter at the nanoscale to create new and unique materials and products with astonishing properties, having enormous potential to change society and standards of living. Nanotechnology has the potential to lead to great breakthroughs and foster information about materials that we never knew before [1]. Novel engineered nano- and biomaterials, nanodevices, and components are fabricated by employing nanotechnology techniques, which explore and fine tune the properties, functions, and responses of living and nonliving matter at the nanoscale. The applications of engineered nanomaterials in optical and magnetic components, electronic and mechanical devices, biotechnology, tissue engineering, wound dressing, quantum computing, energy sectors, and many other fields, with very small features, generally nanosized, are the economically most indispensable segments of nanotechnology today and most probably in the years to come. Nanotechnology-based products are rapidly growing daily and more and more nano-products are available in the international market. The perpetual revolution in nanotechnology is resulting in the fabrication of engineered nanomaterials with

Synthesis and Applications of Electrospun Nanofibers. DOI: https://doi.org/10.1016/B978-0-12-813914-1.00006-7

exotic properties, which will have a positive impact on our lives, as well as in the environment, health, medical, electronic, and energy sectors [1−3].

In the energy generation sector, the conventional fossil fuels (e.g., coal, oil, and natural gas) resources cannot remain the supreme energy source due to the increase in consumption worldwide and emissions and pollutants, therefore alternative sources of energy that are safe, clean, affordable, and environmentally friendly must be promoted. Fossil fuel reserves are finite and are depleting rapidly. To effectively address the depletion of fossil fuels and the serious environmental problems accompanying their combustion, modern society has been searching for a new form of energy that is clean, renewable, cheap, and a viable alternative to fossil fuels [4]. The combustion of fossil fuels (e.g., coal, natural gas, gasoline, and oil) causes various types of pollution, resulting in a number of gas emissions, such as carbon dioxide, carbon monoxide, nitrogen, nitrous oxide, sulfur dioxide, volatile organic compounds, acidic/basic water vapors, and heavy metals. Additionally, the fluctuating and skyrocketing prices of fossil fuels in the international market, rising standards of living, and energy insecurity have caused substantial concerns to the global economy [1].

Innovative fuels, hydrogen, and solar cell technologies that use engineered nanostructure materials, present tremendous technological potential due to their outstanding features, such as large-scale and low-cost production [4]. The worldwide energy demand is growing continuously, and, according to some estimates, it is expected to rise approximately 50% by 2030. Currently, around 80% of energy demand is supplied by fossil fuels. The fossil fuel reserves will not be able to meet the ever-rising energy demand globally in the near future. Therefore, there is an urgent need to identify new sources of energy that are environmentally friendly and able to replace fossil fuels. Polymer batteries, fuel cells, solar cells, wind power, and geothermal power generators are some possible alternatives [3]. Given their high porosity, large surface area, nanosized diameter, interconnecting pores, and tunable surface morphology and porosity, electrospun nanofibers are being considered for polymer batteries, solar cells, and polymer electrolyte membrane fuel cells (PEMFCs) and supercapacitors [4].

6.2 ELECTROSPUN NANOFIBERS FOR BATTERY MEMBRANES

Many advancements have been made in battery technology in the past, through consistent improvement in electrochemical systems and the introduction of the latest battery chemistries; however, the fact of the matter is that there is still no ideal battery that can foster optimum performance under all kinds of operating conditions. Likewise, there is not a single separator or membrane that can be considered ideal for all battery geometries and chemistries [1]. A membrane is a porous structure generally placed between electrodes with opposite polarity, permeable to ionic flow but preventing electric current electrodes [1,2]. A wide variety of polymeric membranes have been developed and utilized in the last few years, which were mainly fabricated from cellulosic papers, cellophane, foams, woven and nonwoven fabrics, and microporous flat-sheet membranes made of different polymeric structures [2]. Since the batteries are becoming more sophisticated, membrane function has becoming more challenging and complex [1]. Separators or membranes play a very important role in battery operation. In order to prevent short circuits, the main function is to keep the positive and negative electrodes apart, and allow the transfer of ionic transport of charge carriers that are essential to complete the circuit during the passage of current in an electrochemical cell [1−4]. The membranes should be very good electronic insulators and possess the ability of conducting ions either by soaking electrolyte or by intrinsic ionic conduction [1].

To date, very little progress has been made towards the development and characterization of new materials for membranes for long-lasting service [1,3,5]. As far as literature surveys are concerned, not much attention has been directed towards publishing articles on battery membranes. Nonetheless, a number of recent publications on the fabrication, characterization, and performances of batteries have appeared, but unfortunately none of them has included detailed discussions about membranes, separators, their performances, and the selection of materials. Generally, polyolefin microporous membranes have been employed as membranes in different Li-ion batteries. Traditional separators have relatively high chemical resistance, suitable thickness, and good mechanical and thermal stabilities [6]. Despite the broad use of membranes, an extensive need still exists for enhancing the performance, increasing the lifespan, and minimizing the cost of membranes. Various techniques are available to prepare the porous polymeric separator/membrane for stable and durable battery manufacturing. These include solvent casting; phase separation using a nonsolvent, plasticizer extraction, and more recently, electrospinning [7]. Considering the distinctive and unique features of membranes that can be fabricated by an electrospinning process, this method could stand out as the method of choice for battery membranes [2].

The fabrication of membranes with high ionic conductivity, chemical, mechanical, electrical, dielectric, and thermal stability is one of the prime concerns for the science and technology of membranes and separators. Therefore, a polymeric material that possesses an appropriate glass transition temperature, and that is thermally and chemically stable could be the ideal candidate for battery membrane applications. The fibrous structure of electrospun nanofibers with high porosity, interconnecting pores, electrochemical stability, and high surface area present an optimum solution for separator materials. Electrospun nanofibers have unique fibrous structures, which are very important for high-electrolyte absorbance and low-ionic resistance, thereby resulting in excellent charge/discharge cycles of a battery throughout the service life [2]. The conventional membrane based on microporous structures possesses many disadvantages, such as poor thermal stability, low porosity, and low wettability in liquid electrolyte [8]. Electrospun nonwoven nanofibers possess high surface area, porosity, permeability, and low weight, better interconnecting pathways for ionic transfer. The stochastic arrangement of the fibers is one of the main advantages of nonwoven fibers compared with other fibers for battery-separator applications [2]. Fig. 6.1 shows a lithium ion battery with its charging and discharging process [9].

When a battery is subjected to charging, Li + ions are confined in the anode. On discharging, the Li + ions move to the cathode. Linag et al. [10] reported the fabrication of polyvinylidene fluoride (PVDF) membrane for lithium-ion batteries employing an electrospinning method and subsequent heat treatment. The membrane exhibited remarkably high ionic conductivity at room temperature, good electrochemical stability, and high cycle performance. These results clearly indicated that heat-treated PVDF fibrous membranes were ideal candidates for high-performance lithium-ion batteries. Moreover, nanofibrous membranes, including polyvinylidenefluoride-co-chlorotrifluoroethylene (PVDF-CTFE) and

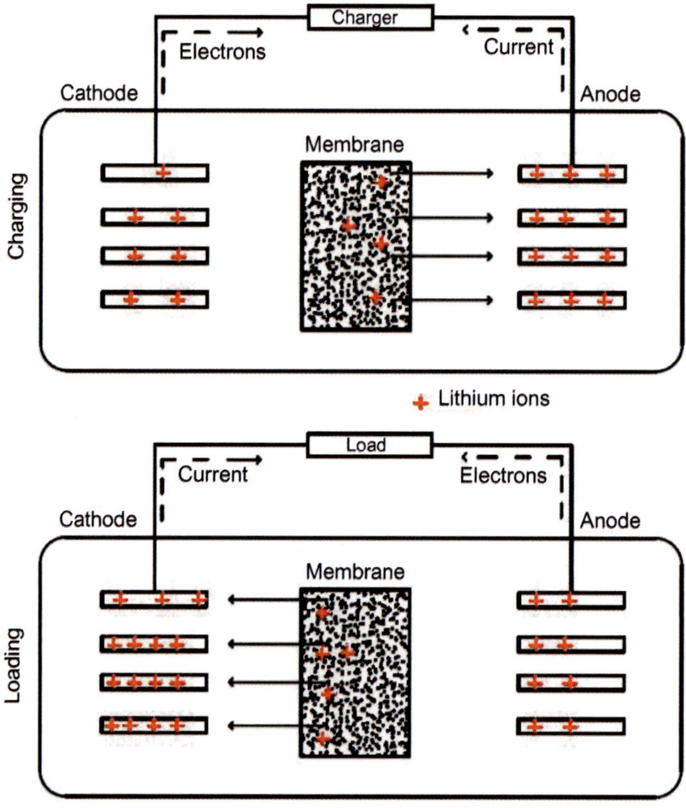

FIGURE 6.1

Loading and charging processes of a lithium-ion battery.

PVDF-CTFE/PVDF-HFP, were also fabricated by means of an electrospinning method for lithium-ion batteries [11].

The PVDF membrane showed very high uptake to electrolyte solution (320%−350%) and high ionic conductivity of approximately 1.7×10^{-3} S/cm at 0°C [12−15]. The porous structure of PVDF membrane favors high uptake of lithium electrolyte, which makes it possible to hold an excessive amount of lithium electrolyte in a smaller space in the battery. The large surface area and interconnecting pathways help in enhancing ionic conductivity, which in turn improves the energy density per unit weight as compared to conventional batteries. Similarly, other polymers, such as polyacrylonitrile (PAN), have also been used as battery membranes. The electrospun PAN nanofibrous membrane showed high ionic conductivity and good electrochemical stability. The PAN membrane with 1 M $LiPF_6$−EC/DMC showed an initial discharge capacity of approximately 145 mAh/g, and around 94% of the initial discharge capacity after 150 cycles at a charge/discharge rate of 0.5 C/0.5 C [16]. Khan et al. [2] used two polymers, that is, polyacrylonitrile (PAN) and poly(methyl methacrylate) (PMMA), and added graphene in varying amounts in both polymers and prepared PAN-based and PMMA-based membranes for lithium-ion batteries. Fig. 6.2 shows the SEM images and photographs of PAN, and PMMA electrospun nanocomposite fibers having 4 wt.% of graphene inclusions prepared for the battery separator

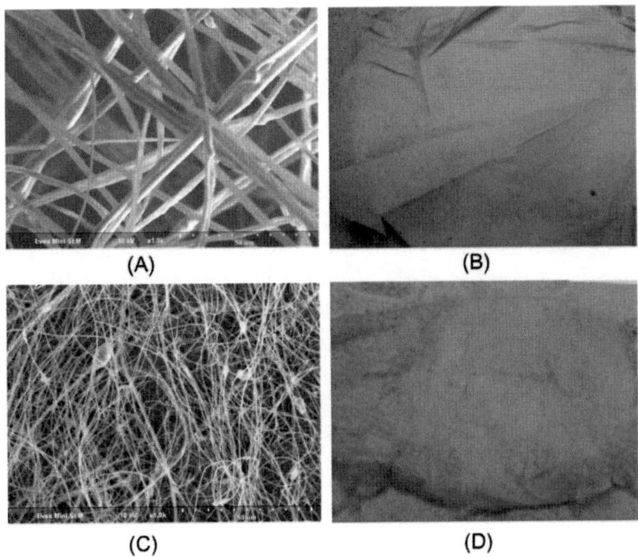

FIGURE 6.2

SEM images and photographs of (A, B) PAN and (C, D) PMMA electrospun nanocomposite fibers having 4 wt.% of graphene inclusions developed for battery separator applications.

applications [2]. The test results exhibited that the PAN-based membrane showed an ionic conductivity of 3.31×10^{-4} S/m at 8 wt.% of graphene and PMMA-based membrane showed an ionic conductivity of 5.52×10^{-4} S/m at 8 wt.% graphene, respectively.

Jayaraman et al. [17] demonstrated the fabrication of gelled electrospun PVDF-HFP nanomembrane for lithium-ion batteries, which exhibited a liquid-like conductivity of approximately 2.9 S/cm at ambient conditions. In the fabrication process, the anode and cathode were electrospun for $TiNb_2O_7$ and $LiMn_2O_4$, respectively. Recently, Yanilmaz et al. [18] prepared a nanofiber membrane via electrospinning SiO_2/nylon 6,6 and evaluated its performance for use in lithium-ion batteries. This membrane showed high liquid electrolyte uptake, low interfacial resistance, and good cycling performance compared to those using commercial microporous polyolefin membrane.

6.3 ELECTROSPUN NANOFIBERS FOR SUPERCAPACITORS

The recent rapidly rising demand for high-speed rechargeable energy storage devices with excellent performance has spurred a lot of researchers around the world to develop new materials for supercapacitors. A supercapacitor is an electrochemical device that stores and releases energy, when required, at an excessively high rate with high power density and long cycle life. Supercapacitors are the most promising new energy storage devices in many areas such as transportation, electricity, communications, defense, consumer electronics, and other applications due to their high power performance, long cycle life, and low maintenance cost [8]. These supercapacitors are outstanding energy storage devices with long cyclic functions and high power densities. They have attracted significant attention due to their low electrical resistivity, large surface area, and superb charging–discharging rate compared to conventional storage devices. They have displayed great performance in a wide range of applications, such as power back-up, in some electrical devices and power sources for hybrid vehicles.

Supercapacitors can be classified into pseudocapacitors (PCs) and electrical double-layer capacitors (EDLCs), depending on the energy storage mechanism [8]. PCs generally store energy based on fast reversible surface redox reactions, whereas EDLCs store energy by employing ion adsorption and desorption at the interface of the electrode and electrolyte [8]. Supercapacitors and EDLCs are commonly classified as energy storage devices that are suitable for rapid charging and discharging. The specific energy of supercapacitors is better than conventional capacitors. They have a higher specific power compared to batteries; however, their specific energy is somewhat lower. Supercapacitors are energy storage devices with intermediate characteristics that are usually between those of batteries and traditional capacitors and they can be charged and discharged a number of times without changing their properties and aging. Supercapacitors work well under hot and cold temperatures [19]. Fig. 6.3 shows a schematic of supercapacitor performance.

Improving the energy density of supercapacitors while maintaining their power density and long life has been the main concern in developing a reliable energy storage system for the future. Nanotechnologies, especially nanofibers, have been receiving tremendous attention from various fields, wherein their excellent properties can contribute to improve product functionality. As is known, electrospinning is perhaps the most facile route to prepare highly porous nanofibers. Electrospinning possesses several advantages over conventional methods, such as easy fabrication, less costly, less time consuming, and capable of mass production. The performance of a supercapacitor depends on the electrolyte/electrode and separator. Various types of materials have been used for separators/electrodes in supercapacitors, such as porous carbon, activated carbon fibers, carbon aerogels, carbon nanotubes, and graphene [19].

Transitional metal oxide nanostructures have gained extensive attention due to their attractive features in energy storage devices, such as supercapacitors, solar cells, and so on. A significant amount of research has been focused on metal oxides due to their outstanding demand in energy storage devices. The metal oxides generally considered for supercapacitor electrode applications are RUO_2, IrO_2, NIO, TiO_2, SnO_2, V_2O_5, and CeO_2 [20,21]. These metal oxides generally exhibit pseudocapacitive behaviors, demonstrating high specific capacitance. However, metal oxide has some drawbacks, such as toxicity, low availability, and high cost, which limits their applications in supercapacitors [22]. Similarly, carbon-based nanocomposites with conducting polymers have been considered for

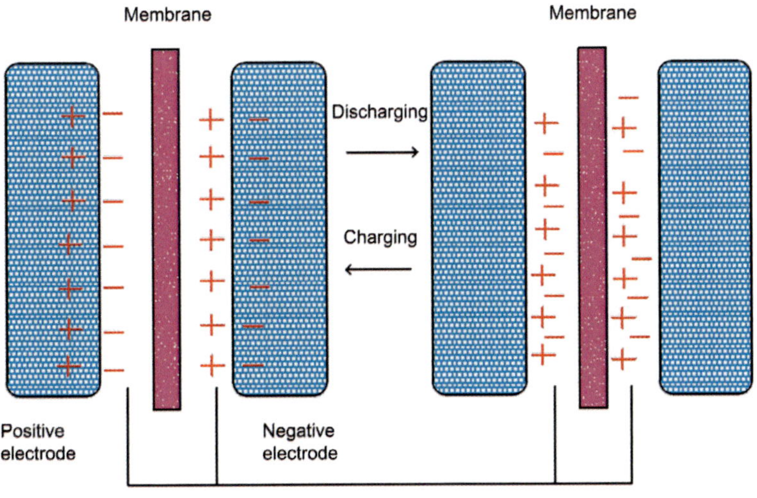

FIGURE 6.3

Schematic of supercapacitor performance and charging/discharging process.

supercapacitor applications. These nanocomposites have high capacitance values due to the fact that their functional groups contain phosphorus, nitrogen, and oxygen, which are termed as pseudocapacitance effects [19].

Electrospinning is considered the most facile and effective process to fabricate woven and nonwoven porous fibers from polymeric solutions. Polymeric fibers fabricated by an electrospinning process have a distinctive texture with micro- and/or nanosize fiber arrangement, high porosities, flexibility, high surface area, and interconnecting pores, and can be an ideal candidate for separators in supercapacitor applications. Therefore, electrospun carbon nanofibers from polymeric solutions of polybenzimidazole (PBI), PAN, and polyimide (PI) can also be used as electrodes for EDLCs after undergoing the process of stabilization, carbonization, and activation, which improves the porosity and surface texture of the nanofibers [8]. Some researchers have investigated the use of $ZnCl_2$, silver, and nickel as additives in polymeric solutions in order to increase the capacitance of nanofiber-based EDLCs [23−25]. The study revealed that the addition of $ZnCl_2$ caused improvement in fiber texture and specific capacitance. The specific capacitance of carbon nanofibers containing 5 wt.% $ZnCl_2$ reaches a highest value of 140 F/g [23−25]. An et al. [26] employed a coaxial electrospinning process to fabricate nanofibers using polyvinylpyrrolidone (PVP)-doped Sn and polyvinylpyrrolidone/polyacrylonitrile mixtures as inner and outer solutions, respectively, combined with a reduction in H_2. The electrochemical test revealed that the capacitance reaches a maximum value with 8% Sn in the polyvinylpyrrolidone solution.

Chaudhari et al. [27] fabricated polyaniline (PANI) nanofiber webs by means of an electrospinning process for electrode material in supercapacitor applications. PANI-webs exhibited higher specific capacitance (~ 267 F/g) than chemically synthesized PANI powder (~ 208 F/g) in IM H_2SO_4. Bhuvanalogini et al. [28] reported the preparation of $LiNi_{0.4}\ Co_{0.6}O_2$ nanofibers by electrospinning followed by the calcination at 450°C. They used $LiNi_{0.4}\ Co_{0.6}O_2$ nanofibers as positive electrode and activated carbon as negative electrode and a porous polypropylene separator in 1 M LiPF6−ethylene carbonate/dimethyl carbonate (LiPF6−EC:DMC) (1:1 v/v) as electrolyte. Cyclic voltammetry studies were conducted in the potential range of 0−3 V displayed the highest specific capacitance of 72.9 F/g. The electrochemical impedance measurements showed the charge transfer resistance was 5.05 Ω and the specific capacitance of the cell was 67.4 F/g [28]. RuO_2 and V_2O_5 are being considered as electrode materials for supercapacitors due to their high electrical conductivity, electrochemical stability, and high capacitance [8]. Lee at al. [29] reported that the $RuO_2−Ag_2O$ composite nanowire electrodes for supercapacitors were produced by an electrospinning process, exhibiting high capacitance of 173.25 F/g at 10 mV/ss and excellent retention of capacitance up to 97% after 300 cycles. The toxicity and high cost are issues with metal oxide, therefore researchers are focusing their attention on conducting polymers such as PANI, polypyrrole (PPy), and poly-p-phenylene (PPP) as electrode materials for pseudocapacitors.

As far as the separators for supercapacitors are concerned, Jabbarnia et al. [19] used a blend of polyvinylidene fluoride (PVDF) and polyvinylpyrrolidone (PVP) polymeric solution and added carbon black power to a prepared separator for a supercapacitor. Their results showed that the capacitance increased as the wt.% of carbon black increased. Recently, some studies have utilized PEDOT-coated electrospun PVP electrodes and PAN electrospun fibers as separators for flexible supercapacitors.

6.4 ELECTROSPUN NANOFIBERS FOR ENERGY CONVERSIONS

6.4.1 ELECTROSPUN NANOFIBERS FOR SOLAR CELLS

Fossil fuels play an important role in the global energy market. The discovery of fossil fuels for energy usage has brought about a revolution in human history. The combustion of fossil fuels (gasoline, coal, natural gas, and oil) has resulted in atmospheric pollution and climate change due to the emission of gases, such as CO, CO_2, nitrogen, nitrous oxide, sulfur oxides, volatile organic compounds, water vapors, and heavy metals [30,31]. Our planet can survive cataclysmic climate changes, if we adopt policies to reduce the usage of energy in our daily life. Fossil fuels are responsible for much of the world's power supply and energy demand [32]. Since 1990, the consumption of fossil fuels has doubled every 20 years, indicating an obvious major problem in worldwide energy dependence [30,31]. Global warming is real and mainly caused by manmade carbon-associated gas emissions, also known as greenhouse gases, resulting in immense climate changes. The major source of carbon emissions is the combustion of fossil fuels in power plants, automobiles, other transportation vehicles, industrial facilities, and other natural and artificial sources. The ever-rising prices of fossil fuels and their depletion at an alarming rate have created major threats to the economy and political stability. The increased consumption of fossil fuels worldwide, global warming, and depletion of oil reserves has motivated researchers to focus on alternative energy sources that are safe, clean, affordable, and environmentally friendly. There is an urgent need to replace fossil fuels with alternative energy resources, such as wind, solar, biomass, and hydrogen cells to mitigate the adverse effects of global warming and oil reserve depletion [31].

Renewable energy systems are of great significance due to their environmentally clean nature and efficiency. Among the renewable energy systems, fuel cells and solar cells are mainly considered these days. Solar cells do not create any emissions or contamination to the atmosphere, while hydrogen fuel cells create only clean water as a byproduct. Solar energy is an inexpensive, decentralized, and inexhaustible natural resource. The magnitude of the available solar power striking on Earth's surface is equal to 130 million 500 MW power plants [33]. Solar radiation is an abundantly available source of energy in almost all parts of the world throughout the year. Despite its availability in such an enormous amount, the energy produced from solar radiation remains at only 0.01% of the total demand [34]. Thus, solar energy that reaches the surface of the Earth is much more than our current needs. Based on some recent estimates, an area of approximately 10^5 m^2 installed with solar cells with operating efficiency of 10% is enough to fulfill our needs without any other alternatives [35]. One of the main problems with solar energy is that the energy conversion and storage rates are relatively low and need improvements.

Over the past few years, solar photovoltaics have shown an annual growth rate of 50% and the trend of using photovoltaics for energy generation has been increasing rapidly [34]. Photovoltaic technology has shown rapid advancements both in terms of materials technology and architecture of devices [34]. Solar cells were developed in 1954 by Chapin et al. [36] with 4.5% conversion efficiency, which open a new era of converting solar energy into electrical energy. Since then, solar technology has gone through different phases, such as polycrystalline silicon solar cells, single-crystalline silicon solar cells, amorphous silicon thin-film solar cells, dye-sensitized solar cells (DSSCs), and hybrid solar cells [34]. The single-crystalline silicon cell can achieve a conversion efficiency of nearly 20%, and this value is larger than other types of solar cells [37]. DSSCs are efficient alternatives to silicon-based solar cells, due to low cost, simple fabrication, and ease of large-scale production [38]. Since the introduction of solar cells, many advances have been made to improve the conversion efficiency. One strategy is the introduction of nanostructured materials, such as electrospun nanofibrous materials.

As an alternative to conventional solar cells, Grätzel and colleagues have developed dye-sensitized solar cells (DSSCs) also known as Grätzel cells. DSSCs are photoelectrochemical cells that generate electricity using wide band-gap mesoporous oxide semiconductor nanoparticles, nanowires, or nanotubes [39]. DSSCs are simple and economical, made from simple relatively cheap materials, such as conductive glass, nanoparticles, blackberry juice (containing natural dye molecules), and an iodine solution [35,40]. The nanoparticles used in DSSC construction include TiO_2, ZnO, SnO_2, and other semiconductor nanoparticles incorporated with indium tin oxide (ITO) nanoparticles, C_{60}, carbon nanotubes (CNTs), and graphene [41].

A typical DSSC is composed of a photoactive n-type semiconductor working electrode (photo anode), a counter electrode made up of either a metal or semiconductor (photo cathode), and an electrolyte [34]. DSSCs consist of a mesoporous photo electrode layer of metal oxide, such as TiO_2, ZnO, SnO_2, CuO, and other semiconductor nanoparticles, on a conducting glass substrate (fluorine-doped tin oxide, $F\text{-}SnO_2$, or FTO). Generally, TiO_2 is used on glass substrate. The metal oxide surface is chemisorbed with a layer of organic dye, which works as a sensitizer. The cell is then completed by sealing both electrodes in the presence of an electrolyte [34]. The photoanode is composed of a transparent conducting glass, which is coated with a porous semiconductor film, and the photosensitizing dye is adsorbed on the surface of the semiconductor film. When sunlight strikes on the surface of the DSSC, it enters through the transparent conductive oxide (TCO) top coat on the glass surface and hits the sensitizing dye molecules coated onto TiO_2 nanoparticles. The photons striking the dye molecules with sufficient energy cause an excitation state of the dye molecules, thereby releasing electrons and transferring excited electrons to the conduction band of the oxide layer and then transferring them out through an external circuit [34,35]. During this process, the maximum light absorption and charge transfer influence the overall photoelectric conversion efficiency. Hence, the photoanode plays a vital role. Fig. 6.4 shows a schematic diagram of an electrospun nanofiber-based DSSC [34].

Electrospun nanofibers, such as TiO_2, CuO, SnO_2, and ZnO are promising alternatives to the mesoporous nanoparticle thin films that are currently used in DSSCs. These nanofiber-based DSSCs have been receiving significant attention recently, both from academia and industries, due to their high efficiency, ease of fabrication, and environmentally friendly nature. However, mostly DSSCs have been based on TiO_2 nanostructured materials owing to the dye adsorption, porosity, charge transfer, and electron transport [34].

Electrospun metal oxide nanofibers as a thin film coated with photoanode are now being considered as materials for anodes due to their high specific surface area and 1D fibrous morphology. The high specific surface area enhances the absorption of photosensitizing dye. The electrospun TiO_2 nanofibers are 1D materials with several advantages, such as large surface area, environmental friendliness, and room temperature processing; therefore, they are ideal candidates in the fabrication of DSSCs [34]. TiO_2 anatase nanofibers prepared by electrospinning have been widely used as photoanode material [8]. Lee et al. [42] reported the

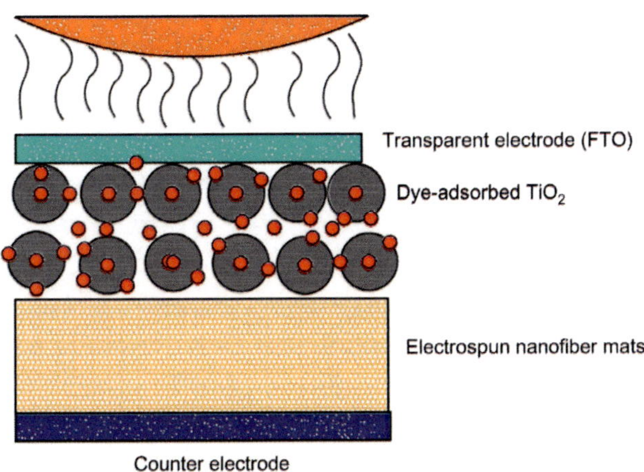

FIGURE 6.4

Schematic diagram of an electrospun nanofiber-based DSSC.

use of electrospun TiO_2 nanorods in DSSC fabrication and achieved a high conversion efficiency. DSSCs are not costly and are easy to manufacture when compared to silicon-based solar cells; however, there are some problems with these cells, such as electrolyte leakage and low conversion efficiency.

An alternative solution is the use of a viscous polymer gel electrode, but it is difficult to infuse a viscous gel into the TiO_2 nanostructure [37]. Song et al. [43] reported the use of TiO_2 nanofiber membranes produced by an electrospinning technique and combined with sol-gel processes to overcome this problem. The viscous polymer gel can easily penetrate into the fibrous nanofiber membrane, thereby enhancing the conversion efficiency to 6.2% [44]. The large controllable pore sizes of electrospun nanofibers and their interconnectivity promote easy penetration of viscous polymer gel electrolyte. These nanofibers cannot adhere to conductive substrate, resulting in lower conversion efficiency. Various methods have been adopted to solve this problem, including conversion of these fibers into nanorods, surface treatment, and hot pressing [8].

Moreover, some researchers have reported the use of nanoparticles incorporated with nanofibers and used them as photoanodes in DSSCs. Wang et al. [45] reported the use of a nonwoven composite of hybrid nanofibrous TiO_2/SiO_2 mat and TiO_2 nanoparticles in the construction of photoanode on FTO/glass substrate and obtained a conversion efficiency of 6.67%. In this composite structure, TiO_2 nanoparticles were used to increase the dye loading, whereas the TiO_2 nanofibers were used to improve electron transport and SiO_2 nanofibers were used to provide mechanical strength [8]. Yang et al. [46] prepared photoanodes for DSSCs by electrospinning TiO_2 nanoparticles/nanofibers with varying diameters of fibers and achieved a conversion efficiency of 8.4%.

In the DSSCs, the counter electrode also plays an important role in transmitting and collecting electrons and photocatalytic activity. The traditional material used in counter electrodes is platinum, due to its high electrocatalytic activity and high photovoltaic performance with redox couples [47]. However, platinum is an expensive metal and its long-term stability is not satisfactory due to the corrosive redox couple [8]. The alternative approach is the use of carbon-based materials, mesoporous carbon, and carbon nanotubes. Joshi et al. [48] reported the preparation of counter electrodes for DSSCs with electrospun carbon nanofibers and determined that the open circuit voltage and current density were in close agreement with the platinum electrode, but the efficiency was not the same. The relatively unsatisfactory performance of carbon nanofiber-based DSSCs could be due to their low fill factor, resulting in high resistance, which could be improved by preparing thinner and porous nanofibers. Noh et al. [49] reported the preparation of counter electrodes by electrospinning carbon fibers incorporated with platinum in varying amounts. DSSCs fabricated with 40 wt.% platinum nanoparticles exhibited better circuit current density, open circuit voltage, fill factor, and a conversion efficiency of 4.47%.

So far, we mentioned the uses of electrospun fibers for electrode materials, although electrolytes also play an important role in cell performance. The power conversion efficiency of DSSC based on liquid electrolyte has been around 12% but electrolyte leakage and volatilization of liquid electrolytes limit its applications. For this reason, organic or inorganic conductors, polymer gel electrolyte, and ionic liquid have been explored as substitutes for liquid electrolyte. The polymer gel electrolyte possesses thermal stability, high permeability, and high ionic conductivity. The commonly used polymer gel electrolytes for DSSCs are polymethylmethacrylate (PMMA), poly(vinylidenefluoride-co-hexafluoropropylene) (PVDF-HFP), and polyacrylonitrile (PAN). The problems associated with polymer gel electrolyte are their complicated fabrication and they generally lower the conversion efficiency [8]. In order to overcome these problems, electrospun polymer gel electrolyte membranes have been investigated. One study demonstrated the fabrication of electrospun poly(vinylidenefluoride-co-hexafluoropropylene) (PVDF-HFP) nanofiber films and spin-coated poly(vinylidenefluoride-co-hexafluoropropylene) (PVDF-HFP) films, and then applied to the polymer matrix in polymer electrolyte for DSSCs. The test results exhibited higher conversion efficiency and higher electrolyte uptake due to the porous structure of the membrane [8].

Asmatulu et al. studied the effects of graphene, carbon nanotubes (CNTs), and C_{60} in electrospun TiO_2 nanofibers on DSSC efficiencies [50]. The TiO_2 nanofibers were produced using chemical mixtures of DMF, PVAC, acetic acid, and titanium (IV) isopropoxide prior to the electrospinning process. Then, the resultant nanofibers were carbonized at 550°C for 2 h to produce TiO_2 nanocomposite fibers. It was reported that adding nanoscale inclusions into the TiO_2 nanofibers significantly increased the DSSC efficiencies (4%–6%), which may be because of the charge/electron transport properties of the inclusions. Fig. 6.5 shows the SEM images of electrospun TiO_2 nanocomposite fibers incorporated with graphene and C_{60} prior to carbonization.

(A)   (B)   (C)

FIGURE 6.5

SEM images of electrospun (A) plain TiO_2 nanocomposite fibers incorporated with 4 wt.%, (B) graphene, and (C) C_{60} prior to the carbonization.

6.4.2 ELECTROSPUN NANOFIBERS FOR FUEL CELLS

Fuel cells are electrochemical devices that convert the chemical energy of a fuel into electricity through an electrochemical reaction in the presence of a catalyst with high efficiency. Methanol, bioethanol, and hydrogen have been used as fuel in fuel cells. Fuel cells are among the most promising electrochemical devices, with their efficiencies reaching approximately 60% and they are quiet in operation and release no pollutants into the atmosphere [51]. The proton exchange membrane fuel cells (PEMFCs) and direct methanol fuel cells (DMFCs) are generally considered to be an ideal choice for commercialization due to their simple operation, low operating temperature, and relatively high power density. A typical single-stack hydrogen polymer electrolyte membrane (PEM) fuel cell can generate approximately 1 A/cm^2 density at an operation voltage of nearly 0.7 V [51]. When a number of fuel cells are stacked together, they can generate enough power to operate automobiles, charge cellphones, laptops, digital cameras, and power other electronic devices [51,52].

Fuel cells are similar to batteries in construction. They are composed of an anode (negative electrode), cathode (positive electrode), and an electrolyte, which can be liquid or solid. Precious nanomaterials, such as platinum and gold, are used as catalysts in most fuel cells. In a typical hydrogen fuel cell, hydrogen gas enters the cell at an anode and splits into protons (positive charge particles) and electrons (negative charge particles), in the presence of catalyst, while oxygen is fed through the cathode [53]. The separator allow protons to travel from the anode to the cathode through the electrolyte, where they combine with oxygen to produce water. Electrons follow a disparate path through an external circuit, thereby generating electricity to power electronic devices. Fig. 6.6 shows a schematic view of a simple hydrogen fuel cell [53].

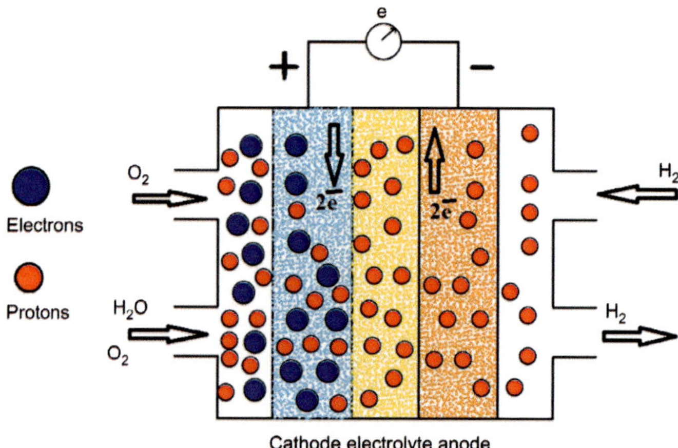

FIGURE 6.6

Schematic view of a simple hydrogen fuel cell.

The use of a high amount of platinum catalyst and the high cost of platinum in fuel cells limit its potential commercial applications. In order to overcome this problem, various carbon materials, such as activated carbon nanofibers, carbon nanotubes, carbon nanofibers, and mesocarbon microbeads for catalyst support in DMFCs have been explored due to their good electronic conductivity and low cost [8,34]. Electrospinning has been recognized as a novel synthesis technique for carbon materials with porous nanostructures and controlled diameters. Polyacrylonitrile, polyaniline, and polyimide have been excessively used to prepare carbon nanofibers or carbon tubes as supporting materials and a possible substitute for platinum catalyst due to their high surface area, high porosity, and nanostructure, which promote uniform dispersion of platinum nanoparticles [8]. It has been reported by some researchers that the carbon nanofibers produced through electrospinning followed by heat treatment exhibit higher conductivity than other types of carbon nanofibers [34]. Park et al. [54] found that platinum catalyst utilization increased to approximately 69%, when that catalyst was deposited onto electrospun carbon fibers, compared to around 35% utilization on carbon black. This clearly shows that carbon nanofibers are excellent support materials for platinum catalyst. Generally, platinum catalyst supported by carbon electrospun nanofibers shows better electrocatalyst activity, better stability, high exchange current, and low charge transfer resistance.

Li et al. [55] reported the deposition of platinum clusters onto electrospun carbon fibers mats by means of electrodeposition using the multicycle CV method for DMFC applications and found that carbon fiber mats enhance the catalyst peak current of methanol, thus indicating that the carbon mats promote catalyst performance. Zhao et al. [56] reported the fabrication of platinum nanospheres and platinum nanocubes on electrospun WO_3 and found that platinum nanocubes were more electrocatalyst than platinum nanospheres and electrocatalyst activity was far better than commercially available WO_3. Xuyen et al. [57] deposited platinum catalyst on polyimide nanofibers by a nucleation process followed by heat treatment and used them as an electrode for a fuel cell. Palladium is a possible substitute for platinum in alcoholic oxidation due to its excellent electrocatalyst activity and it is abundantly available in nature. Su et al. [58] used electroless plating on TiO_2 nanofibers and found excellent electro-oxidation behavior towards glycerol in an alkaline medium.

Proton exchange membrane (PEM) is another very important component of a fuel cell that has been under investigation for a couple of decades. The factors that influence the energy conversion efficiency of DMFC are proton conductivity and crossover. Currently, DMFC has achieved an efficiency of 40%. Therefore, researchers have been focusing their attention on the development of an efficient membrane that will impede methanol crossover. Nafion is one of the most widely used commercial polymer electrolyte membranes in fuel cells and has been used for several decades. Nafion exhibits relatively high proton exchange conductivity [52]. The electricity generation in PEMFCs is via a chemical reaction of hydrogen (fuel) at the anode and oxygen at the cathode (Fig. 6.6). Protons are transmitted through an electrolyte membrane (electrospun Nafion fiber membrane), which contains distilled water, to the cathode, while electrons are transmitted to the cathode through an external circuit. Fig. 6.7 shows the workings of a fuel cell with electrospun electrolyte membrane [52,53].

The basic properties of electrolyte membranes are high proton conductivity and shielding of electron transmission through the membrane. Since the membrane needs to hold distilled water for proton conductivity, water retention of the membrane is also critical. Nafion, a perfluorosulfonic acid polymer film, has been excessively used so far. However, Nafion membranes are expensive to use. For the same membrane area, electrospun Nafion fiber membranes require less material than conventional Nafion fuel cell membranes, thereby reducing cost. Porous nanofiber membranes are capable of holding distilled water, thereby increasing proton conductivity. Therefore, such nanofiber membranes have the potential to be used in PEMFCs.

Many researchers have attempted to modify Naifon with nanoparticles in order to decrease methanol crossover and swelling to increase proton conductivity [34]. In PEMFCs, the handling of hydrogen fuel is a critical issue. For achieving high proton conduction, water management is very important. Various alternatives have been proposed to address this issue, such as using DMFCs as an alternative to PEMFCs in electronic devices. As mentioned earlier, electrospinning Nafion perfluorosulfonic acid (PFSA) has spurred interest in using Nafion incorporated with some polymer to fabricate membrane for fuel cells. Naifon is insoluble in common organic solvents and requires a high-molecular-weight carrier polymer, such as poly(vinyl alcohol) (PVA), poly(acrylic acid) (PAA), or polyethylene oxide (PEO) [8]. Martwiset et al. [59] fabricated a water-stable membrane by electrospinning poly(vinyl alcohol) (PVA) and poly(4-styrenesulfonic acid). Choi et al. [60] fabricated perfluorosulfonic acid (PESA) nanofibers and used

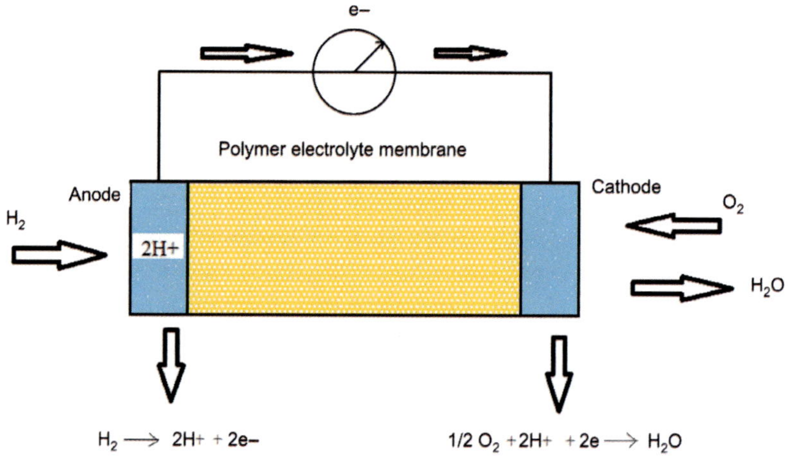

FIGURE 6.7

Fuel cell with electrospun electrolyte membrane.

polyethylene oxide (PEO) as the carrier polymer, along with an adhesive. PEO was then removed and fibers were crosslinked using adhesive chemistry. They found that the proton conductivity was increased to 0.16 S/cm.

It has been found that the addition of an excessive amount of carrier polymer reduces the proton conductivity. Therefore, recent studies have focused on minimizing the content of carrier polymer in order to maintain proton conductivity. Dong et al. [61] used only 0.1 wt.% carrier polymer to electrospun Nafion nanofibers and reported that the proton conductivity was an order of magnitude higher than the bulk Naifon film. Tamura et al. [62] reported the synthesis of a composite membrane for PEMFCs by electrospinning sulfonated polyimide nanofibers in the presence of sulfonated polyimide and found that polyimides were completely oriented within nanofibers and proton conductivity in the parallel direction was significantly high. Mollá et al. [63] fabricated a composite membrane by electrospinning an aqueous solution of PVA and infiltrated Nafion into a porous mat. This membrane exhibited low proton conductivity, due to the fact that an excessive amount of carrier polymer was used. However, other properties such as mechanical stability and methanol permeability were significantly improved. Asmatulu et al. investigated the transport properties of graphene-based thin Nafion membrane for polymer electrolyte membrane fuel cells [52]. The studies indicated that the graphene substantially increased the proton conductivity of Nafion membrane at lower graphene concentrations.

Salahuddin et al. reported the effects of the superhydrophobic PAN nanofibers for gas diffusion layers of proton exchange membrane fuel cells for cathodic water management after surface modifications [64]. Fig. 6.8 shows the SEM images of the electrospun PAN nanofibers before and after carbonization at 850°C for 60 min. The carbonized PAN nanofibers were then treated with hydrophobizing agents in different patterns to make some part of the surface of the carbonized nanofibers superhydrophobic (water contact angle is 162 degrees) while remaining hydrophilic (water contact angle is less than 10 degrees). This study indicated that the proposed approach had great success in the field [64].

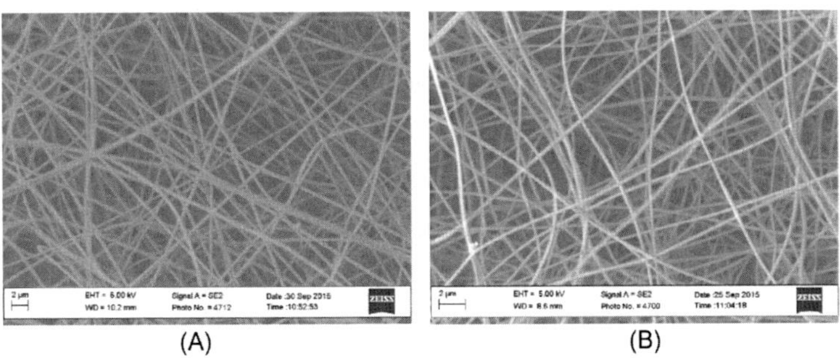

(A)   (B)

FIGURE 6.8

SEM images of the electrospun PAN nanofibers (A) before and (B) after carbonization at 850°C for 60 min.

6.4.3 SOLID OXIDE FUEL CELLS BASED ON ELECTROSPUN METAL OXIDES

Some studies have reported the use of electrospun nanofibers as electrode materials in solid oxide fuel cell (SOFC) applications. The nanofibers of Ni/yttria-stablized zirconia, prepared by electrospinning and electroless plating, tested as the anodes for a commercial half-cell, exhibited twice the peak power density compared to conventional ball-milled powders of the same material [65,66]. Recently, electrospinning has been used to prepare $Ba_{0.5}Sr_{0.5}Fe_{0.8}Cu_{0.2}O_{3-\delta}$ fibers with high porosity and surface area as cathode material for a proton-conducting ceramic fuel cell [67]. A study reported carbon nanofiber preparation by gas-assisted electrospinning and solution blowing process for use in microbial fuel cells [21,22,68,69]. The results showed improved anode current density. When lanthanum strontium cobalt ferrite (LSCF) nanofibers were used as the cathode in a fuel cell, they exhibited a power density of 0.9 W/cm^2 at 1.9 A/cm^2 and 750°C, which then improved more by incorporating 20 wt.% of gadolinia-doped ceria (GDC) into LSCF nanofiber cathodes. It showed a power density of 1.07 W/cm^2 at 1.9 A/cm^2 and 750°C [34]. This enhancement in performance is due to the porous structure, high thermal stability, and interconnecting pores of nanofibers that provide continuous pathways for charge transport.

6.5 CONCLUSIONS

Electrospinning is encountering revitalized immense interest and current advances stipulate a high potential of electrospun materials in energy-related applications. Electrospinning has been recognized as a versatile, straightforward, efficient, and cost-effective technique for fabricating nanofibers from all kinds of polymeric solutions or melts. The 1D nanofibers prepared by electrospinning have shown tremendous advantages for applications in energy devices, such as solar cells, fuel cells, lithium-ion batteries, and supercapacitors. As far as their applications in solar cells are concerned, electrospun nanofibers have exhibited high photoelectric conversion efficiency due to efficient charge separation and transport and maximum light absorption, which are mainly attributed to the high specific surface areas and high porosity. For the application of nanofibers in supercapacitors, the electrode prepared by electrospun nanofibers has demonstrated high specific capacitance and better cycling stability due to its unique fiber morphology, including large surface area to volume ratio and small diameter. Electrospun nanofibers have been shown to be ideal substitute materials for supporting catalysts in fuel cell electrodes, as well as acting as membranes in PEMFCs and SOFCs. The fibrous structure of electrospun nanofibers with high porosity, interconnecting pores, electrochemical stability, and high surface area is ideal for battery membranes.

REFERENCES

[1] K.W. Guo, Green nanotechnology of trends in future energy: a review, Int. J. Energy Res. 36 (2012) 1−17.

[2] W.S. Khan, R. Asmatulu, V. Rodriguez, M. Ceylan, Enhancing thermal and ionic conductivities of electrospun PAN and PMMA nanofibers by graphene nanoflake additions for battery-separator applications, Int. J. Energy Res. 38 (2014) 2044−2051.

[3] D. Linden, T.B. Reddy, Handbook of Batteries, Third ed., McGraw-Hill, New York, 2002.

[4] L. Zang, Energy Efficiency and Renewable Energy through Nanotechnology, Springer, New York, 2011.

[5] A. Pankaj, Z. Zhengming, Battery separator, Chem. Rev. 104 (2004) 4419−4462.

[6] N. Nuraje, R. Asmatulu, G. Mul, November Green Photo-Active Nanomaterials: Sustainable Energy and Environmental Remediation, RSC Publishing, Cambridge, England, 2015.

[7] V. Chabot, S. Farhad, Z. Chen, A.S. Fung, A. Yu, F. Hamdullahpur, Effect of electrode physical and chemical properties on lithium-ion battery performance, Int. J. Energy Res. 37 (2013) 1723−1736.

[8] S. Xiaomin, Z. Weiping, M. Delong, M. Qian, B. Denzel, M. Ying, et al., Electrospinning of nanofibers and their applications for energy devices, J. Nanomater. 2015 (2015). Article ID 140716, 20 pages.

[9] J. Fang, H.T. Niu, T. Lin, X.G. Wang, Applications of electrospun nanofibers, Chin. Sci. Bull. 53 (15) (2008) 2265−2286.

[10] Y. Liang, S. Cheng, J. Zhao, C. Zhang, S. Sun, N. Zhou, et al., Heat treatment of electrospun polyvinylidene fluoride fibrous membrane separators for rechargeable lithium-ion batteries, J. Power Sources 240 (2013) 204−211.

[11] H. Lee, M. Alcoutlabi, J.V. Watson, X. Zhang, Electrospun nanofiber-coated separator membranes for lithium-ion rechargeable batteries, J. Appl. Polym. Sci. 129 (4) (2013) 1939−1951.

[12] S.W. Choi, S.M. Jo, W.S. Lee, Y.R. Kim, An electrospun poly (vinylidene fluoride) nanofibrous membrane and its battery applications, Adv. Mater. 15 (23) (2003) 2027−2032.

[13] J.R. Kim, S.W. Choi, S.M. Jo, W.S. Lee, B.C. Kim, Electrospun PVdF-based fibrous polymer electrolytes for lithium ion polymer batteries, Electrochim. Acta 50 (1) (2004) 69−75.

[14] S.S. Choi, Y.S. Lee, C.W. Joo, S.G. Lee, J.K. Park, K.S. Han, Electrospun PVDF nanofiber web as polymer electrolyte or separator, Electrochim. Acta 50 (2-3) (2004) 339−343.

[15] K. Gao, X. Hu, C. Dia, T. Yi, Crystal structures of electrospun PVDF membranes and its separator application for rechargeable lithium metal cells, Mater. Sci. Eng. B 131 (1-3) (2006) 100−105.

[16] S.W. Choi, J.R. Kim, S.M. Jo, W.S. Lee, Y.R. Kim, Electrochemical and spectroscopic properties of electrospun PAN-based fibrous polymer electrolytes, J. Electrochem. Soc. 152 (5) (2005) A898−A995.

[17] S. Jayaraman, V. Aravindan, P.S. Kumar, W.C. Ling, S. Ramakrishna, S. Madhavi, Exceptional performance of $TiNb_2O_7$ anode in all one-dimensional architecture by electrospinning, ACS Appl. Mater. Interfaces 6 (11) (2014) 8660–8666.

[18] M. Yanilmaz, M. Dirican, X. Zhang, Evaluation of electrospun SiO_2/nylon 6, 6 nanofiber membranes as a thermally-stable separator for lithium-ion batteries, Electrochim. Acta 133 (2014) 501–508.

[19] A. Jabbarnia, W.S. Khan, A. Ghazinezami, R. Asmatulu, Investigating the thermal, mechanical and electrochemical properties of PVdF/PVP nanofibrous membranes for supercapacitor applications, J. Appl. Polym. Sci. 133 (30) (2016) 43707–43717.

[20] G. Binitha, M.S. Soumya, A.A. Madhavan, P. Parveen, A. Balakrishnan, K.R.V. Subramanian, et al., Electrospun α- Fe_2O_3 nanostructures for supercapacitor Applications, J. Mater. Chem. A 1 (2013) 11698–11704.

[21] M. Moniruddin, B. Ilyassov, X. Zhao, E. Smith, T. Serikov, N. Ibrayev, et al., Recent progress on perovskite nanomaterials in photovoltaic and water splitting applications, Mater. Today Energy 6 (2017) 1–14.

[22] R. Asmatulu, Nanotechnology Safety, Elsevier, Amsterdam, The Nederland, 2013.

[23] C. Kim, B.T.N. Ngoc, K.S. Yang, K. Kojima, Y.A. Kim, A.J. Kim, et al., Self-sustained thin Webs consisting of porous carbon nanofibers for supercapacitors via the electrospinning of polyacrylonitrile solutions containing zinc chloride, Adv. Mater. 19 (17) (2007) 2341–2346.

[24] E. Han, D. Wu, S. Qi, G. Tian, H. Niu, G. Shang, et al., Incorporation of silver nanoparticles into the bulk of the electrospun ultrafine polyimide nanofibers via a direct ion exchange self-metallization process, ACS Appl. Mater. Interfaces 4 (5) (2012) 2583–2590.

[25] J. Li, E.-H. Liu, W. Li, X.-Y. Meng, S.-T. Tan, Nickel/carbon nanofibers composite electrodes as supercapacitors prepared by electrospinning, J. Alloys Compd. 478 (1-2) (2009) 371–374.

[26] G.H. An, H.J. Ahn, Activated porous carbon nanofibers using Sn segregation for high-performance electrochemical capacitors, Carbon 65 (2013) 87–96.

[27] S. Chaudhari, Y. Sharma, P.S. Archana, R. Jose, S. Ramakvishna, S. Mhaisalkar, et al., Electrospun polyaniline nanofibers web electrodes for supercapacitors, J. Appl. Polym. Sci. 129 (4) (2013) 1660–1668.

[28] G. Bhuvanalogini, N. Muruganathem, V. Shobana, A. Subramania, Preparation, characterization, and evaluation of $LiNi_{0.4} Co_{0.6} O_2$ nanofibers for supercapacitor appkications, J. Solid State Electrochem. 18 (2014) 2387–2392.

[29] J.B. Lee, S.Y. Jeong, W.J. Moon, T.Y. Seong, H.J. Ahn, Preparation and characterization of electro-spun RuO_2-Ag_2O composite nanowires for electrochemical capacitors, J. Alloys Compd. 509 (11) (2011) 4336–4340.

[30] A. Alharbi, I.M. Alarifi, W.S. Khan, A. Swindle, R. Asmatulu, Synthesis and characterization of electrospun polyacrylonitrile/graphene nanofibers embedded with SrTiO3/NiO nanoparticles for water splitting, J. Nanosci. Nanotechnol. 17 (2017) 1–9.

[31] A. Alharbi, I.M. Alarifi, W.S. Khan, R. Asmatulu, Synthesis and analysis of electrospun $SrTiO_3$ nanofibers with NiO_x nanoparticles shells as photocatalysts for water splitting, Macromol. Symp. 365 (2016) 246–257.

[32] D.S. Rupali, Development of Solar Sensitive Thin Film for Water Splitting and Water Heating Using Solar Concentrator, PhD dissertation, University of Trento, Italy, 2010.

[33] G.W. Michael, L.W. Emily, M.R. James, B.E. Shannon, Q. MI, A.S. Elizabeth, et al., Solar water splitting cells, Chem. Rev. 110 (11) (2010) 6446−6473.

[34] S.K. Palaniswamy, J. Sundaramurthy, S. Sundarrajan, V.J. Babu, G. Singh, S.I. Allakhverdiev, et al., Hierarchical electrospun nanofibers for energy harvesting, production and environmental remediation, Energy Environ. Sci. 7 (10) (2014) 3192−3222.

[35] A. Alharbi, I.M. Ibrahim, W.S. Khan, R. Asmatulu, "Semiconductor Nanomaterials for Water Splitting," Advances in Material Science Research, Vol 21, Nova Science Publishers, 2015, ISBN: 978-1-63483-547.

[36] D.M. Chapin, C.S. Fuller, G.L. Pearson, A new silicon p-n junction photocell for converting solar radiation into electrical power, J. Appl. Phys. 25 (5) (1954) 676−677.

[37] S. Ramakrishna, K. Fujihara, W.E. Teo, T. Yong, Z. Ma, R. Ramaseshan, Electrospun nanofibers: solving global issues, Mater. Today 9 (3) (2006) 40−50.

[38] S. Chuangchote, T. Sagawa, S. Yoshikawa, Efficient dye-sensitized solar cells using electrospun TiO_2 nanofibers as a light harvesting layer, Appl. Phys. Lett. 93 (2008) 033310−033313.

[39] S.S. Sun, N.S. Sariciftci, Organic Photovoltaics: Mechanisms, Materials, and Devices, Taylor and Francis, Boca Raton, FL, USA, 2005.

[40] R. Asmatulu, A. Karthikeyan, D.C. Bell, S. Ramanathan, M.J. Aziz, Synthesis and variable temperature electrical conductivity studies of highly ordered TiO_2 nanotubes, J. Mater. Sci. 44 (2009) 4613−4616.

[41] H.M. Haynes, Production Alternatives to Screen Printing for Dye Sensitized Solar Cells in Laboratory Testing, M.S. thesis, Wichita State University, Wichita, KS, USA, December, 2010.

[42] B.H. Lee, M.Y. Song, S.Y. Jang, S.M. Jo, S.Y. Kwak, D.Y. Kim, Charge transport characteristics high efficiency dye-sensitized solar cells based on electrospun TiO_2 nanorod photoelectrodes, J. Phys. Chem. C 113 (51) (2009) 21453−21457.

[43] M.Y. Song, D.K. Kim, K.J. Ihn, S.M. Jo, D.Y. Kim, Electrospun TiO_2 electrodes for dye-sensitized solar cells, Nanotechnology 15 (12) (2004) 1861.

[44] M.Y. Song, Y.R. Ahn, S.M. JO, D.Y. Kim, TiO_2 single- crystalline nanorod electrode for quasi-solid-state dye-sensitized solar cells, Appl. Phys. Lett. 87 (11) (2005) 113113.

[45] X. Wang, M. Xi, H. Fong, Z. Zhu, Flexible, transferable, and thermal-durable dye-sensitized solar cell photoanode consisting of TiO_2 nanoparticles and electrospun TiO_2/SiO_2 nanofibers, ACS Appl. Mater. Interfaces 6 (18) (2014) 15925−15932.

[46] L. Yang, W.W.F. Leung, Application of a bilayer TiO_2 nanofiber photoanode for optimization of dye-sensitized solar cells, Adv. Mater. 23 (39) (2011) 4559−4562.

[47] M. Grätzel, Dye-sensitized solar cells, J. Photochem. Photobiol. C Photochem. Rev. 4 (2) (2003) 145−153.

[48] P. Joshi, L. Zhang, Q. Chen, D. Galipeau, H. Fong, Q. Qiao, Electrospun carbon nanofibers as low-cost counter electrode for dye-sensitized solar cells, ACS Appl. Mater. Interfaces 2 (12) (2010) 3572−3577.

[49] S.I. Noh, T.-Y. Seong, H.J. Ahn, Carbon nanofibers combined with Pt nanoparticles for use as counter electrodes in dye-sensitized solar cells, J. Ceram. Process. Res. 13 (4) (2012) 491−494.

[50] R. Asmatulu, M.A. Shinde, A. Alharbi, I.M. Alarifi, Integrating graphene and C_{60} into TiO_2 nanofibers via electrospinning process for the enhanced energy conversion efficiencies, Macromol. Sympos. 365 (2016) 128−139.

[51] R. Asmatulu, W.S. Khan, "Nanotechnology Safety in the Energy Industry," in Nanotechnology Safety, Elsevier, ed. R. Asmatulu, ISBN: 978-0-444-59438-9.

[52] R. Asmatulu, A. Khan, V.K. Adigoppula, G. Hwang, Enhanced transport properties of graphene-based, thin Nafion® membrane for polymer electrolyte membrane fuel cells, Int. J. Energy Res. (2017). Available from: https://doi.org/10.1002/er.3834.

[53] F. Barbir, PEM Fuel Cells: Theory and Practice, Elsevier Academic Press, New York, NY, USA, 2005.

[54] J.H. Park, Y.W. Ju, S.H. Park, H.R. Jung, K.S. Yang, W.J. Lee, Effects of electrospun polyacrylonitrile-based carbon nanofibers as catalyst support in PEMFC, J. Appl. Electrochem. 39 (8) (2009) 1229−1236.

[55] M. Li, G. Han, B. Yang, Fabrication of the catalytic electrodes for methanol oxidation on electrospinning-derived carbon fibrous mats, Electrochem. Commun. 10 (6) (2008) 880−883.

[56] Z.G. Zhao, J.Z. Yao, J. Zhnag, R. Zhu, Y. Jin, Q.W. Li, Rational design of galvanically replaced pt- anchored electrospun WO_3 nanofibers as efficient electrode materials for methanol oxidation, J. Mater. Chem. 22 (32) (2012) 16514−16519.

[57] N.T. Xuyen, H.K. Jeong, G. Kim, K.P. So, K.H. An, Y.H. Lee, Hydrolysis-induced immobilization of Pt $(acac)_2$ on polyimide-based carbon nanofiber mat and formation of Pt nanoparticles, J. Mater. Chem. 19 (9) (2009) 1283−1288.

[58] A. Su, W. Jia, A. Schempf, Y. Lei, Palladium/titanium dioxide nanofibers for glycerol electrooxidation in alkaline medium, Electrochem. Commun. 11 (11) (2009) 2199−2202.

[59] S. Martwiset, K. Jaroensuk, Electrospinning of poly (vinyl alcohol) and poly (4-styrenesulfonic acid) for fuel cell applications, J. Appl. Polym. Sci. 124 (3) (2012) 2594−2600.

[60] J. Choi, K.K.M. Lee, R. Wycisk, P.N. Pintaura, P.T. Mather, Nanofiber composite membranes with low equivalent weight perfluorosulfonic acid polymers, J. Mater. Chem. 20 (2010) 6282−6290.

[61] B. Dong, L. Gwee, D.S.D. La Cruz, K.I. Winey, Y.A. Elabd, Super proton conductive high-purity nafion nanofibers, Nano Lett. 10 (9) (2010) 3785−3790.

[62] T. Tamura, H. Kawakami, Aligned electrospun nanofiber composite membranes for fuel cell electrolytes, Nano Lett. 10 (4) (2010) 1324−1328.

[63] S. Mollá, V. Compañ, Polyvinyl alcohol nanofiber reinforced Nafion membranes for fuel cell applications, J. Membr. Sci. 372 (1-2) (2011) 191−200.

[64] M. Salahuddin, M.N. Uddin, G. Hwang, R. Asmatulu, Superhydrophobic PAN nanofibers for gas diffusion layers of proton exchange membrane fuel cells for cathodic water management, Int. J. Hydrog. Energy (2017). Available from: http://dx.doi.org/10.1016/j.ijhydene.2017.07.229.

[65] S. Cavaliere, S. Subianto, I. Savych, D.J. Jones, J. Roziere, Electrospinning: designed architectures for energy conversion and storage devices, Energy Environ. Sci. R. Soc. Chem. 4 (2011) 4761−4785.

[66] L. Li, P. Zhang, R. Liu, S.M. Guo, Preparation of fibrous Ni-coated-YSZ anodes for solid oxide fuel cells, J. Power Sources 196 (3) (2011) 1242−1247.

[67] S. Shahgaldi, Z. Yaakob, D.J. Khadem, M. Ahmadrezaei, W.R.W. Daud, Synthesis and characterization of cobalt-free $Ba_{0.5}Sr_{0.5}Fe_{0.8}Cu_{0.2}O_{3-\delta}$ perovskite oxide cathode nanofibers, J. Alloys Compd. 509 (37) (2011) 9005−9009.

[68] S.S. Ray, A.L. Yarin, B. Pouredeyhimi, The production of 100/400 nm inner/outer diameter carbon tubes by solution blowing and carbonization of core−shell nanofibers, Carbon 48 (12) (2010) 3575−3578.

[69] S. Chen, H. Hou, F. Harnisch, S.A. Patil, A.A. Carmona-Mertinez, S. Agarwal, et al., Electrospun and solution blown three-dimensional carbon fiber nonwovens for application as electrodes in microbial fuel cells, Energy Environ. Sci. 4 (2010) 1417−1421.

Electrospun nanofibers for filtration applications

7

CHAPTER OUTLINE

7.1 NANOTECHNOLOGY IN FILTRATION

Nanotechnology, the science and art of manipulating matter at the nanoscale, presents the potential of engineered nanomaterials for filtration applications. Nanotechnology is one of the fastest-growing scientific disciplines due to its immense potential in creating novel materials that have advanced domestic and industrial applications. Nanotechnology has tremendously impacted many different science and engineering disciplines, such as chemistry, biology, physics, electronics, materials science, and engineering. Electrospun nanofibers, due to their high surface area to volume ratio, flexibility, and porosity, find applications as filter media, adsorption layers in protective clothing, environmental and industrial uses, and so on. Scientists have discovered that materials at nanoscale can have significantly improved most of their properties when compared to their bulk counterparts. Therefore, there are numerous possibilities for improved devices, structures, components, and materials if we understand this difference and learn how to control the assembly of small structures or nanostructures [1].

The impact of nanotechnology in the advancement of filtration will be more promising in the near future. Nanotechnology has a lot of potentials for separation science and technology for a wide range of applications in food, beverage, environmental, and biomedical fields. In recent years, nanoscale membranes made from engineering techniques have become competitive when compared to many other conventional techniques, such as ion exchangers, sand filters, and adsorption [1]. The distinctive advantage of membrane technology is its high efficiency, low power consumption, and minimum investment. Electrospun nanofibrous membrane possesses a high surface area to volume ratio, high porosity, tunable pore size, and ease in fabrication, which make them ideal candidates in filtration and membrane science.

Synthesis and Applications of Electrospun Nanofibers. DOI: https://doi.org/10.1016/B978-0-12-813914-1.00007-9

Nanotechnology has the potential of playing a vital role in wet and dry filtration for solid—liquid separation, dewatering, and dust control. Polymeric nanofibrous membranes have been used in a wide variety of commercial applications recently, and these nanomembranes hold promise for technological exploitation in expanding the advanced filtration applications. Nanotechnology presents the potential for treatment of surface water, groundwater, wastewater, and air by toxic particles, organic and inorganic substances, and microorganisms.

The separation of suspended solid particles, microorganisms, and droplets from a liquid or gas by means of a porous medium which retain the solid particles, microorganisms, and droplets of liquid or gas and allows the liquid to pass through, is called filtration. The filter medium is permeable to the liquid, so it retains the suspended solid particles and permits only the liquid to pass through, collecting the filtrate on the membranes. The volume of filtrate collected per unit time is generally referred to as the filtration rate. Filtration has been extensively used in both households and industries for removing suspended solid particles from air and liquid. As far as the application of filtration in the military is concerned, it is used in isolated bags, respirators, and uniform garments, to disinfect aerosol dusts, bacteria, viruses, fungi, and molds [2]. Medical applications need fabrics to be free from bacteria and viruses.

In a fiber-based filter, the particle-removal process is determined by different mechanisms. Large particles are generally blocked on the surface of filter media due to molecular sieve effects. However, the particles that are smaller than the pore size can still be collected by the fiber media, through a different mechanism, such as interception, impaction, or static electrical attraction [2]. Very fine suspended particles can also be captured by the Brownian motion effect [2]. The filtration efficiency is an important parameter in the filtration process. The filtration efficiency depends on the pore size, thickness of filter media, and surface properties of the media. Besides the filtration efficiency, the other parameters, such as pressure drop and flux resistance, are also important factors to be evaluated [2]. The chemical and biological contaminants present in air and water pose a serious threat to human health and they must be eliminated by means of an appropriate filtration technology.

The research and development on polymeric nanofibers have gained much interest in recent years due to their wide range of applications in engineering and medical fields. When compared to conventional microfibers, polymeric nanofibers possess high filtration efficiency and separation capability of micro- and nano-scale objects because of their higher inertia impaction and interception. Additionally, due to slip flow at the nanofiber surface, drag force and pressure drop significantly decrease. Slip flow causes more contaminants trapped near the surface of nanofibers and inertia impaction, interception efficiency, and filtration efficiency increases for the same pressure drop as compared to conventional membranes. These benefits, besides high surface area, high porosity, and flexibility, facilitate adsorption of contaminants from air are the primary reasons for increasing attention towards polymeric nanofibrous membrane for filtration

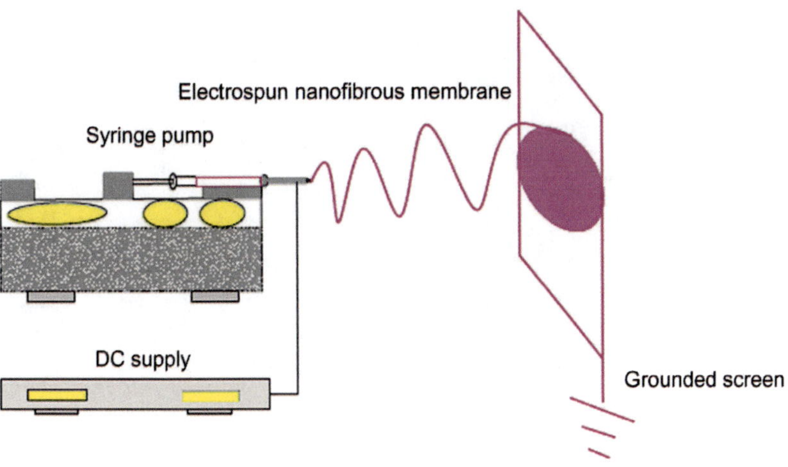

FIGURE 7.1

General view of an electrospinning process.

applications. A number of processing methods have been available for polymeric nanofiber fabrication. However, electrospinning stands in a unique position due to its low investment and easy fabrication. Filtration technology has been improved by the induction of electrospun nanofibers since they possess small pore size and higher surface area than regular fibers. Electrospinning is a novel process that produces continuous ultrafine polymer fibers through the action of an external electrostatic field applied on a polymeric solution or melt. Fig. 7.1 shows a general view of an electrospinning process.

Electrospun nanofibers possess high surface area to volume ratio, high porosity, nanosized diameter, low basic molecular weight, high permeability (based on hydrophilic nature of the polymers), ability to control composition, and well-connected pore structures [3]. Electrospun nanofibers with high filtration efficiency, small pore size, and low cost are materials of choice for many filtration applications. These outstanding properties of nanofibers are most suitable for filtration media (or nanomembranes), such as air/dust filtration and water/liquid/oil filtration. Although polymeric nanofibrous membranes are being currently employed for commercial applications in air filtration and dust control, their applications in water/liquid filtration are yet to be exploited. The fibrous membrane thus fabricated via electrospinning possesses flexibility, high surface area, and porous structure, which allow much higher sites for separation processes, and are generally referred to as electrospun nanofibrous membranes (ENMs) [4−7]. Fig. 7.2 shows an SEM image of electrospun nanomembranes made of PAN nanofibers.

The polymers generally used for fabricating nanoporous membranes are cellulose nitrate, polyvinylchloride, polyacrylonitrile, cellulose acetate, aromatic

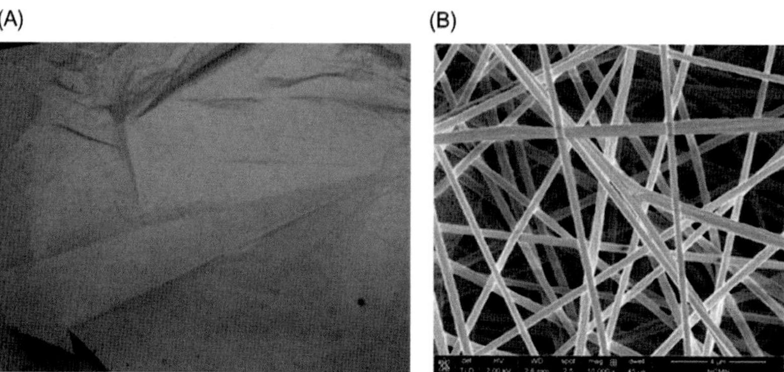

FIGURE 7.2

(A) Photograph and (B) SEM image of electrospun nanomembranes made of PAN nanofibers.

polyamide, aliphatic polyamide, polysulfone, polycarbonate, polytetrafluoroethylene, polyvinylidene fluoride, polydimethylsiloxane, polypropylene, polyvinylidene difluoride, etc. ENMs possess high flux and low transmembrane pressure. The problem with ENMs is that the membrane possesses electrostatic charge since an electrostatic field has been used during electrospinning and the charge intensity would intensify as the thickness of the membrane increases. Additionally, ENMs require additional support in order to provide mechanical strength. Therefore, nowadays, most of the applications of ENMs in membrane separation technology are generally based on a hybrid system [8]. In this system, nanofibers are generally placed on a support (substrate) or sandwiched between different layers or blended together with micro-sized fibers [8]. ENMs are an effective means of treating wastewater. However, they experience fouling and biofouling during the service, which substantially reduce the filtration rates. Fouling is the accumulation of solutes on the membrane surface resulting in the resistance in mass-transfer, thereby reducing the membrane productivity. Surface modification is one of the methods to alleviate fouling. Interfacial polymerization, surface coating, blending, and grafting are the processes generally used for surface modification. Plasma-induced graft polymerization is another very efficient method for applying a selective layer of polymer on the top surface of a membrane. The number of pores is reduced by this method; however, the porosity remains the same. Surface modification plays an important role in enhancing the performance of membranes.

ENMs have shown promising results in the laboratory testing stage; nevertheless, their readiness for large-scale commercialization is still facing some technical challenges, such as compatibility with the existing infrastructures, environmental and health risk, potential degradation of polymer with time, and initial cost. However, some companies such as Donaldson (US) have been producing nanofiber-based filters for domestic and industrial applications for a long time [2]. Recently, AMSOIL developed a nanofibrous-based filter for automobile applications. Similarly, DuPont has been producing electrospun fabric products for HVAC, automotive and liquid filtration, bedding protection, and apparel applications [2]. ENMs provide a significant increase in filtration efficiency in comparison to conventional filters at the same pressure drop. ENMs with ultrafine diameters have much higher capability to collect fine suspended particles, since slip flow around the fibers increases the diffusion, interception, and inertia impact efficiency [9].

Both experimental studies and theoretical calculations clearly reveal that electrospun nanofiber mats are extremely efficient at trapping airborne small particles compared to conventional filters. A thin layer of electrospun fibers on a porous substrate is sufficient to eliminate particle penetration [10]. Fibrous membranes possess advantages, such as high filtration efficiency and low fluid resistance. Electrospun nanofiber layers present minimum impedance to moisture vapor diffusion, which is an important parameter in protecting cloths in decontamination applications [10]. ENMs can also remove tiny liquid droplets within a liquid—liquid immiscible system. ENMs are being used as a supporting scaffold for infiltration of oil/water emulsion separation. The possibility of incorporating a variety of polymers, biological agents, and particulate nanofibers during the electrospinning process will lead to the development of new sets of nanocomposite/hybrid nanofibrous membranes with high efficiency and a much broader range of environmental and health applications [11]. Fig. 7.3 shows SEM images of electrospun PVC nanofibers incorporated with 4 wt.% of PVP obtained at 25 KV DC voltage, 2 mL/h pump speed, and 25 cm separation distance [12].

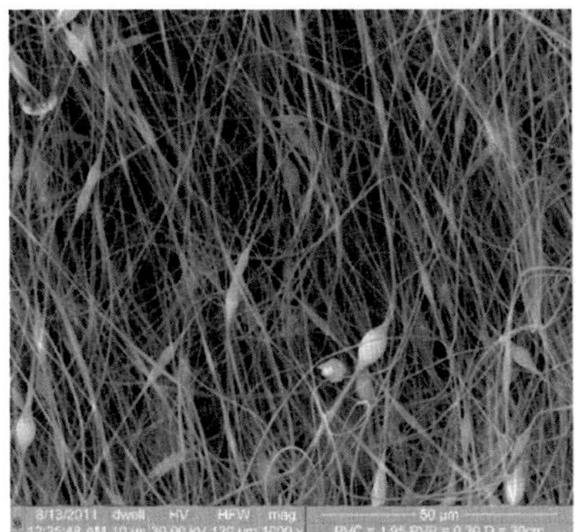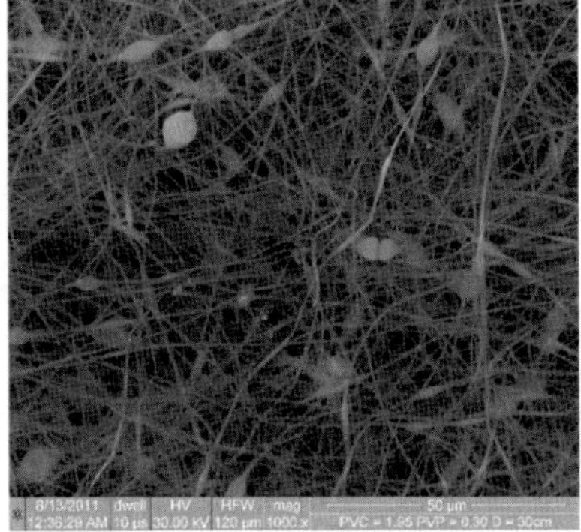

FIGURE 7.3

SEM images showing electrospun PVC nanofiber membranes incorporated with 4 wt.% of PVP obtained at 25 KV DC voltage, 2 mL/h pump speed, and 25 cm separation distance [12].

There are two factors that define the functionality of membrane: flux and selectivity. Selectivity refers to the surface properties of a membrane that distinguish the types of species that can pass through the membrane. Selectivity in removing contaminants, mechanical strength, and porosity can be modified in ENMs to enhance their efficiency [13]. Surface modification of ENMs enhances the efficiency of filtration. Flux means the rate at which the species are transferred across the membrane. These two factors depend on parameters such as pore size, wettability, porosity, tortuosity, pressure drop across the membrane, and thickness of the membrane [14]. The pore size of the ENM is determined by the bubble-point method. The process involves the measurement of pressure needed to blow air through a liquid-filled membrane [14]. The membrane is placed in the supporting cell of distilled water and connected to a bubble-flow meter. The pressure is applied to the membrane base and at each pressure, and the corresponding bubble flow rate is measured. The relationship between the pore size and the corresponding pressure is given by the Young–Laplace equation [14]:

$$R = \frac{2\gamma}{\Delta P}\cos\theta \tag{7.1}$$

where "R" is the radius of the pore, "ΔP" is the differential pressure, "γ" is the surface tension, and "θ" is the contact angle. In recent years, a number of research articles have been published on nanofibrous membranes for water and air filtration.

7.2 ELECTROSPUN NANOFIBERS FOR WET FILTRATION

Wet filtration encompasses a wide range of methods employed to separate solid particles, gas and liquid droplets, and microorganisms from liquid. Wet filtration is generally used in petroleum, chemical industries, pigments, metallurgy, pharmacy, food, paper mills, and coal and water treatment. ENMs are replacing woven fabrics in an increasing number of wet filtration applications, due to their higher filtration efficiency and an ability to retain very fine particles. Electrospun nanofibers have very high length (hundreds of kilometers long), so they can never become airborne and diffuse into the human body.

Water pollution is an important concern in the recent era. Water filtration/purification with the latest technology is an urgent need in today's economy. There are approximately 1 billion people in the world that have inadequate access to potable water and around 2.6 billion people lack access to sanitation [15]. Millions of people die every year due to maladies transmitted through unclean water and human excreta. With the increase in human population and associated environmental degradation, the scarcity of clean and potable water supply will constitute a major problem to the present state of the world's water reserve. This water crisis is growing annually and will continue to grow in the coming years, since water scarcity is occurring rapidly due to droughts, urbanization, population growth, and industrialization [12].

The release of wastewater from industries, commercial establishments, and domestic usage causes a catastrophic scenario. The presence of arsenic in water can cause major health concerns, such as bladder, liver, lung, kidney, and skin cancers. Dioxins are released during the combustion process, such as waste, forest fires, and oil and coal burning. They can destroy water sources. Adding fluoride to drinking water is a standard practice but it is a neurotoxin and an endocrine disruptor. It can harm the thyroid gland and calcify the pineal gland. It is so toxic that several countries have banned water fluoridation. Water contains lead, which is toxic to almost all organs. Mercury is very toxic and causes brain damage, blindness, nerve damage, cognitive disability, headaches, weakness, muscle atrophy, tremors, mood swings, memory loss, and skin rashes. Polychlorinated biphenyls (PCBs) are used in industries for insulation, oils, paints and adhesives, electronics, and fluorescent lights. Experimental tests show that PCBs cause cancer, as well as affecting the immune system, nervous system, and endocrine system. Perchlorate, a chemical compound is used in explosive and rocket fuel causes thyroid cancer.

Membrane technology is playing a vital role in water treatment because of the fact that conventional processes of water treatments, such as sedimentation, flocculation, coagulation, and active carbon are incapable of eliminating fine-size organic pollutants to specific limits. Water treatment processes consist of using several types of membranes, such as microfiltration (MF), ultrafiltration (UF), reverse osmosis (RO), and nanofiltration (NF) [16]. RO can provide the purest water; however, NF membrane has been recognized as a novel process in water treatment. NF can swiftly and economically address the major portion of the total dissolved solids from surface and ground water, pathogens (e.g., bacteria, viruses, molds, and fungi), monovalent and multivalent anions and cations (water-softening agents), minerals, salts, and other suspended micro- and nanoscale particles [16]. ENMs are a possible solution for providing potable water with minimum investment. Engineers and scientists have established a critical relationship between turbidity and human diseases. According to this relation, more turbidity means more toxic substances, which may increase the chances of diseases in human. A number of studies have shown that ENMs reduce turbidity in water to a significant level. ENMs can rapidly and economically eliminate total dissolved solids (TDSs), pathogens, monovalent and multivalent anions and cations, salts, minerals, and other suspended nanomaterials [12]. ENMs can effectively remove protozoa (*Ascryptosporidium* and *Giardia*) from water sources. Fig. 7.4 shows a schematic of the ENM filtration process.

As can be seen in Fig. 7.4, filtration is a process of removing or eliminating suspended particulates from raw water by applying pressure to drive the water through porous permeable media. The polymers used in ENM fabrication are mechanically and chemically stable with a hydrophilic surface for the use as filter media. ENM filtration can remove pathogenic microorganisms in wastewater and

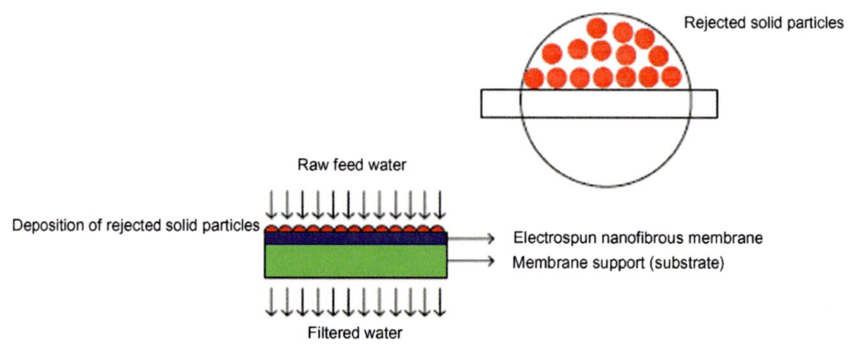

FIGURE 7.4

Schematic diagram of the ENM filtration process.

water to prevent waterborne diseases. Many research studies have reported the applications of ENMs for eliminating bacteria, particles, or dye from water [17]. Bacteria are generally in microns, whereas viruses are in the range of tens of nanometers. An ENM can easily remove bacteria; however, for virus removal the size of the pore should be extremely small (50−500 nm). By controlling the process parameter, electrospun membrane can be fabricated with nanosized pores. The only problem with small pore size is that the water flux is reduced. Sato et al. [18] used cellulose fine fibers infused on a PAN ENM nonwoven substrate to fabricate a composite membrane for removing bacteria and viruses. In order to remove viruses, they charge the membrane (polyacrylonitrile ENM membrane) by applying a cellulose fiber layer on the top surface, so the positively charged cellulose fibers attract and trap the negatively charged viruses. They achieved 99.99% efficiency in removing *Escherichia coli* (*E. coli*), which is in close agreement to 2 cu/mL set up by National Sanitation Foundation Standard [18].

Additionally, ENMs eliminate most bacteria such as, coliform, *Salmonella*, *Cryptosporidium*, *Giardia*, and other waterborne microorganisms. Some studies have reported embedding silver nanoparticles (3−6 nm) in polyacrylonitrile electrospun membrane and testing for Gram-positive *Bacillus cereus* and Gram-negative *E. coli* microorganisms [19]. The membrane showed excellent antibacterial activity. The methodology of amidoxime-functionalized PAN was used in those studies. Tests were conducted for finding microbes such as *S. aureus* and *E. coli* for Ag + and its reduction to Ag nanoparticles. The ASFPAN-3, which were amidoxime-functionalized nanofibers, after immersion for 20 min in NH_4OH showed complete (log 7) reductions (killing all unwanted objects) [19]. A similar trend was observed for $AgNO_3$ solution, in which polyacrylonitrile nanofibers were dipped in solution for 30 min and Ag nanoparticles/PAN nanofibers displayed seven orders of magnitude bacteria reduction from the media [19]. The ENMs can virtually eliminate all the possible bacteria that can cause many diseases, including typhoid fever, infectious hepatitis, flu, polio, tetanus, cholera, dysentery, meningitis, and respiratory diseases.

Wastewater also contains some traces of heavy metals in water sources. ENMs can eliminate heavy metals from wastewater. Chromium is a toxic pollutant in wastewater, which causes cancer. ENMs display excellent performance in chromium removal. Teha et al. [20] reported the removal of chromium (VI) with the removal rate reaching up to 19.45 mg/g by means of functionalized cellulose acetate/silica composite membrane. The same group used poly(vinyl alcohol) polymer matrix for removing Cr (II) up to 97 mg/g [21]. Lead and copper can be removed by chitosan nanofiber membrane [19]. Aliabadi et al. [22] reported the removal of other metal ions, including cadmium, copper, nickel, and lead by employing ENMs. Generally, the important parameters that must be taken into account for measuring the quality of filtered water are turbidity, total dissolved solids (TDSs), chemical oxygen demand (COD), pH, and biochemical oxygen demand (BOD). ENMs reduced these parameters to acceptable limits. Alharbi et al. [16] reported the fabrication of electrospun nanofibrous membrane with a blend of polyacrylonitrile (PAN) and polyvinylpyrrolidone (PVP) polymeric solutions incorporated with gentamicin sulfate to filter wastewater and dam water. Their results revealed that turbidity, TSS, COD, and BOD were reduced to acceptable limits. Fig. 7.5 shows SEM images of electrospun PAN nanofibers incorporated with 5 wt.% PVP and 5 wt.% gentamicin (antibacterial agent) before and after the filtration of wastewater. The PAN nanofibers were produced using 25 KV voltage, 3 mL/h pump speed, and 25 cm tip-to-collator distance. Fig. 7.6 also shows images of ENMs before and after dam water filtration, while Fig. 7.7 shows images of ENMs before and after wastewater filtration.

The filtration of oil and water is a challenging issue due to increased amount of oily industrial wastewater and polluted ocean water. ENMs are capable of filtering oily wastewater. Oil pollution coming from textile, petrochemical, food industries, and frequent oil spill accidents during offshore oil production and marine transportation, cause great loss of energy sources and damage the

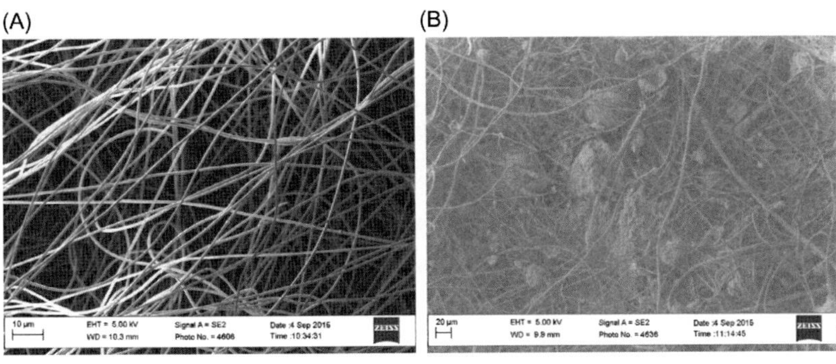

FIGURE 7.5

SEM images of electrospun PAN nanofibers incorporated with 5 wt.% PVP and 5 wt.% gentamicin (antibacterial agent) (A) before and (B) after filtration of wastewater.

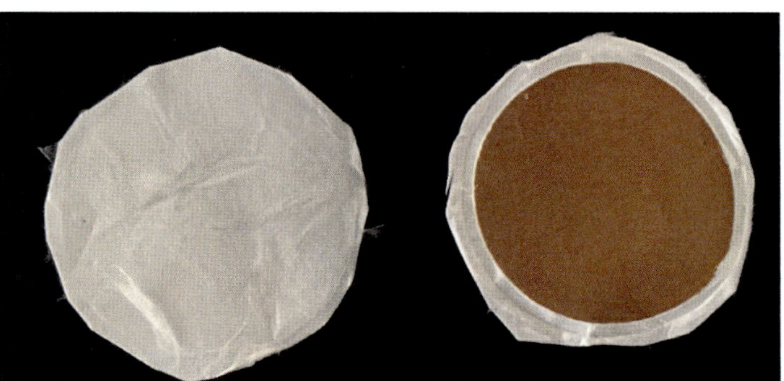

FIGURE 7.6

PAN/PVP membrane with gentamicin sulfate before and after dam water filtration.

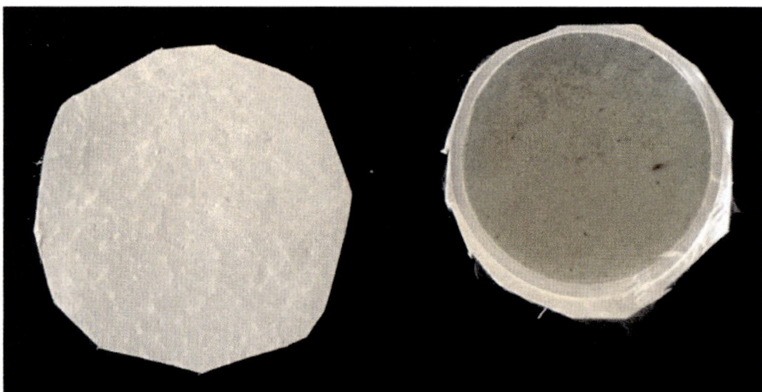

FIGURE 7.7

PAN/PVP membrane with gentamicin sulfate before and after wastewater filtration.

ecological environment [23]. Conventional methods, such as flotation, ultrasonic separation, and skimming are marred by low separation efficiency, high energy cost, and secondary pollution and are incapable of separating oil/water emulsion [23]. Therefore, an efficient and cost-effective novel functional membrane is urgently needed to address this issue. Electrospinning provides an ideal solution for fabricating a wettable surface for effective oil/water separation. The conventional approach for oil spill cleanup is mechanical extraction by sorbents, such as nonwoven polypropylene fibrous mats, which have been widely used in oil spill cleanup. However, they suffer from a low sorption capacity (<30 g/g) [23]. ENMs have shown great promise in this regard.

The sorption capacity of electrospun nanofibers is significantly higher than conventional sorbents since electrospun nanofibrous mats can drive the oil not only into the voids between fibers but also into its pores when the surfaces are highly hydrophobic and oleophilic. The membrane technology is the most promising technology for oily wastewater separation. Fabrication of a fibrous membrane with good wetting properties can be achieved by manipulating the chemical composition and surface geometry [23]. These membranes can be classified into three types: oil-removing, water-removing, and smart separation membranes. The oil-removing membrane (superhydrophobic and superoleophilic) with superwettable features can repel water and allows oil to flow through, thereby achieving high efficiency and selectivity [24]. In the context of wettability of solid surfaces, the Wenzel and Cassie−Baxter model described that the introduction of roughness could make a hydrophobic surface superhydrophobic and an oleophilic surface superoleophilic [25]. Therefore, a membrane that exhibits both superhydrophobic and superoleophilic properties can be fabricated with high surface roughness and low surface energy [26]. The prepared nanoscale membranes can be effectively employed for oil/water separation and other filtration purposes.

Wang et al. [23] reported the use of an in situ polymerization process for the synthesis of superhydrophobic and superoleophilic nanofibrous membrane for filtration and oil/water separation. They prepared a nanofibrous membrane by combining electrospun nanofibers and an in situ polymerized fluorinated polybenzoxazine (F-PBZ) functional layers embedded with nanoparticles (SiO_2 and Al_2O_3). The electrospun nanofibers used can be cellulose acetate, poly(m-phenylene isophthalamide) or poly(m-phenylene isophthalamide). By using the F-PBZ/nanoparticles modification, the surface of the membrane displayed superhydrophobicity with an average water contact angle of 161 degrees, and superoleophilicity with an average water contact angle of 3 degrees [23]. This membrane exhibited swift and efficient oil−water separation under gravitational force. Some studies have shown the development of oil-removing nanofibrous membrane by using Ag nanocluster or hydrophobic nanosilica on the membrane surface [27,28]. Superhydrophobic and superoleophilic electrospun nanofibrous membranes with efficient oil−water separation performance were fabricated by combining the amination of polyacrylonitrile nanofibers and the electroless plating-immobilization of Ag nanoclusters and subsequent surface modification. This membrane displayed excellent separation efficiency and good recyclability, which would certainly make them an ideal candidate in industrial oil−water separation and marine oil spill cleanup [27]. The oil-removing membranes can be fouled or sometimes blocked by thick oil due to their oleophilic property, which affects their efficiency.

The water-removing membrane is a new development in membrane technology having superhydrophilicity and superoleophobicity characteristics, rendered an alternative method to address the fouling issue in oil−water separation [29]. A superhydrophilic and superoleophobic membrane can be fabricated by using $CaCO_3$-based coatings on poly(acrylic acid) grafted polypropylene nanofibrous membranes [23]. The mineral coating traps excessive amounts of water in an aqueous environment and forms a hydrated layer on the membrane surface, thereby endowing the membrane with underwater superoleophobicity. The membrane can separate the oil/water mixture and emulsion with high separation efficiency and high water flux, under pressure or under the influence of gravity [23]. Recently, a superoleophobic cellulose PVDF-co-hexafluoropropylene (PVDF-HFP) nanofibrous membrane has been developed by an electrospinning process that can separate water from oil with an efficiency of 99.9% [30].

ENMs are considered a very good candidate for water filtration with high permeability and lower energy consumption than conventional methods and they can be used for oil/water separation after surface modification. However, the low-flux and fouling problems can be overcome by depositing a thin layer of hydrophilic materials, such as poly(vinyl alcohol) (PVA) [31,32], chitosan [33], or polyamide [34] onto the surface of the membrane by physical absorption or interfacial polymerization. Chu et al. [31−33] demonstrated the surface modification of a hydrophilic layer on nanofibrous membrane to attain high flux in oil/water separation. A recent advancement in membrane technology is the development of membrane with stimuli-responsive wettability. The controllable wettability in such a membrane can be achieved by applying an external stimulus, such as light, pH, temperature, and an electric field [23]. Based on wettability switch, smart nanofibrous membranes have been developed with controllable oil/water wettability, which can treat oil-contaminated water [23]. Those polymers that have acidic or basic functional groups generally contain pH-responsive wetting features since their conformation and charges are influenced by different pH solutions. There are many ways of fabricating a membrane with pH-responsive wettability. However, a new process has been mentioned in the literature, wherein a smart fibrous membrane can be fabricated by depositing poly(methyl methacrylate)-block-poly(4-vinylpyridine) (PMMA-b-P4VP) fibers on a stainless steel mesh through electrospinning process [23]. The pH-responsive P4VP and the underwater oleophilic/hydrophobic PMMA displayed that the membrane possesses switchable wettability towards oil and water [23]. Oil passes through the membrane easily and after wetting the membrane with acidic water having pH=3, the reverse separation can be achieved [23].

7.3 ELECTROSPUN NANOFIBERS FOR DRY FILTRATION

Nanofibers are fibers with diameter less than 500 nm. Generally, electrospinning can produce fibers with diameters from 3 nm to 10,000 nm. Polymeric nanofibers have been used in a wide variety of commercial air filtrations for the last two decades. These fibers have shown great promise in expanding their filtration applications. Air pollution has becoming a more and more serious issue, especially haze pollutions [35−37]. Haze is caused by suspended fine particles in air that absorb and scatter light, before it reaches us [38]. The airborne fine particles with aerodynamic radius of 2.5 microns can cause many diseases [35,39]. In our modern society, transportation, construction, pharmaceuticals, and manufacturing industries continuously release large quantities of dust, dirt, soot, and smoke into the atmosphere (suspended micro- and nanoparticles), which cause heavy air pollution, leading to serious health-hazardous conditions. Some of the emissions and particles come from natural sources, such as volcanoes and wind dust storms [40]. Other sources of emissions are chimneys, car exhausts, and mining [40]. The introduction of chemicals, particulates, and biological matters into the atmosphere and some other known and unknown sources can bring about many diseases and in some cases death or permanent body/organ damage to humans. Airborne fine particles are a major source of cardiovascular and respiratory diseases in many developed and developing countries. Moreover, the presence of volatile organic compounds, ozone, sulfur dioxide, and nitrogen dioxide is extremely harmful for individuals suffering from asthma. Air pollution also causes allergies and skin diseases. Therefore, there is an urgent need for the development of advanced filtration technologies to address this issue.

Conventionally, various materials such as activated carbon and glass fibers and various other methods, including sediment deposition, adsorption, and ion exchange have been used to eliminate contaminants and other pollutants [41]. Activated carbon is used to eliminate toxic chemicals by means of an adsorption process, and high-efficiency particulate air (HEPA) filters are used to eliminate lint and other debris from the air [42]. Usually, charcoal impregnated with metals and their oxides, such as Ca, Zn, Cu, Ag, and Mo along with trietylene diamine (TEDA) and other compounds has been effectively employed in protective clothing and face mask applications [41]. Although these methods and materials are convenient and efficient, they have some drawbacks, such as heavy moisture adsorption and their disposal after use. In order to find suitable materials for filtration and separation, many researchers have explored various types of materials and methods for such applications. Among them, electrospun nanofibers are the material of choice due to their wide range of applications and ease of fabrication. High permeability, small pore size, high surface area, and small fiber diameter makes nanofibers ideal in filtration applications. The capability of nanoscale fibrous membranes and structures to permeate moisture and eliminate chemical vapors make them an ideal candidate in filtration and separation applications, including protective clothing, wastewater treatment, chemical and biological contaminants, and in the textile industry [41].

ENM provides high filtration efficiency since nanofibers with small fiber diameter can capture fine particles easily. The small diameters of the fibers cause slip flows at the fiber surface, thereby increasing the interception and inertia efficiency of ENMs. The small diameters of fibers lead to higher filtration efficiency. ENMs have been tested for adsorption of volatile organic compounds present in the air by many researchers. Scholten et al. [43] found that the adsorption and desorption of volatile organic compounds by ENMs were much faster than conventionally used activated carbon. Cyclodextrins are a family of compounds generally made up of sugar molecules, that possess the ability to form complexes with various molecules, such as hazardous substances and pollutants. Uyar et al. [44] incorporated cyclodextrins into poly(methyl methacrylate) to produce ENMs and reported that volatile organic compounds such as aniline, styrene, and toluene can be eliminated by employing cyclodextrin-based poly(methyl methacrylate) electrospun membrane.

Xu et al. [45] reported adsorption of volatile organic compounds in biotreatment of municipal sewage by ENMs. The incorporation of nanoparticles such as TiO_2, Al_2O_3, and MgO into nanofibers to decontaminate toxic gases, chemical contaminants, biological contaminants (e.g., bacteria, viruses, fungi, and molds), and pesticides for air filtration has been reported in many studies [41]. Podgorski et al. [46] determined the fractional efficiency of nanofibrous membrane and found that a decrease in fiber diameter causes an increase in fractional efficiency. The nanomaterials can be incorporated with textile products to provide antimicrobial properties, decrease luster, and protect against UV radiation, oxidation, and moisture. Metal oxide nanoparticles can decrease luster or provide protection against UV radiation and other harmful effects. Alkoxysilane-modified TiO_2 nanoparticles can absorb UV rays. Some nanoparticles, such as Ag, TiO_2, and ZnO provide antimicrobial and UV properties [41]. Monazite can provide thermal protection for spacecraft reentry applications [41]. Lala et al. [47] determined the antimicrobial (*E. coli* and *Pseudomonas aeruginosa*) activity of poly(vinyl chloride), cellulose acetate, and polyacrylonitrile nanofibrous membrane containing Ag nanoparticles. The functionalization of nanofibers can be beneficial in attracting fine particles and thereby increasing the filtration efficiency. For the filtration of chemical and biological weapon agents, dusts, and debris, ENM nanofibers are made of some specific polymers or coated with some selective agents.

The potential applications of ENMs are enormous because they can be used as filter media in clean air applications in offices, factories, transportation vehicles, schools, hospitals, homes, and other buildings in the near future. Ahn et al. [48] reported that the filtration efficiency of nylon-6 nanofibrous membrane is better than commercially available HEPA filters. The nanofiber filters can be used in hospitals and buildings, wherein the contaminated air (bacteria and pathogens) can be filtered before entering into rooms, as well as other applications [40,48−50].

7.4 CONCLUSIONS

Physical, chemical, and biological contaminants present in environments, such as in air, water, and soil sources pose a great threat to human health, agriculture, and wildlife. For the betterment of human life prior to the consumption of air and water, these pollutants must be eliminated from their sources using various methods. The polymeric nanofibers produced via electrospinning have extremely high surface area, porous, flexibility, and permeable structure with good pore-connectivity plus functionalities and surface chemistry, which make them ideal candidates for a range of advanced separation applications. These properties are suitable for filtering media of water and air that may have been contaminated. Electrospun nanofibrous membranes (ENMs) are a cutting-edge membrane technology that offers substantial high flux and high rejection rate compared to conventional membranes. ENMs present a breakthrough in water and wastewater treatment by offering a lightweight, cost-effective, and less energy-consuming process, when compared to conventional membranes. The tendency of nanofiber membranes to permeate moisture and capture chemical vapors makes them ideal candidates in filtration and separation applications, such as protective clothing, and in the prevention of physical, chemical, and biological contaminants. ENMs incorporated with metal oxide nanoparticles can decontaminate toxic gases, chemical contaminants, biological contaminants, insecticides, and pesticides from the atmosphere for a number of different domestic and industrial applications.

REFERENCES

[1] R. Balamurugan, S. Sundarrajan, S. Ramakrishna, Recent trends in nanofibrous membranes and their suitability for air and water filtrations, Membranes 1 (2011) 232–248.

[2] J. Fang, X. wang, T. Lin, Nanotechnology and Nanomaterials "Nanofibers-Production, Properties and Functional Applications," Intech. 2011.

[3] C. Wang, E. Yan, Z. Huang, Q. Zhao, Y. Xin, Fabrication of highly photoluminescent TiO2/PPV hybrid nanoparticle-polymer fibers by electrospinning, Macromol. Rapid Commun. 28 (2007) 205–209.

[4] B. Ghorani, N. Tucker, M. Yoshikawa, Approaches for the assembly of molecularly imprinted electrospun nanofiber membranes and consequent use in selected target recognition, Food Res. J. 78 (2015) 448–464.

[5] J. Isezaki, M. Yoshikawa, Molecularly imprinted nanofiber membranes: localization of molecular recognition sites on the surface of nanofiber, J. Membr. Sep. Technol. 3 (2014) 119–126.

[6] I.M. Alarifi, A. Alharbi, W. Khan, R. Asmatulu, Carbonized electrospun polyacrylonitrile nanofibers as highly sensitive sensors in structural health monitoring of composite structures, J. Appl. Polym. Sci. 133 (13) (2016) 43235–43241.

[7] I.M. Alarifi, A. Alharbi, W.S. Khan, A. Swindle, R. Asmatulu, Thermal, electrical and surface hydrophobic properties of electrospun polyacrylonitrile nanofibers for structural health monitoring, Materials 8 (2015) 7017–7031.

[8] C. Shin, G.G. Chase, Water-in-oil coalescence in micro-nanofiber composite filters, AIChE J. 50 (2) (2004) 343–350.

[9] K. Kosmider, J. Scott, Polymeric nanofibers exhibit an enhanced air filtration performance, Filtrat. Sep. 39 (6) (2002) 20–22.

[10] J. Fang, T. Niu, T. Lin, G. Wang, Applications of electrospun nanofibers, Chinese Science Bulletin 53 (15) (2008) 2265–2286.

[11] S. Homaeigohar, M. Elbahri, Nanocomposite electrospun nanofiber membranes for environmental remediation, Materials 7 (2014) 1017–1045.

[12] R. Asmatulu, H. Muppalla, Z. Veisi, W.S. Khan, A. Asaduzzaman, N. Nuraje, Study of hydrophilic electrospun nanofiber membranes for filtration of micro and nanosize suspended particles, Membranes 3 (4) (2013) 375–388.

[13] S.S. Homaeigohan, Functional Electrospun Nanofibrous Membrane for water Filtration, PhD Dissertation, der Christian-Albrechts-Universität zu Kiel, 2011.

[14] R. Gopal, S. Kaur, Z. Ma, C. Chan, S. Ramakrishna, T. Matsuura, Electrospun nanofibrous filtration membrane, J. Membr. Sci. 28 (1–2) (2006) 581–586.

[15] World Health Organization (WHO), Meeting the MDG drinking water and sanitation target: A mid–term assessment of progress. Geneva, 2004.

[16] A. Alharbi, I.M. Alarifi, W.S. Khan, R. Asmatulu, Highly hydrophilic electrospun polyacrylonitrile/polyvinylpyrrolidone nanofibers incorporated with gentamicin as filter medium for dam water and wastewater treatment, J. Membr. Sep. Technol. 5 (2016) 38–56.

[17] Al. Baz Ismail, R. Otterpohl, C. Wendland, Efficient Management of Wastewater, Springer-Verlag, Berlin Heidelberg, 2008.

[18] A. Sato, R. Wang, H. Ma, B.S. Hsiao, B. Chu, Novel nanofibrous scaffolds for water filtration with bacteria and virus removal capability, J. Electron Microsc. 60 (3) (2011) 201–209.

[19] S.A.A.N. Nasreen, S. Sundarrajan, S.A.S. Nizar, R. Balamurugan, S. Ramakrishna, Advancement in electrospun nanofibrous membranes modification and their application in water treatment, Membranes 3 (2013) 266–284.

[20] T.A. Ahmed, Wu Yi-na, W. Hongtao, Li F, Preparation and application of functionalized cellulose acetate/silica composite nanofibrous membrane via electrospinning for Cr (VI) ion removal from aqueous solution, J. Environ. Manag. 12 (15) (2012) 10–16.

[21] T.A. Ahmed, J. Qiao, F. Li, B. Zhang, Preparation and application of amino functionalized mesoporous nanofiber membrane via electrospinning for adsorption of Cr^{3+} from aqueous solution, J. Environ. Sci. 24 (4) (2012) 610–616.

[22] M. Aliabadi, M. Irani, J. Ismaeili, H. Piri, M.J. Parnian, Electrospun nanofiber membrane of PEO/Chitosan for the adsorption of nickel, cadmium, lead and copper ions from aqueous solution, Chem. Eng. J. 220 (2013) 237–243.

[23] X. Wang, J. Yu, G. Sun, B. Ding, Electrospun nanofibrous materials: a versatile medium for effective oil/water separation, Mater. Today 19 (7) (2016) 403–414.

[24] Z. Xue, Y. Cao, Na Liu, L. Feng, L. Jiang, Special wettable materials for oil/water separation, J. Mater. Chem. A 2 (2014) 2445–2460.

[25] R.N. Wenzel, Resistance of solid surfaces to wetting by water, Ind. Eng. Chem. 28 (8) (1936) 988–994.

[26] M.H. Tai, P. Gao, B.Y.L. Tan, D.D. Sun, J.O. Leckie, Highly efficient flexible electrospun carbon-silica nanofibrous membrane for ultrafast gravity-driven oil-water separation, ACS Appl. Mater. Interfaces 6 (12) (2014) 9393–9401.

[27] X. Li, M. Wang, Ce Wang, C. Cheng, X. Wang, Facile immobilization of Ag nanofibrous membrane for oil/water separation, ACS Appl. Mater. Interfaces 6 (17) (2014) 15272–15282.

[28] L. Wang, Y. Wang, X.G. Sun, Z.Y. Pan, J.Q. He, Y. Zhou, et al., Microstructure and surface residual stress of plasma sprayed nanostructured and conventional ZrO_2 - 8 wt. % Y_2O_3 thermal barrier coatings, Surf. Interface Anal. 43 (5) (2011) 869–880.

[29] X. Gao, Li Xu, Z. Xue, L. Feng, J. Peng, Y. wen, et al., Dual-scaled porous nitrocellulose membranes with underwater superoleophobicity for highly efficient oil/water separation, Adv. Mater. 26 (11) (2014) 1771–1775.

[30] F.E. Ahmed, B.S. Lalia, N. Hilal, R. Hashaikeh, Underwater Superoleophobic cellulose/electrospun PVDF-HFP membranes for efficient oil/water separation, Desalination 344 (2014) 48−54.

[31] X. Wang, X. Chen, K. Yoon, D. Fang, B.S. Hsiao, B. Chu, High flux filtration medium based on nanofibrous substrate with hydrophilic nanocomposite coating, Environ. Sci. Technol. 39 (19) (2005) 7684−7691.

[32] X.X. Wang, L. Chen, Manufacture and characteristic of fibrous composite Z-pins, Appl. Mech. Mater. 33 (2010) 110−113.

[33] K. Yoon, K. Kim, X. Wang, D. Fang, B.S. Hsiao, B. Chu, High flux ultrafiltration membranes based on electrospun nanofibrous PAN scaffolds and chitosan coating, Polymer 47 (7) (2006) 2434−2441.

[34] M. Obaid, N.A.M. Barakat, O.A. Fadali, M. Motlak, A.A. Almajid, K.A. Khalil, Effective and reusable oil/water separation membranes based on modified polysulfone electrospun nanofiber mats, Chem. Eng. J. 259 (1) (2015) 449−456.

[35] J. Li, F. Gao, L.Q. Liu, Z. Zhang, Needleless electrospinning nanofibers used for filtration of small particles, eXpress Polym. Lett. 7 (8) (2013) 683−689.

[36] D. Ji, Y. Wang, L. Wang, L. Chen, B. Hu, G. Tang, et al., Analysis of heavy metals episodes in selected cities of north china, Atmos. Environ. 50 (2012) 338−348.

[37] J. Ma, X. Xu, C. Zhao, P. Yan, A review of atmospheric chemistry research in china: photochemical smog, haze pollution, and gas-aerosol interactions, Adv. Atmos. Sci. 29 (2012) 1006−1026.

[38] G.J. Watson, Visibility: science and regulation, J. Air Waste Manag. Assoc. 52 (2002) 628−713.

[39] A. Wei, Z. Meng, Evaluation of micronucleus induction of sand dust storm fine particles (PM 2.5) in human blood lymphocytes, Environ. Toxicol. Pharmacol. 22 (2006) 292−297.

[40] R. Asmatulu, E. Asmatulu, Importance of recycling education: a curriculum development at Wichita State University, J. Mater. Cycles Waste Manag. 13 (2011) 131−138.

[41] R. Balamurugan, S. Sundarrajan, S. Ramkrishna, Recent trends in nanofibrous mmebranes and their suitability for air and water filtrations, Membrane 1 (2011) 232−248.

[42] S. Sundarrajan, K.L. Tan, S.H. Lim, S. Ramakrishna, Electrospun nanofibers for air filtration applications, Proc. Eng. 75 (2014) 159−163.

[43] E. Scholten, L. Bromberg, G.C. Rutledge, T.A. Hatton, Electrospun polyurethane fibers for absorption of volatile organic compounds from air, ACS Appl. Mater. Interfaces 3 (10) (2011) 3902−3909.

[44] T. Uyar, R. Havelund, Y. Nur, A. Balan, J. Hacaloglu, L. Toppare, et al., Cyclodextrin functionalized poly (methyl methacrylate) (PMMA) electrospun nanofibers for organic vapors waste treatment, J. Membr. Sci. 365 (1−2) (2010) 409−417.

[45] Z. Xu, Q. Gu, H. Hu, F. Li, A novel electrospun polysulfone fiber membrane: application to advanced treatment of secondary bio-treatment sewage, Environ. Technol. 29 (1) (2008) 13−21.

[46] A. Podgorski, A. Balazy, L. Gradon, Application of nanofibers to improve the filtration efficiency of the most penetrating aerosol particles in fibrous filters, Chem. Eng. Sci. 61 (20) (2006) 6804−6815.

[47] N.L. Lala, R. Ramaseshan, B. Li, S. Sundarrajan, R.S. Barhate, L. Ying-Jun, et al., Fabrication of nanofibers with antimicrobial functionality used as filters-protection against bacterial contaminants, Biotechnol. Bioeng. 97 (6) (2007) 1357−1365.

[48] Y.C. Ahn, S.K. Park, G.T. Kim, Y.J. Hwang, C.G. Lee, H.S. Shin, et al., Development of high efficiency nanofilters made of nanofibers, Curr. Appl. Phys. 6 (6) (2006) 1030−1035.

[49] A. Jabbarnia, W.S. Khan, A. Ghazinezami, R. Asmatulu, Investigating the thermal, mechanical and electrochemical properties of PVdF/PVP nanofibrous membranes for supercapacitor applications, J. Appl. Polym. Sci. (2016). Available from: https://doi. org/10.1002/a43707.

[50] W.S. Khan, R. Asmatulu, V. Rodriguez, M. Ceylan, Enhancing thermal and ionic conductivities of electrospun PAN and PMMA nanofibers by graphene nanoflake additions for battery-separator applications, Int. J. Energy Res. 38 (2014) 2044−2051.

Electrospun nanofibers for catalyst applications

8.1 INTRODUCTION

The world's population has been increasing and will likely continue to increase in the future. All these new inhabitants will require clean water, food, transportation, health care, shelter, and consumer products, all of which necessitate the need for inexpensive and clean energy, either directly or indirectly. With the increase in population and expeditious advancements in many developing countries, the demand for energy will more than double by mid-century [1]. The increase in population is not the only factor for higher energy demand, our luxurious standard of living and speedy industrialization, especially in developing countries, are also responsible for the higher energy demand. Modern society heavily depends on fossil fuels for electricity generation, transportation, and industries. Fossil fuels are an important segment in the energy market of the world and fulfill much of the world's energy demand [2]. However, their resources are limited, and in the near future they will not be able to meet the ever-rising energy demand. The other problem with fossil fuels is the emission of polluting gases such as CO, CO_2, nitrogen oxides, and sulfur dioxide, as well as volatile organic compounds, suspended particulates, and heavy metals into the atmosphere, which are bringing unacceptable changes to the Earth's climate. The major problem of greenhouse gas emissions is global warming, and the influence of global warming is far greater than the increase in the temperature on the Earth's surface. Some of the other impacts of global warming include primary and secondary pollutants,

Synthesis and Applications of Electrospun Nanofibers. DOI: https://doi.org/10.1016/B978-0-12-813914-1.00008-0

coastal erosion, melting glaciers, excessive drought or flooding, and some infectious diseases. Almost all resources of green energy are subjected to seasonal limitations, and regional variations. The urgent energy and environmental problems, which we are facing now, are due to the use of fossil fuels. There is a compelling demand to address the energy issue swiftly by lessening our dependence on fossil fuels and at the same time searching for alternative energy resources that will not cause environmental problems.

In order to address some of the issues of rapid depletion of fossil fuels and detrimental environmental concerns accompanying their combustion, scientists and engineers have been exploring new forms of energies that are renewable, inexpensive, clean, environmentally friendly, and viable alternatives to fossil fuels. Recently, much attention has been paid to hydrogen generation via solar means as a next-generation energy source. Energy harvesting from sunlight is an alternative approach to solve the energy crisis with minimum environmental effects [3]. Solar energy is a decentralized, inexhaustible, and economically feasible natural resource, with the magnitude of the available solar power striking the Earth's surface equal to 130 million 500 MW power plant, which far exceeds our needs [3]. However, some important factors must be addressed before utilizing solar energy in order to meet the rising power demand. First, the means for solar energy conversion, storage, and distribution should be environmentally friendly. The next important factor is to provide a stable and constant energy flux throughout the year [3,4]. Due to the variations in the intensity of sunlight from region to region and country to country, energy harvesting from sunlight is strenuous, especially in those countries where the Sun appears only for a few hours. Despite these challenging factors, the most important issue is whether energy harvesting is economically feasible on a large scale and that will meet the demand [4].

Hydrogen has been recognized as being an efficient, cheap, and environmentally friendly fuel, which can be easily transported [5]. Hydrogen is the simplest and most abundantly available element in the universe. It is always found combined with other elements, such as water and organic compounds. Most of the hydrogen production needs fossil fuels. The most important aspect of hydrogen fuel production via solar means is that hydrogen fuel can be easily produced by direct decomposition of water, without any polluting byproducts, such as greenhouse gases, which makes it an ideal candidate for future energy demand [6]. Solar water splitting is the process by which energy in solar photons is used to break down liquid water into molecules of hydrogen and oxygen gases. Hydrogen has a high heat of combustion and its combustion product is water vapors and droplets, instead of carbon dioxide. However, there is no natural reserve of hydrogen gas and renewable hydrogen production is still not popular due to its high cost. Hydrogen and oxygen used in fuel cells to generate electricity produce water as a byproduct [7,8]. Solar energy can be converted and stored in a chemical bond, as shown in Eq. (8.1). Hydrogen produced from fossil fuels, as shown in Eq. (8.2), results in the emission of carbon dioxide and this process is not environmentally friendly.

$$H_2O \xrightarrow[\text{catalyst}]{h\nu} H_2 + \tfrac{1}{2}O_2 \tag{8.1}$$

$$CH_4 + 2H_2O \rightarrow 4H_2 + CO_2 \quad \Delta G^\circ = 131 \text{ KJ mol}^{-1} \tag{8.2}$$

Overall water splitting, as shown in Eq. (8.1), is a thermodynamically unfavorable process and has a positive Gibbs free energy change ($\Delta G = +237.2$ kJ mol^{-1}, 2.46 eV per molecule). Some of the semiconductor photocatalysts can absorb photons and generate electrons and holes on their surfaces by absorbing solar/photon energy. The photogenerated electron and hole pairs are able to drive the reduction and oxidization reactions, respectively, of the water molecules. The ideal way to produce hydrogen is artificial photosynthesis. Heterogeneous photocatalysis for overall water splitting has been recognized as a potential method for hydrogen production [4]. Heterogeneous photocatalysis has been studied extensively since the discovery of photoactivated water splitting using titanium dioxide as an electrode. Heterogeneous photocatalysis is a process in which a photocatalyst reaction is carried out using sunlight. Fujishima and coworkers discovered that water could be splitted into hydrogen and oxygen through a photocatalyst reaction in 1972 [9,10]. Therefore, their early studies were focused on the production of hydrogen fuel from water using solar energy [9]. Further studies in this connection revealed that various irradiated semiconductor particles could be used as a photocatalyst.

Among the various semiconductors, including CdS, SnO$_2$, WO$_3$, SiO$_2$, ZrO$_2$, ZnO, Nb$_2$O$_3$, Fe$_2$O$_3$, and SrTiO$_3$ that were studied for a photocatalytic reaction under solar energy, titanium dioxide (TiO$_2$) remains as the benchmark among other semiconductors [9]. TiO$_2$ is an excellent photocatalyst due to its high resistance to photocorrosion, desirable bad-gap energy, availability, chemical inertness, nontoxicity, and durability [9]. Generally, a photoexcited electron−hole pair can be generated by irradiation with light having energy greater than the band gap energy of TiO$_2$ (3.2 eV for anatase) [11]. Despite these advantages, there are some drawbacks of using TiO$_2$ as a photocatalyst: (1) the photogenerated electron−hole pairs can combine at a faster rate, which affects the photocatalyst efficiency, and (2) TiO$_2$ can be excited with ultraviolet (UV) light only, which is approximately less than 5% of solar radiation, because of its wide energy band gap [11]. These drawbacks of TiO$_2$ result in low photocatalyst activity in practical applications [11]. Therefore, numerous studies have focused on modifying TiO$_2$ band gap and enhance its photocatalytic activity under visible light conditions (usually between 390 and 700 nm wavelength) [11,12].

Water splitting into hydrogen and oxygen using a photocatalyst under the UV solar radiations was initiated by the pioneering work of Fujishima and Honda [4,10]. They found that overall water splitting could be achieved by using a photoelectrochemical (PEC) cell containing titania (TiO$_2$) as a photoanode and platinum as a counter-electrode under UV irradiation and an external bias. This work stimulated the exploration of new materials for both anodes and cathodes and integrated a configuration that uses photovoltaic cell junctions to enhance the output voltage for a single-band or dual-band gap device [3]. Following Fujishima and Honda's work, a large number of these PEC cells were developed for efficiently utilizing solar energy. Nevertheless, their development was marred by the fact that suitable photoelectrode materials with appropriate band gaps and stability are relatively difficult to find in natural forms. The maximum efficiency obtained during overall water splitting was approximately 5.9%; however, this figure is much less than the requirement for practical applications (10%) [7].

Many research studies were focused on modifying TiO_2 in order to increase its photocatalyst efficiency in the visible light range. Researchers have doped TiO_2 with metals such as platinum, iron, palladium, silver, and ruthenium [9−14] and nonmetals such as nitrogen, boron, sulfur, and carbon as dopants for enhancing photocatalyst efficiency [6]. TiO_2 can be incorporated with semiconductors such as NiO, ZnO, CdS, $SrTiO_3$, Fe_2O_3, and SnO_2 to improve the photo responding range and facilitate charge separation [6]. Recently, p-n junction photocatalysts such as NiO/TiO_2, NiO/ZnO, and $NiO/InVO_4$ have been receiving attention because of their high charge separation. NiO is a p-type semiconductor incorporated with many different n-type semiconductors and is used as a cocatalyst because of its high mobility, cost-effectiveness, and high p-concentration ability [6].

8.2 HYDROGEN PRODUCTION

Hydrogen is one of the most important fuels for the future energy demand because it is clean, energy-efficient, and abundant in nature. Hydrogen is a chemical element with the symbol H and atomic number 1. Hydrogen is the lightest element in the periodic table, having a atomic weight of 1.008. Its monatomic form (H) is the most abundant chemical substance in the Earth, constituting roughly 75% of all baryonic mass. The enthalpy of combustion of hydrogen gas is 286 kg/mol. Hydrogen is a good energy source for the following reasons: (1) it is available on the Earth and it exists in both water and biomass form; (2) it has a better energy yield of around 122 KJ/g compared to gasoline (40 KJ/g); and (3) it is environmentally friendlier since its end use does not produce greenhouse gases [15]. Hydrogen can be stored in gaseous, liquid, or metal hydride forms and can be distributed over long distances through pipelines or by means of oil tankers in compressed gas or cryogenic liquid [15].

Most of the hydrogen is produced by a process called "steam reforming" from hydrocarbon fuels at elevated temperatures [16]. In this process, methane is used as a fuel source since it contains the highest hydrogen-to-carbon ratio among hydrocarbons and has the lowest carbon footprint to lower the greenhouse effects. The steam methane reformer is generally used to produce hydrogen. Generally, there are two major steps in the steam reforming process: (1) in the first step methane is mixed with steam and passed over a metal-based catalyst (nickel) at high temperature (700−900°C) and high pressure (1.5−3 MPa) to produce a mixture of hydrogen and carbon monoxide (CO) as shown in Eq. (8.4); and (2) in the second step additional hydrogen can be recovered by a lower-temperature gas-shift reaction with the carbon monoxide produced, in the presence of a copper or iron catalyst. The second step is the shift reaction in which CO from the first step reacts with additional steam to produce CO_2 and more free hydrogen gas (Eq. 8.4) [15].

$$CH_4 + H_2O \rightarrow CO + 3H_2 \qquad (8.3)$$

$$CO + H_2O \rightarrow CO_2 + H_2 \qquad (8.4)$$

The other process that is used to produce hydrogen gas is coal gasification. Coal gasification is a process in which coal undergoes partial oxidation at higher temperatures and pressures with the help of oxygen and steam to produce a mixture consisting of CH_4, CO_2, CO, H_2, and water vapor. At a temperature of $1000°C$ and pressure of nearly 1 bar, mostly CO and hydrogen remain in the system. The chemical reactions that take place during coal gasification are represented by the following equations [15].

$$C + \frac{1}{2}O_2 \rightarrow CO \qquad (8.5)$$

$$C + H_2O \rightarrow CO + H_2 \qquad (8.6)$$

Biomass of plants, crops, seeds, algae, animal waste, and recycled organic compounds can be used to generate hydrogen gas via thermochemical and biological processes. Pyrolysis and gasification processes are generally employed for hydrogen production, whereas biophotolysis, biological gas shift reaction, and fermentation are other excellent biological processes that have been under investigation for a while [15]. In a typical pyrolysis process, biomass is generally heated to an elevated temperature in the absence of oxygen to produce hydrogen, CH_4, CO, CO_2, carbon, and other compounds, depending on the type of the biomass. Generally, a temperature of around $600°C$ and a pressure of around 0.5 MPa are used in this process [15]. In the electrolysis process, electric current is used to split water into hydrogen at the cathode (+) and oxygen at the anode (−). Steam electrolysis (a variation of conventional electrolysis) uses heat, instead of electricity, to provide some of the energy needed to split water, making the process more energy-efficient. Thermochemical water splitting uses chemicals and heat in multiple steps to split water into its component parts. Photobiological systems use microorganisms to split water using sunlight and organics. Biological systems use microbes to break down a variety of biomass feedstocks into hydrogen (anaerobic bacteria, viruses, etc.).

8.2.1 HYDROGEN PRODUCTION BY SOLAR ENERGY

Hydrogen production using solar water splitting can be classified into three major steps: (1) thermochemical water splitting; (2) photobiological water splitting; and

(3) photocatalytic water splitting [15]. Solar water splitting is the process by which energy in solar photons is used to break down liquid water into molecules of hydrogen and oxygen gas. Hydrogen produced through solar water does not emit carbon into the atmosphere. A thermochemical hydrogen production process of water splitting is one in which water, thermal energy, or heat are used as the input, and the output of the process is hydrogen and oxygen and possibly (or probably) some waste heat. In this process, high temperature from concentrated solar power or from the waste heat of nuclear power reactions and chemical reactions is used to produce hydrogen and oxygen from water. Generally, heat from sunlight, which typically can reach 2000 °C, is used in the presence of a catalyst. Large-scale concentrator systems are necessary to obtain high temperatures; therefore, this process is not cost-effective [15].

Photobiological water splitting involves sunlight, a biological component, catalysts, and an engineered system. Specific organisms, algae and bacteria, produce hydrogen as a byproduct of their metabolic reaction processes. These organisms generally live in water, and therefore can biologically split water into its component elements. Currently, this technology is still in the research and development stage and the theoretical sunlight conversion efficiencies have been estimated at 24%. Photocatalyst water splitting is another promising technique to produce clean hydrogen from water and other sources. It has the following advantages: (1) reasonable solar-to-hydrogen efficiency with the photocatalyst; (2) low manufacturing cost; (3) ability to achieve separate hydrogen and oxygen gases throughout the chemical reaction; and (4) small reactor systems suitable for household applications, thus providing for a huge market potential for large-scale productions in both urban and rural areas. The following sections briefly talk about the overview of hydrogen generation by photocatalytic water splitting [15].

8.2.2 HYDROGEN PRODUCTION BY PHOTOCATALYTIC ACTIVITY

Recently, extensive studies have been conducted on hydrogen as a potential energy source using photocatalyst technology. Numerous attempts have also been made to develop new photocatalysts that can work not only under UV light but also under visible light in order to enhance photocatalyst efficiency. Solar energy is an everlasting, inexhaustible, and clean energy source; meanwhile, hydrogen energy is an efficient, clean, and environmentally friendly energy source. Sunlight is an inexpensive, abundant, nonemitting, and nonpolluting renewable source that is available throughout the year in most parts of the world [16]. The Sun is a major source of energy with a magnitude of 3.0×10^{24} J/year ($\sim 10^5$ terawatts) [7]. The energy consumption of the world is approximately 4×10^{20} J/year (~ 12 terawatts) corresponding to around 0.01% of the solar energy reaching the Earth's surface [7]. Thus, solar energy that reaches the Earth is far greater than our needs. The amount of energy from sunlight that strikes the Earth's surface yearly is approximately 10,000 times the total energy consumed on Earth;

therefore, the conversion of solar energy into a usable form has spurred interest in recent years [16].

However, a major disadvantage of utilizing solar energy is that the Sun does not appear in all parts of the world regularly. In some parts of the world sunlight is available for just a few months in a year. The amount of sunlight a region receives depends upon geographical location, seasons, weather conditions, and time of day. Sustainable hydrogen production is a major target in the development of alternative energy systems of the future for providing a clean, affordable, and environmentally friendly energy supply. The conversion of solar energy into hydrogen via photocatalyst water splitting using a photosemiconductor catalyst is one of the most reliable technologies for the future because large quantities of hydrogen can potentially be generated in a sustainable manner. Unequivocally, the conversion of solar energy into clean hydrogen fuel under ambient conditions is one of the greatest challenges being faced by researchers in this era. Many semiconductor materials and electrospun nanofibers can be great options for water splitting [17,18].

8.3 BASIS OF PHOTOCATALYTIC WATER SPLITTING

Photocatalysis works through the photon-emitting process by sunlight as an energy source in the splitting of water into hydrogen and oxygen. Photocatalysis is a chemical reaction induced by photoirradiation in the presence of a photocatalyst to carry out a redox reaction [4,6,15,17,19,20]. A photocatalyst is a substance that is capable to produce by absorption of sunlight a chemical reaction or chemical transformations of reaction participants without being consumed, oxidized, or transformed. Fig. 8.1 shows the basic principle of water splitting using a semiconductor photocatalyst. Thermodynamically, water splitting is an uphill reaction with a large positive change in Gibbs free energy (238 kJ/mol) [4]. When energy from sunlight that is equivalent to or greater than the band gap energy of a semiconductor photocatalyst, electrons (e^-) and holes (h^+) are generated in the conduction band and valence band, respectively [4,6]. The photocatalyst uses one proton to excite an electron from the valence band to the conduction band, thereby resulting in an excited state [19]. Due to the incident light, the electrons (e^-) are excited into a conduction band, leaving behind holes (h^+) in the valence band. These electrons (e^-) and holes (h^+) cause a reduction reaction and an oxidation reaction, respectively [4,6]. These theories can be generalized for most semiconductor materials, including electrospun fibers. As shown in Eqs. (8.7), (8.8), and (8.9), the decomposition of water into hydrogen and oxygen by means of an electrochemical cell is a two-electron stepwise process:

$$\text{Oxidation: } H_2O + 2h^+ \rightarrow \tfrac{1}{2} O_2 + 2H^+ \text{ (HER)} \tag{8.7}$$

$$\text{Reduction: } 2H^+ + 2e^- \rightarrow H_2 \text{ (ORE)} \tag{8.8}$$

$$\text{Overall reaction: } H_2O \rightarrow \tfrac{1}{2} O_2 + H_2 \quad \Delta G = +237.2 \text{ kJ/mol} \tag{8.9}$$

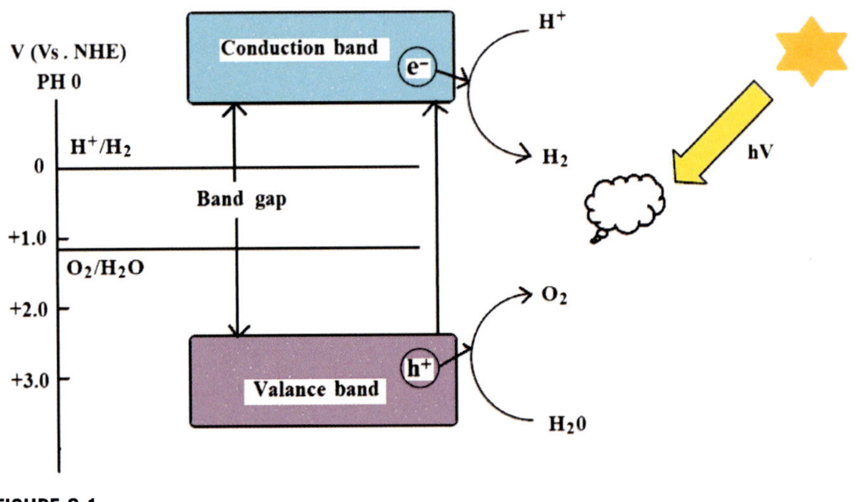

FIGURE 8.1

Basic principle of overall water splitting using heterogeneous photocatalyst [6,15].

Photocatalysts having different semiconductive properties can absorb photons, and generate electrons and holes on their surfaces by absorbing sunlight. The structure of the photocatalyst remains the same if an equal number of electrons and holes are generated in a chemical reaction [4,6]. In order to achieve an overall water-splitting process in an efficient way, the bottoms of the conduction bands must be located at a more negative potential than the reduction potential of H^+ to H_2 (0 V vs. NHE at pH 0), while the tops of the valence bands must be positioned more positively than the oxidation potential of H_2O to O_2 (1.23 V vs. NHE). It was stated that the top of the valence band must be located at a higher positive potential when compared to the oxidation potential of H_2O to O_2 (1.23 V vs. NHE) [4,6]. The minimum photon energy that is needed to carry out the splitting reaction is 1.23 eV, corresponding to a wavelength of ca. 1000 nm in the infrared region of the solar energy [4]. If the entire spectrum of visible light is used, the photocatalyst efficiency will be higher. However, there is a barrier in the charge-transfer process known as the activation barrier between water molecules and the photocatalyst, requiring photon energy that is higher than the band gap of the photocatalyst in order to carry out the overall water-splitting reaction [4].

Additionally, water formation during the backward reaction must be strictly inhibited, and the photocatalysts themselves must be strongly stable in the backward reaction; otherwise, a short circuit may take place and overall hydrogen evaluation reactions will end up here. Furthermore, there are numerous materials that possess suitable band gap energy for water splitting; conversely, there are very few materials that can work as an efficient photocatalyst for overall water splitting due to some prerequisite factors, as mentioned below [4,6]. As can be seen in Fig. 8.2, water splitting occurs in three steps: (1) the photocatalyst absorbs

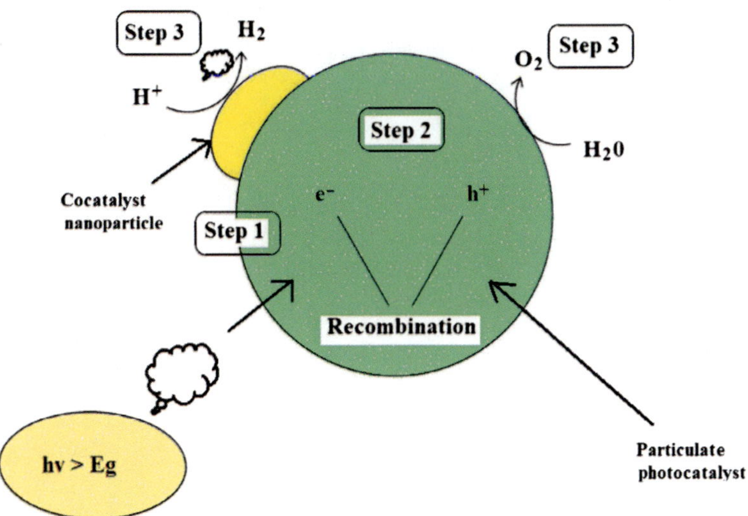

FIGURE 8.2

Process of overall water splitting using heterogeneous photocatalyst.

photon energy from sunlight that is equal to or greater than the band gap energy of the photocatalyst, thereby generating electron–hole pairs in the bulk; (2) the photoexcited electron–hole pairs migrate to the surface; and (3) the absorbed species are subjected to reduction and oxidation by the electron–hole pairs to produce hydrogen and oxygen. The first two steps depend on the electronic properties and structure of the photocatalyst [6,15].

Generally, the crystallinity of photocatalyst has a beneficial effect on photocatalyst activity since the density of defects, which works as a recombination, centers between photogenerated electrons and holes, decreased with increasing crystallinity [6,15]. In the third step, decomposition reactions of the water molecules take place on the surface of the cocatalyst, which is typically a noble metal (e.g., Pt, Rh) or metal oxide (e.g., NiO, RuO$_2$), and is loaded onto the photocatalyst surface as a dispersion of nanoparticles in order to activate the sites and reduce the activation energy for gas evolution [6,15]. Therefore, the surface properties of cocatalyst and structures of photocatalyst must be developed in such a way as to favor photocatalyst activity [15].

Generally, photocatalyst splits water in two ways: (1) by using photoelectrochemical cells that consist of a photoanode and a metal counter-electrode. In a PEC cell, a semiconductor electrode is illuminated in an appropriate electrolyte, causing an electrochemical reaction at both electrodes. The second electrode is used as a reference electrode. The main component of the PEC cell is the semiconductor, which converts incident photons to electron–hole pairs. These electrons and holes are spatially separated from each other due to the presence of an electric field inside the semiconductor [15]. The excited electrons percolate

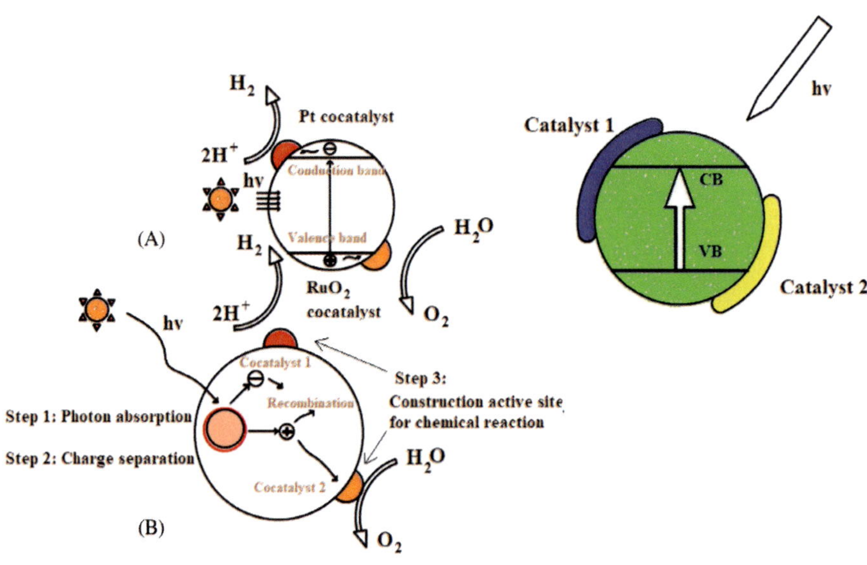

FIGURE 8.3

(A) Schematic of overall water-splitting reaction using solid photocatalyst; and
(B) schematic of electron transport during the water-splitting reaction [10].

through the semiconductor reaching the counter-electrode, via the external circuit to promote water reduction at its surface. The photogenerated holes are swept towards the interface of the semiconductor/electrolyte, where they form water in an oxidation process; and (2) by utilizing a photocatalytic system. The photocatalyst is modified with a suitable cocatalyst to provide an active redox site in visible light spectrum. Fig. 8.3 shows a particulate photocatalyst system for overall water splitting [10].

From the point of view of semiconductor photochemistry, the function of a photocatalyst is to initiate reduction and oxidation (overall redox) reactions in the presence of an irradiate semiconductor. During the photocatalyst process, the light absorption and subsequent illumination of semiconductor photocatalyst cause photoexcitation of electron−hole pairs, when the energy of incident photons matches or exceeds the bad gap energy of the semiconductor photocatalyst. When the intensity of photons (sunlight) is equal to or greater than the band gap energy of photocatalyst semiconductor, the photocatalyst semiconductor activates and a redox reaction take place, an electron is excited in the conduction band, thus reducing the water molecules into hydrogen gas. As the electron moves to the conduction band, the holes in the valence band move to the active site of the photocatalyst and oxidize the water molecules into oxygen gas by means of a cocatalyst. The electron−hole pairs migrate to the surface of the semiconductor without recombination, they can participate in oxidation and reduction reactions with adsorbed species such as water, oxygen, and other organic or inorganic species. These oxidation and reduction reactions are the basic mechanisms of photocatalytic hydrogen production.

8.3.1 PHOTOELECTROLYSIS

The Gibbs free energy change, $\Delta G = +237.2\,kJ/mol$, corresponds to a photon with a wavelength of 1000 nm in overall water splitting for converting one molecule of H_2O to H_2 and $1/2\ O_2$. Thus, a photon from the visible range of the solar spectrum has sufficient energy to photolyze water. The energy required to photolyze water must be 1.23 eV per electron transferred, according to the Nernst equation [3]. Therefore, semiconductor photocatalyst that is used in overall water splitting must be able to absorb sunlight photon energy that is greater than 1.23 eV (wavelengths of ~ 1000 nm and shorter) and convert this energy into hydrogen and oxygen [21]. This process generates two electron–hole pairs per molecule of H_2 (1.23 eV $\times$ 2 = 2.46 eV) or four electron–hole pairs per molecule of O_2 (1.23 eV $\times$ 4 = 4.92 eV) [21]. A single semiconductor having a band gap energy E_g sufficient or large enough to split water as well as possess a conduction band energy E_{cb} and valence band energy E_{vb} that straddles the electrochemical potentials $E^{o}(H^+/H_2)$ and $E^{o}(O_2/H_2O)$ can carry out both the oxygen evolution reaction (OER) and the hydrogen evolution reaction (HER) using the electron–hole pairs produced under photon energy from sunlight, as shown in Fig. 8.4 [3].

In order to carry out an oxygen evolution reaction and hydrogen evolution reaction for overall water splitting without recombination, photoinduced free

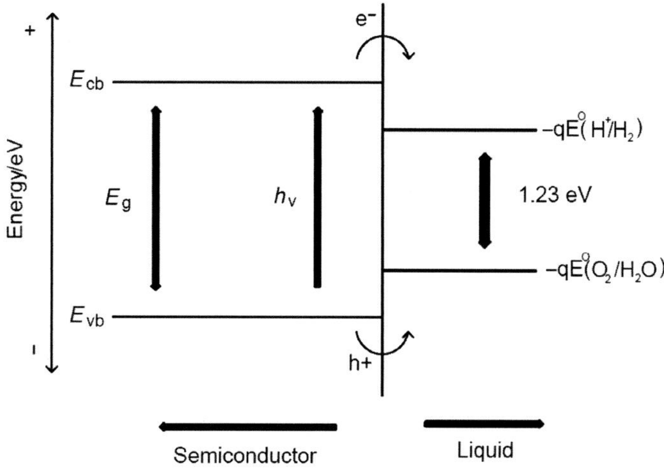

FIGURE 8.4

Oxygen evolution reaction (OER) and hydrogen evolution reaction (HER) for overall water splitting.

charge carriers (electrons and holes) in the semiconductor have to travel to a liquid junction, and must react only with solution species directly at the semiconductor surface. The electron-transfer processes at semiconductor/liquid junctions incur some losses due to the concentration and kinetic overpotentials needed to drive the HER and the OER. The energy needed for photoelectrolysis at a semiconductor photoelectrode has been found to be $1.6-2.4$ eV per electron$-$hole pair generated in order to account for losses [3].

8.4 ELECTROSPUN FIBERS EMBEDDED WITH NANOCATALYST FOR WATER SPLITTING

Electrospinning is a versatile, novel, and straightforward technique based on an electrohydrodynamic process for producing continuous thin fibers with diameters ranging from a few micrometers to a few nanometers [21]. The ultrafine fibers produced via electrospinning methods, generally called "nanofibers," are unique nanomaterials because of their nanoscaled dimensions in cross-sectional direction and macroscopic length of the fiber axis. These ultrafine fibers have the advantages of functionality due to their nanoscaled dimension in cross-section and ease of manipulation due to their macroscopic length. Nearly all naturally occurring and synthetic polymers, polymer blends, conjugated polymers, and polymers loaded with chromophores, nanoparticles, additives, metals, and ceramics have been electrospun [22,23]. Polymers can be modified chemically and doped/ blended with additives/elements, such as B, P, In, and As, as well as carbon-black particles, enzymes, bacteria, viruses, metallic nanoparticles, and nanostructured semiconductor materials [22]. The dispersion of additives into a polymer solution before electrospinning is of significance due to the excellent properties of nanocomposite fibers and also the continuously increasing demand for miniaturization of electronic components, sensors, optical detectors, and devices [24,25]. Electrospun nanocatalyst-based membranes are found to be more effective in photocatalysis for overall water splitting than conventional membranes. These membrane have high surface-area-to-volume ratio, and their porous structure offers higher surface active sites for efficient photocatalysis [24]. Fig. 8.5 shows a schematic view of an electrospinning process.

The electrospinning process has been explained in Chapters 1 and 2 in detail, and briefly in other chapters. Since the pioneering work by Fujishima and Honda, TiO_2 has been extensively investigated for water splitting due to its efficient photocatalyst activity, stability, lower cost, and environmentally friendly nature [5,6]. Despite these advantages, TiO_2 possesses some drawbacks, such as its wide band gap energy (3.2 eV) and higher recombination rate of charge carriers, which have impeded its popularity (efficiency). In order to overcome these problems, a number of attempts have been made to modify TiO_2 by fabricating its nanoscaled structure in such a way as to enhance its photoefficiency [5,6]. For example,

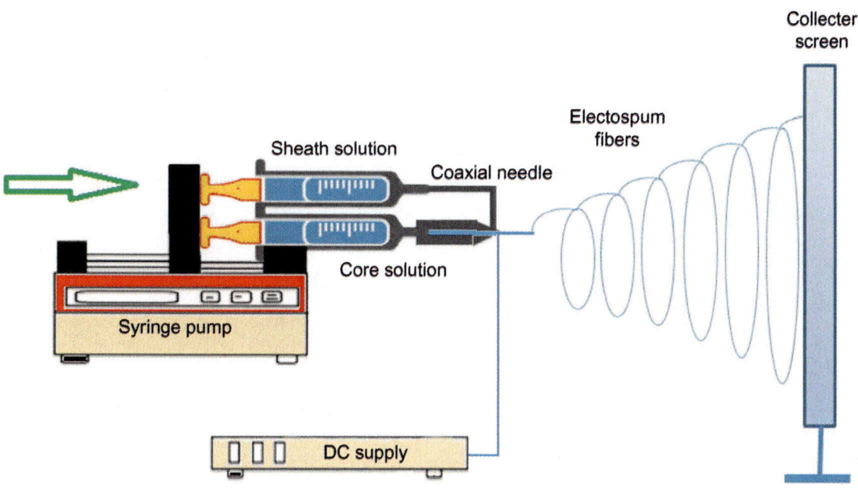

FIGURE 8.5

Schematic of an electrospinning process.

Cheng et al. reported the preparation of TiO_2 nanorods array photoelectrodes and studied the effect of annealing on the PEC performance and rhodamine B degradation [26]. Wolcot et al. used dense and aligned TiO_2 nanorod photoanodes and studied PEC water splitting [27]. Cowan et al. achieved improved electron−hole separation with nanostructured photoelectrodes prepared from TiO_2 colloid paste containing 15-nm particles [28]. Zhang et al. [29] reported higher PEC performance by fabricating crystalline TiO_2 nanotubular arrays with intrinsic p-n junction properties, attained through potentiostatic anodization without post-annealing. Yun et al. reported the preparation of TiO_2 nanospheres and nanorods doped with platinum in order to study the effects of these nanostructured materials on hydrogen evolution [30]. Their study showed that the nanorods exhibited better hydrogen evolution than nanospheres due to the reduction in the e^-/h^+ recombination rate in the rods.

Recently, the electrospinning technique has been used for fabricating photocatalyst fibers or polymer fibers embedded with a photocatalyst for water splitting to generate hydrogen. In electrospinning, it is very easy to control the morphology, diameter, and length of fibers; therefore, some researchers have prepared TiO_2 nanofibers for water-splitting applications [5,6,18]. Indium tin oxide (ITO) has also been added to TiO_2 in order to improve the photocatalyst efficiency. Also, TiO_2 nanofibers have been generated via electrospinning using polyvinyl acetate (PVAc), dimethylformamide (DMF), acetic acid (96%), and titanium (IV) isopropoxide, and a small amount of ITO has been added in order to improve the photocatalyst efficiency. Since photocatalytic water splitting necessitates a minimum energy band gap of 1.23 eV, ITO possesses a large band gap of 4 eV and is mostly transparent in the visible part of the Sun spectrum. Therefore, ITO-doped

TiO_2 nanofiber can be used as a catalyst, absorbing sunlight in the visible spectrum and converting it into solar energy, which is required for water splitting.

The photocatalyst water splitting is a promising technique to convert energy from sunlight into highly pure chemical fuel (hydrogen) [31]. According to some estimates, the heterogeneous photocatalyst running at 10% solar energy conversion efficiency can produce hydrogen fuel at a cost of $1.63/kg hydrogen [31]. This is much cheaper than hydrogen produced by PEC cells or from the combination photovoltaic cells with water electrolytes [31]. Inorganic nanomaterials are small in size, and therefore, full sunlight penetration can be achieved, thus becoming dominant photocatalysts in water-splitting applications. Kudo and coworkers determined that $LiNbO_3$ nanowires modified with RuO_2 cocatalyst can split water with a quantum yield of 0.7% [31,32]. Yan et al. studied overall water splitting with RuO_2-modified Zn_2GeO_4 nanorods with UV light [33]. Domen's group determined that overall water splitting could be possible with NiO-loaded $NaTaO_3$ nanocrystals around 40 nm in size; however, deep penetration of UV light was required for catalyst activity [34]. These results were highly encouraging, but higher photocatalyst efficiency and better sunlight absorption in the visible spectrum are needed before these systems can be used for practical applications in hydrogen fuel production.

Some researchers have reported a nano-NiO-$SrTiO_3$ (NiO-STO) system, i.e., nanoscale titanate photocatalyst for overall water splitting having a 3.2 eV band gap. This system allows better light absorption than germinates, tantalates, and niobates [31]. Townsend et al. described the primary relationships between nanoscale NiO and $SrTiO_3$ (NiO-STO) for overall water splitting [31]. With a 3.2 eV band gap STO allows significantly higher light absorption than germinates, tantalates, and niobates. A number of different ternary titanates have been developed for overall water splitting under UV light sources. Shibata et al. employed layered structures of $Na_2Ti_3O_7$, $K_2Ti_2O_5$, and $K_2Ti_4O_9$ in aqueous methanol solutions using a Pt cocatalyst for hydrogen evolution [35]. It was also reported that the quantum yield of $K_2Ti_2O_5$ materials obtained 10% efficiency. These fibers have substantially high surface area and their porous structure allows more active sites for catalytic activity under sunlight. Recently, much attention has been paid to strontium titanate ($SrTiO_3$) in overall water splitting. Similar to anatase TiO_2, strontium titanate ($SrTiO_3$) has a band gap energy level suitable for photocatalyst water splitting due to the fact that its valence band and conduction band are lower and higher than the oxidation−reduction levels of water, respectively [36]. Additionally, the perovskite structure of $SrTiO_3$ makes it suitable for doping for electronic modifications [36−38]. Alharbi et al. employed a coaxial electrospinning process to produce NiO-STO (NiO-$SrTiO_3$) nanocomposite fibers for obtaining a higher photocatalyst efficiency [39]. Fig. 8.6 shows a schematic view of the coaxial electrospinning process. NiO_x−$SrTiO_3$ photocatalyst acts as three components in water splitting, such as Ni−STO−NiO catalyst, where STO absorbs light, Ni produces protons, and NiO oxidizes water [5,17].

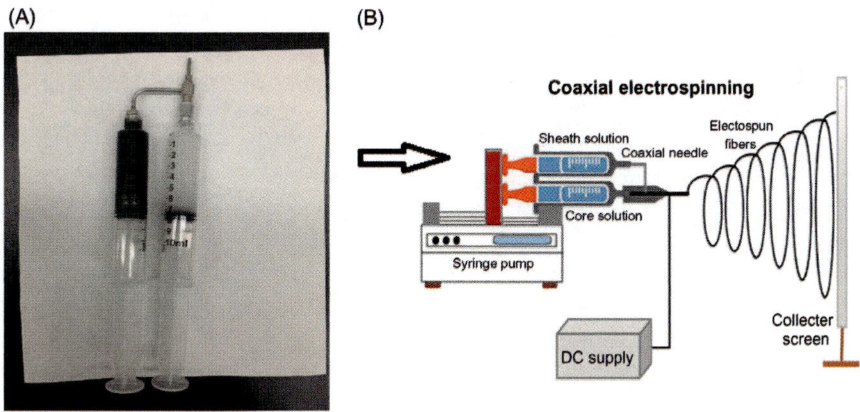

FIGURE 8.6

(A) Coaxial capillary tube; (B) schematic of coaxial electrospinning.

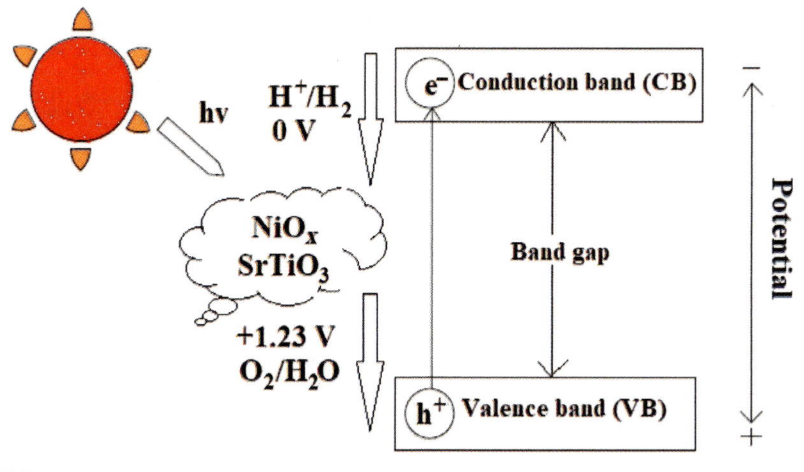

FIGURE 8.7

Process of overall water splitting using homogeneous Ni−STO−NiO photocatalyst.

In the NiO−STO system, two catalysts—NiO and $SrTiO_3$—are used. Nickel oxide (NiO) is a p-type semiconductor material with an energy band gap of $E_{gb} = 3.85$ eV, while strontium titanate ($SrTiO_3$) has a band gap of $E_{gb} = 3.25$ eV. The band gap of the NiO−STO system should be between 3.85 and 3.25 eV. In some literature reviews, the band gap of the NiO−STO system in fibrous form is 3.65 eV, while in solid form it is 3.2 eV. As can be seen in Fig. 8.7, water splitting takes place in three major steps: (1) Ni−STO−NiO catalyst possesses greater bandgap energy when compared to the photon energy from sunlight, efficiently absorbs sunlight, resulting in the generation of an electron—hole pair for water splitting; (2) these electron—hole pairs are transferred to the surface; and (3) these electron—hole pairs reduce and oxidize water on the surface of photocatalyst to produce hydrogen and oxygen [5,17]. Strontium titanate is an oxide of strontium and titanium elements having typical perovskite structures at room temperature. Strontium titanate has a band gap energy level (3.25 eV), so it is very suitable for photocatalyst water splitting due to the fact that its valence band and conduction band are lower and higher than the oxidation—reduction levels of water, correspondingly [17]. Asmatulu et al. integrated some of the targeted nanoscale inclusions (e.g., graphene and C_{60} into TiO_2) in nanofibers using an electrospinning process for the enhanced transportation and energy conversion efficiencies [18]. The dye-sensitized solar cell studies indicated that these nanocomposite fibers can be a great way of improving the overall efficiency of these solar cells.

8.5 ELECTROSPUN NANOFIBERS IN HYDROGEN STORAGE

Hydrogen is an outstanding energy alternative to fossil fuels since the only byproduct in its conversion process is water [40,41]. However, the major hurdle lies in finding suitable materials that possess the advantage of durability, cheap, and safety in storage of large amounts of hydrogen and at the same time fulfilling the necessary needs of vehicular transport applications. Therefore, efficient and durable hydrogen storage is essential to develop and utilize hydrogen energy. Many researchers have focused their attention on this critical issue. It has been found that nanostructured materials are ideal candidates due to their unique advantages such as high specific surface area and small pore size, which can facilitate hydrogen storage. Also, carbon materials such as carbon nanotubes, graphite, and fullerene have spurred attention due to their unique properties including safety, low mass density, and high reliability. Among carbon materials, activated carbon and carbon fibers present cost-effective storage media with reasonable storage capacity, although more hydrogen can be absorbed on carbon nanotubes.

Hydrogen adsorption on carbon materials generally takes place due to van der Waals and other weak interactions. This adsorption depends on the surface area, pressure, and temperature. The surface area can easily be extended by introducing nanoscale pores on the surface of carbon materials [41]. Single-walled carbon nanotubes and mesoporous carbon possess high surface areas of ~ 1300 and ~ 2000 m^2/g, respectively. These high surface area carbon materials are generally synthesized by vapor deposition or template processes, thereby making them unbefitting for cost-effective large-scale production, which prevents them from being used for hydrogen storage. The electrospun porous carbon fibers are capable of hydrogen storage of as much as 2.5 wt.% hydrogen at 300K and 30 MPa [41,42].

Kurban used ammonia borane and polystyrene to fabricate core-shell composite nanofibers employing an electrospinning process [43]. In this process, polystyrene was selected as the shell material due to the fact that it has a good H_2 permeability value (23.8 barrers) and a melting point of approximately 240°C. The results revealed a significant improvement in the dehydrogenation speed of hydrogen due to the nanostructure effects. Many other researchers have focused on improving hydrogen storage properties, such as alteration of the diffusion pathway, lower dehydrogenation, and even enhanced reversibility using templates to modified thermodynamics and/or kinetics [44–46]. Xia et al. reported the preparation of Li_3N nanofibers via an electrospinning process for hydrogen storage [47]. Their results showed that Li_3N porous nanofibers displayed better hydrogen performance with stable reversibility over 10 cycles of de/rehydrogenation at a temperature of 250°C.

Alipour et al. prepared PMMA/ammonia borane (AB) nanofiber composites via an electrospinning process to study the synergetic nanoconfinement effects of nanofibers on dehydrogenation temperature and the removal of unwanted byproducts of AB [48]. The results revealed that PMMA/AB nanofiber composites can decrease the enthalpy of exothermic decomposition due to the interaction between AB molecules and PMMA nanofiber structures and strongly reduce the emission of byproduct impurities by decreasing the loss of AB weight. Kim et al. fabricated polyacrylonitrile-derived carbon fibers after carbonization at 800°C in argon and water [41]. Water vapors were used during carbonization in order to promote pore structure. Palladium salt solution was electrosprayed during the electrospinning process. The results showed that the surface area of palladium-coated carbon fibers was $815.6 \, m^2/g$ and hydrogen adsorption capacity was determined as 0.35 wt.% at 298K and 0.1 MPa. Generally speaking, electrospinning can prepare nanomaterials with desired pore size, nanoporous structure, and morphology for hydrogen storage by optimizing process parameters.

Overall, the electrospinning technique is a novel, versatile, efficient, and low-cost method for fabricating nanofibers from all naturally occurring and synthetic polymers. The one-dimensional nanomaterials prepared by electrospinning have displayed huge advantages for energy applications, such as hydrogen generation/splitting, solar cells, fuel cells, lithium ion batteries, and supercapacitors. Electrospun nanofibers incorporated with nanophotocatalyst have shown enhanced photocatalytic activity in water splitting for hydrogen generation because of their large surface area and improved crystallinity. The electrospun nanofibers have exhibited high photoelectric conversion efficiency due to their efficient charge separation and transport and the maximum light absorption rate, which is mainly attributed to their high specific surface areas, flexibility, and high porosity. The efficiency of photocatalytic water splitting for hydrogen generation is still unsatisfactory due to the relatively low activity of photocatalysts. Therefore, how to prepare novel photocatalysts with high catalytic activity by electrospinning is another problem that needs to be addressed. With respect to this issue, fabricating electrospun composite organometal halide perovskite/TiO$_2$ NFs or composite organometal halide perovskite/graphene nanofibers with diversified morphologies as photocatalysts may be a feasible and promising solution. The proposed electrospun nanofibers are highly functional, easy to apply, and safe to handle [18,49,50].

Overall, the electrospinning technique has been used for fabricating photocatalyst micro- and nanoscale fibers embedded with photocatalysts for water splitting and hydrogen generation under sunlight. Nanofibers, also called ultrafine fibers, possess a high surface area and porous structure, which allow more sites for photocatalytic activity. In electrospinning, it is relatively easy to control fiber morphology and diameter and incorporate catalysts into the fibers. Therefore, some researchers have prepared TiO$_2$ fibers via electrospinning and incorporated indium tin oxide into TiO$_2$ fibers to enhance the photocatalytic efficiency. Some studies also produced NiO−STO (NiO−SrTiO$_3$) nanocomposite fibers using the coaxial electrospinning technique in order to obtain higher photocatalytic efficiency. Research is still in progress in this area, and scientists have been experimenting with all kind of materials and techniques in order to find a suitable catalyst that has high photocatalytic efficiency and can also function in both visible and UV light.

8.6 CONCLUSIONS

Hydrogen is an outstanding energy alternative to fossil fuels since the only byproduct in its conversion process is water. However, the major hurdle lies in finding suitable materials that possess the advantage of durability, cheap, and safety in storage of large amounts of hydrogen. Many research studies have focused their attention on improving hydrogen storage properties such as alteration of the diffusion pathway, lower dehydrogenation, and even enhanced reversibility using templates to modified thermodynamics and/or kinetics. Electrospun nanofibers, due to their unique features, can be the materials of choices for hydrogen storage. Some researchers reported the preparation of nanofibers via an electrospinning process for solar energy capture, hydrogen splitting, and hydrogen storage. The studies showed that Li$_3$N porous nanofibers have displayed better hydrogen performance with stable reversibility over 10 cycles of de/rehydrogenation at a temperature of 250°C. Other studies fabricated polyacrylonitrile-derived carbon fibers after carbonization at 800°C in argon and water vapor and electrosprayed palladium during electrospinning. Their results showed that the surface area of palladium-coated carbon fibers was 815.6 m^2/g and hydrogen adsorption capacity was determined as 0.35 wt.% at 298 K and 0.1 MPa. Broadly speaking, electrospinning can prepare nanomaterials with desired pore size, shape, nanoporous structure, and morphology for hydrogen production and storage by optimizing system and process parameters.

REFERENCES

[1] F.E. Osterloh, B.A. Parkinson, Recent developments in solar water-splitting photocatalysis, MRS Bull. 36 (1) (2011) 17−22.

[2] R.S. Dholam, "Development of Solar Sensitive Thin Film for Water Splitting and Water Heating Using Solar Concentrator," PhD dissertation, University of Trento, Italy, 2010.

[3] M.G. Walter, E.L. Warren, J.R. McKone, S.W. Boettcher, Q. Mi, E.A. Santori, et al., Solar water splitting cells, Chem. Rev. 110 (2010) 6446−6473.

[4] K. Maeda, K. Domen, New non-oxide photocatalysts designed for overall water splitting under visible light, J. Phys. Chem. C 111 (2007) 7851−7861.

[5] M.G. Mali, S. An, M. Liou, S.S. Al-Deyab, S.S. Yoon, Photoelectrochemical Solar water splitting using electrospun TiO_2 nanofibers, Appl. Surf. Sci. 328 (2015) 104−114.

[6] A.R. Alharbi, I.M. Alarifi, W.S. Khan, R. Asmatulu, Semiconductor nanomaterials for water splitting, Advances in Material Science Research, 21, Nova Science Publishers, 2015. ISBN: 978-1-63483-547.

[7] N. Nuraje, R. Asmatulu, S. Kudaibergenov, Metal oxide-based functional materials for solar energy conversion: a review, Curr. Inorg. Chem. 2 (2) (2012) 23.

[8] A. Kudo, Y. Miseki, Heterogeneous photocatalyst materials for water splitting, Chem. Soc. Rev. 38 (1) (2009) 253−278.

[9] L.C. Sim, K.W. Ng, S. Ibrahim, P. Saravanan, "Preparation of Improved p-n Junction NiO/TiO2 Nanotubes for Solar-Energy-Driven Light Photocatalysis," Vol. 2013, 2013 Article ID 659013, 10 pages.

[10] J. Wang., Z. Lin, Dye-sensitized TiO_2 nanotube solar cells with markedly enhanced performance via rational surface engineering, Chem. Mater. 22 (2) (2010) 579−584.

[11] B. Wawrzyniak, A.W. Morawski, B. Tryba, Preparation of TiO_2-nitrogen-doped photocatalyst active under visible light, Int. J. Photoenergy (2006). Hindawi Publishing Corporation, 1−8, Article ID 68248.

[12] L. Wen, B. Liu, X. Zhao, K. Nakata, T. Murakami, A. Fujishima, Synthesis, characterization, and photocatalysis of Fe-doped: a combined experimental and theoretical study, Int. J. Photoenergy (2012). 10 pages, Article ID 368750.

[13] X. Feng, D.J. Sloppy, T.J. LaTempa, M. Paulose, S. Komarneni, N. Bao, et al., Synthesis and deposition of ultrafine Pt nanoparticles within high aspect ratio TiO_2 nanotube arrays: application to the photocatalytic reduction of carbon dioxide, Journal of Mater. Chem. 21 (2011) 13429−13433.

[14] N. Sasirekha, S.J.S. Basha, K. Shanthi, Photocatalytic performance of Ru doped anatase mounted on silica for reduction of carbon dioxide, Appl. Catal. B 62 (1−2) (2006) 169−180.

[15] C.H. Liao, C.W. Huang, C.S.W. Jeffrey, Hydrogen production from semiconductor-based photocatalysis via water splitting, Catalysts 2 (2012) 490−516.

[16] H.M. Chen, C.K. Chen, R.S. Liu, L. Zhang, J. Zhang, D.P. Wilkinson, Nano-architecture and material designs for water splitting photoelectrodes, Chem. Soc. Rev. 41 (2012) 5654−5671.

[17] A. Alharbi, I.M. Alarifi, W.S. Khan, R. Asmatulu, Synthesis and analysis of electrospun $SrTiO_3$ nanofibers with NiO nanoparticles shells as photocatalysts for water splitting, Macromol. Symp. 365 (2016) 246−257.

[18] R. Asmatulu, M.A. Shinde, A. Alharbi, I.M. Alarifi, Integrating graphene and C_{60} into TiO_2 nanofibers via electrospinning process for the enhanced energy conversion efficiencies, Macromol. Symp. 365 (2016) 128−139.

[19] A. Alharbi, I.M. Alarifi, W.S. Khan, R. Asmatulu, "Comparative Studies on different Nanofiber photocatalysts for Water Splitting," Proceedings of SPIE, Smart Materials and Nondestructive Evaluation for Energy systems, Las Vegas, Nevada, USA, 2016.

[20] A. Alharbi, I.M. Alarifi, W.S. Khan, R. Asmatulu, "Co-axial electrospinning of strontium titanata nanofibers associated with nickel oxide nanoparticles for water splitting," CAMX Conference, Dallas, TX October 27−29, 2015, 13p.

[21] W.S. Khan, R. Asmatulu, M. Ceylan, A. Jabbarnia, Recent progress on conventional and non-conventional electrospinning processes, Fibers Polym. 14 (2013) 1235−1247.

[22] A. Greiner, J.H. Wendorf., Electrospinning: a fascinating method for the preparation of ultrathin fibers, Angew. Chem. Int. Ed. 46 (2007) 5670−5703.

[23] W.S. Khan, R. Asmatulu, M. Ceylan, A. Jabbarnia, Recent progress on conventional and non-conventional electrospinning processes, Fibers Polym. 14 (8) (2013) 1235−1247.

[24] S. Mishra, S.P. Ahrenkiel, Synthesis and characterization of electrospun nanocomposite TiO_2 nanofibers with Ag nanoparticles for photocatalysis applications, J. Nanomater. 16 (2012).

[25] Y. Wang, Y. Li, Y.G. Sun, G. Zhang, H. Liu, J. Du, Q. Yang, Fabrication of Au/PVP nanofiber composites by electrospinning, J. Appl. Polym. Sci. 105 (6) (2007) 3618−3622.

[26] X. Cheng, G. Pan, X. Yu, T. Zheng, A. Umar, Q. Wang, Effect of post-annealing treatment on photocatalytic and photoelectrocatalytic performances of TiO_2 nanotube arrays photoelectrode, J. Nanosci. Nanotechnol. 13 (2013) 5580−5585.

[27] A. Wolcott, W.A. Smith, T.R. Kuykendall, Y. Zhao, J.Z. Zhang, Photoelectrochemical water splitting using dense and aligned TiO_2 nanorod arrays, Small 5 (1) (2009) 104−111.

[28] A.J. Cowan, W. Leng, P.R.F. Barnes, D.R. Klug, J.R. Durrant, Charge carrier separation in nanostructured TiO_2 photoelectrodes for water splitting, Phys. Chem. Chem. Phys. 15 (201) (2013) 8772−8778.

[29] J. Zhang, X. Tang, D. Li, One-step formation of crystalline TiO_2 nanotubular arrays with intrinsic p-n junctions, J. Phys. Chem. C 115 (2011) 21529−21534.

[30] H.J. Yun, H. Lee, J.B. Joo, N.D. Kim, J. Yi, Effect of TiO_2 nanoparticle shape on hydrogen evolution via water splitting, J. Nanosci. Nanotechnol. 11 (2011) 1−4.

[31] T.K. Townsend, N.D. Browning, F.E. Osterloh, Nanoscale strontium titanate photocatalysts for overall water splitting, ACS Nano 6 (8) (2012) 7420−7426.

[32] K. Saito, K. Koga, A. Kudo, Lithium niobate nanowires for photocatalytic water splitting, Dalton Trans. 40 (2011) 3909−3913.

[33] S. Yan, L. Wan, Z. Li, Z. Zou, Facile temperature-controlled synthesis of hexagonal Zn_2GeO_4 nanorods with different aspect ratios toward improved photocatalytic activity for overall water splitting and photoreduction of CO_2, Chem. Commun. 47 (19) (2011) 5632−5634.

[34] T. Yokoi, J. Sakuma, K. Maeda, K. Domen, K.T. Tatsumi, J.N. Kondo, Preparation of a colloidal array of $NaTaO_3$ nanoparticles via a confined space synthesis route and its photocatalytic application, Phys. Chem. Chem. Phys. 13 (7) (2011) 2563−2570.

[35] M. Shibata, A. Kudo, T. Akira, D. Kazunari, M. Kenichi, T. Takharu, Photocatalytic activities of layered titanium compounds and their derivatives for H_2 evolution from aqueous methanol solution, Chem. Lett. 16 (6) (1987) 1017−1018.

[36] L. Macaraig, S. Chuangchote, T. Sagawa, Electrospun $SrTiO_3$ nanofibers for photocatalytic hydrogen generation, J. Mater. Res. 29 (1) (2014) 123−130.

[37] R.U.E. Lam, L.G.J. de Haart, A.W. Wiersma, G. Blasse, A.H.A. Tinnemans, A. Mackor., The sensitization of $SrTiO3$ photoanodes by doping with various transition metal ions, Mater. Res. Bull. 16 (12) (1981) 1593−1600.

[38] X. Zhou, J. Shi, C. Li, Effect of metal doping on electronic structure and visible light absorption of $SrTiO_3$ and $NaTaO3$ (Metal = Mn, Fe, and Co), J. Phys. Chem. C 115 (16) (2011) 8305−8311.

[39] A. Alharbi, M.I. Alarifi, W.S. Khan, R. Asmatulu, "Electrospun Strontium Titanata incorporated with Nickel Oxide Nanoparticles for Improved Photocatalytic Activities," SPIE Smart Structures/Non-Destructive Evaluation Conference, San Diego, CA, March 8−12, 2015, 8p.

[40] S. Xiaomin, Z. Weiping, M. Delong, M. Qian, B. Denzel, M. Ying, et al., Electrospinning of nanofibers and their applications for energy devices, J. Nanomater. 2015 (2015). Article ID 140716, 20p.

[41] H. Kim, D. Lee, J. Moon, Co-electrospun Pd-coated porous carbon nanofibers for hydrogen storage applications, Int. J. Hydrog. Energy 36 (2011) 3566−5373.

[42] M. Rzepka, P. Lamp, M.A. De la Casa-lillo, Physisorption of hydrogen on microporous carbon and carbon nanotubes, J. Phys. Chem. B 102 (52) (1998) 10894−10898.

[43] Z. Kurban, Electrospun nanostructured composite fibers for hydrogen storage applications, PhD Dissertation, Department of Physics and Astronomy University College, London, UK, 2011.

[44] P.E. de Jongh, P. Adelhelm, Nanosizing and nanoconfinement: new strategies towards meeting hydrogen storage goals, ChemSusChem 3 (12) (2010) 1332−1348.

[45] M. Fichtner, Nanoconfinement effects in energy storage materials, Phys. Chem. Chem. Phys. 13 (48) (2011) 21186−21195.

[46] T.K. Nielsen, F. Besenbacher, T.R. Jensen, Nanoconfined hydrides for energy storage, Nanoscale 3 (2011) 2086−2098.

[47] G. Xia, D. Li, X. Chen, Y. Tan, Z. Tang, Z. Guo, et al., Carbon-coated Li_3N nanofibers for advanced hydrogen storage, Adv. Mater. 25 (43) (2013) 6238−6244.

[48] J. Alipour, A.M. Shoushtari, A. Kaflou, Electrospun PMMA/AB nanofiber composites for hydrogen storage applications, e-Polymers 14 (5) (2014) 305−311.

[49] N. Nuraje, R. Asmatulu, G. Mul, Green Photo-Active Nanomaterials: Sustainable Energy and Environmental Remediation, RSC Publishing, Cambridge, England, 2015.

[50] R. Asmatulu, Nanotechnology Safety, Elsevier, Amsterdam, The Nederland, 2013.

Electrospun nanofibers for nanosensor and biosensor applications

CHAPTER OUTLINE

9.1 ELECTROSPUN NANOFIBERS FOR NANOSENSORS

9.1.1 GENERAL BACKGROUND

Recently, the importance of control and observation on commercial products used in environmental protection, food business, clinical diagnoses, hygiene, drug development, and forensics have been drastically increased for a couple of decades [1−3]. However, most of the currents sensors have had some concerns in terms of reliability, accuracy, selectivity, and response rates. Thus, there is a great need for reliable analytical devices to measure contaminants, pathogens, and other chemical and biological threats in both accurately and quickly. Nanosensors are new generations of sensors that can sense many chemicals, compounds, and microorganisms at very low levels with high accuracy and rapid phase [4,5]. A nanosensor is a nanometer-sized device to detect and analyze information at near atomic/molecular levels. Nanosensors can be made of electrospun nanofibers, as well as nanowires, nanoprobes, nanoelectromechanical systems (NEMS), carbon nanotubes (CNTs), nanoporous materials, etc. Some of these sensors can be fabricated using electrospinning, top-down lithography, bottom-up assembly, and molecular self-assembly techniques. The new generation of nanosensors is commonly utilized due to their performance, functionality, cost, and efficiency [5−8].

Synthesis and Applications of Electrospun Nanofibers. DOI: https://doi.org/10.1016/B978-0-12-813914-1.00009-2

CNTs are one of the most studied nanomaterials because of their small-scale tubes made of graphitic carbon atoms and excellent physical and chemical properties, such as mechanical, thermal, electrical, dielectrical, and optical. They are currently being used in different types of applications, including nanosensors, scanning probes, nanoelectronics, electrochemical sensing systems, nanoactuators, storage devices, and nanoelectromechanical systems. In general, these sensors are widely employed in monitoring environmental contamination, biomarkers, temperature, pressure, biomedicine, and hormonal and molecular disorders [9]. CNTs can be incorporated with electrospun nanofibers to enhance the overall properties of new sensors.

Nanofibers and nanowires can be in the forms of metals and alloys, polymers, composites, and ceramics based on the selected materials and methods. The miniature size, shape, and morphology of these nanostructured materials can be utilized in sensing devices, electrochemical actuators, artificial muscles, pressure, temperature, dielectric, and electric resistive sensors for a number of industrial applications [10]. Some of these nanosensors are used in measuring molecular depositions on the surfaces with high resolution and estimation are externally monitored in real time. When compared to other devices, nanofibers and nanowires are preferred since they are designed in many ways that the agents and elements cannot enter into the porous cavities when they are hydrophobic or can deposit into the structure of nanofibers when they are completely hydrophilic [9−11]. Any surface changes on nanofibers and nanowires can be sensed by the sensing devices. Fig. 9.1 shows comparisons of label-free sensor technologies for standard and nanomaterial sensors [2,13].

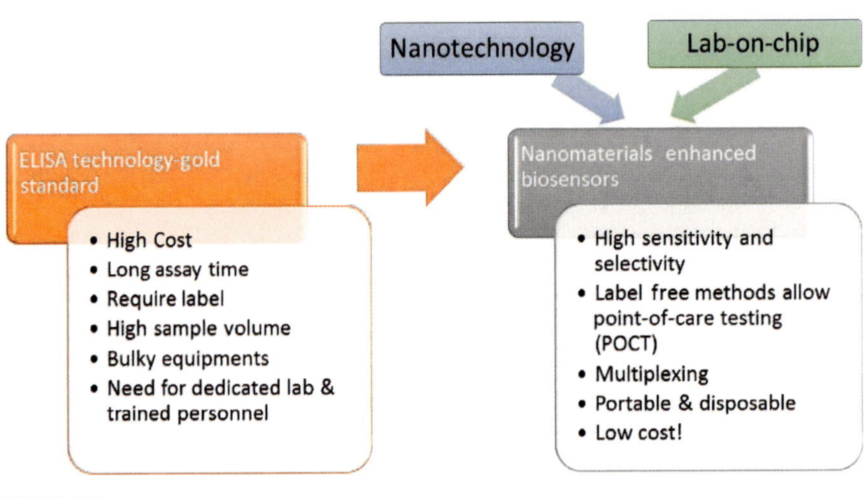

FIGURE 9.1

Comparisons of label-free sensor technologies for standard and nanomaterial sensors.

Nanofibers can be used to sense any changes in resonance, impedance, plasmon resonance, light visibility and intensity, pressure, pH, magnetic, electric, dielectric, optical, and thermal properties [9]. Many sensor applications prefer to use resonance-based sensing devices as they offer appropriate output for the specific requirements. These types of sensors usually work depending on their resonant frequency and resonant mass [12−14]. The working principle of nanosensors is that the resonant frequency is directly related to the resonant mass in which any changes between micromole and picomole levels can be easily detected [12]. The resonant mass is usually the combination of the mass of the resonator and the mass of the attached material deposited prior to the measurements. In other words, any changes in mass of the attached materials on the surfaces of nanofibers and nanowires will result in changes to the resonant frequencies. Depending on the amount of substances interacted with the surfaces, the magnitude of frequencies will be drastically changed/shifted. In this way, these nanomaterials can be utilized in toxic gas detection, nanoparticle concentrations, structural health monitoring (SHM), air, water, and soil contaminations, UV intensity, humidity, vibration, velocity, fluid flow, as well as DNA, enzyme, and protein detection. A number of different industries, such as manufacturing, healthcare, defense, agriculture, food, environmental, and industrial monitoring have been using sensors for a long time.

9.1.2 ELECTROSPINNING FOR NANOFIBERS

Electrospinning is a unique process for producing micro- and nanoscale fibers in woven and nonwoven forms for different sensor applications. These nanofibers have some specific features including improved mechanical properties, high surface functionality, high porosity, and high permeability because of their small diameters and structures. Fig. 9.2 shows the basic set up of the electrospinning process [12−14]. The system (e.g., viscosity, polymer and solvent types and structures, conductivity and surface tension of polymers) and process (e.g.,

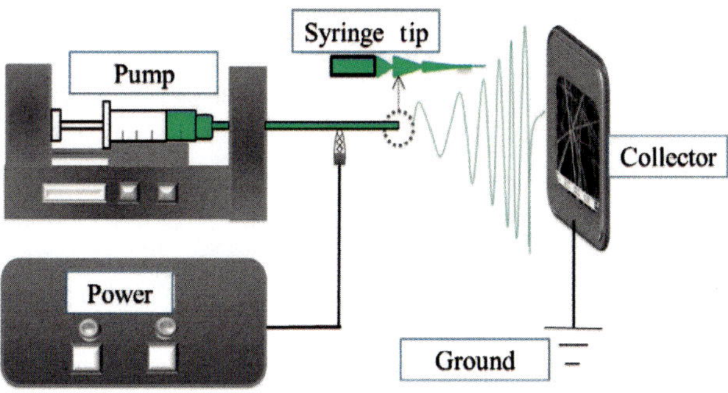

FIGURE 9.2

The basic set up for the electrospinning process.

electric potential, flow rate and polymer concentration, temperature, distance between the capillary and screen, humidity, air velocity effects in the chamber, conductive, and inclusions) parameters can be adjusted to produce different micro- and nanoscale fibers for nano- and biosensor applications. Table 9.1 gives some of the frequently used polymers, solvents, and their testing conditions. Fig. 9.3 also shows SEM images of PVC and polystyrene (PS) nanofibers.

9.1.3 ELECTROSPUN NANOFIBERS FOR NANOSENSORS

Nanotechnology and nanosensors have been significantly increasing for a few decades, and this trend will continue in the near future. Integration of nanotechnology into nanosensors is mainly through novel approaches in the field to advance new features that cannot be achieved using conventionally used technologies in classic sensors. Various nanofibers, nanowires, and their mats/fabrics can be integrated into the sensing elements, so these nanosensors can be utilized as alternatives to conventional sensors. The nanofibers have some promising properties, such as lower production costs, high efficiency, selectivity, small quantities of sample requirements, lower detection limits, higher stability and sensitivity, reduced power consumption, small dimensions, direct analyte (chemical species) detection, and multifunctionality [12]. These achievements are mainly because of the novel properties of nanofibers, including small dimensions, flexibility, higher specific surface area, and variable conduction properties, which make them ideal materials for different industrial applications. By investigating the system and process parameters of nanofibers, the surface functionality and dimensions of nanofibers can be tailored to increase the diffusion and absorption rates for rapid signal transferring and faster response timing. These properties can also improve the sensitivity, dynamic performance, and label-free detection level at very small quantities of samples [1,12].

Recent studies were aimed at developing new classes of electrospun nanofibers as strain gages in structural health monitoring (SHM) of composite aircraft structures [14−19]. The composite aircraft industry has been growing rapidly because of the distinct advantages of fiber-reinforced composites when compared to conventional metals and alloys, which are considerably heavier and corrosive in harsh environments. Fatigue and creep strengths of metal-based aircraft and their parts are other major concerns, which may cause safety concerns in long service life [14]. Because of the easy manufacturing options and high performances, carbon-based piezoresistive sensors, including carbon nanofibers, graphite, carbon nanotubes, and graphene, have been developed as alternatives to the traditional silicon/metal-based microelectromechanical system (MEMS) and NEMS for different industries [20]. These nanosensors can be employed for prognostic and diagnostic purposes due to their small size and high sensitivity to small forces. Real-time detection of small flaws and monitoring of composite structures on a continuous and routine basis is the primary motivation of these SHM sensor devices. Nanofibers play a crucial role in developing nanotechnology-based

Table 9.1 The Frequently Used Polymers, Solvents, and Testing Parameters

Polymers	Solvents	Concentration of polymer (%)	Applied DC voltage (kV)	Tip-to-collector distance (cm)	Potential applications
Naylon 6,6	Formic acid	10	15–25	15–30	Protective coating
Polyurethanes	Dimethylformamide	10	15–25	15–30	Protective coating
Polycarbonate	Dichloromethane	15	15–25	15–30	Sensor, filter
Polylactic acid	Dichloromethane	14	15–25	15–30	Drug Delivery systems
Polyacrylonitrile	Dimethylformamide and acetone	10–20	15–25	15–30	Carbonized fibers, sensor
Polyvinyl chloride	Dimethylacetamide	10–20	15–25	15–30	Filter, sensor

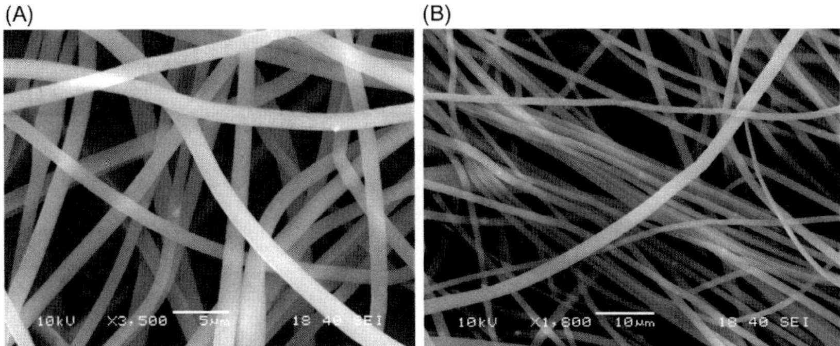

FIGURE 9.3

SEM images showing (A) PVC and (B) PS nanofibers. The processing parameters of PVC fibers including DMAC solvent (80:20), 30 cm distance, 25KV DC voltage, and 1.5 mL/h pump speed, while that of PS fibers including DMF solvent (80:20), 30 cm distance, 25 KV DC voltage, and 2.5 mL/h pump speed.

sensors, and carbon nanofibers are ideal candidates owing to their high strength and stiffness as well as other unique mechanical, thermal, and electrical properties in these applications.

Alarifi et al. reported that electrospun PAN nanofibers could be produced using various electrospinning conditions (e.g., 25 KV DC voltage, 25 cm tip-to-collector distance, and 2 mL/h pump speed) after dissolving in DMF solvent at an 80:20 ratio. The produced nanofibers were stabilized at 260 °C in ambient conditions for 60 min, and carbonized at 600−900 °C in argon purge for an additional 60 min. A characterization test showed that these nanofibers were mainly crystalline, highly conductive, defect free, and in submicron and nanoscale ranges. The prepared nanofiber films/mats were then placed on the top layer of pre-preg carbon fiber laminate composites, and cocured in a vacuum oven following the curing kinetics of pre-preg composites. These lightweight and cost-effective PAN-derived electrospun fibers were used for SHM applications of composites and tested for durability under different conditions. The carbonized SHM sensors were sensitive to any small changes in strain and stress on the composite structures [14].

Fig. 9.4 shows SEM images of electrospun PAN nanofibers before and after carbonization at 850 °C. As shown in Fig. 9.4A, the average diameter of the PAN fibers was around 550 ± 50 nm before carbonization; however, after carbonization, the diameter was reduced to 400 ± 50 nm (Fig. 9.4B), retaining their shape and morphology [20]. The shrinkage is primarily because of the mass reduction and densification of the PAN nanofibers. Fig. 9.5 shows schematic images of SHM testing setups: (A) layup fabrication scheme of carbon fiber composite incorporated with carbonized PAN nanofibers; and (B) specimen used in strain sensing [20].

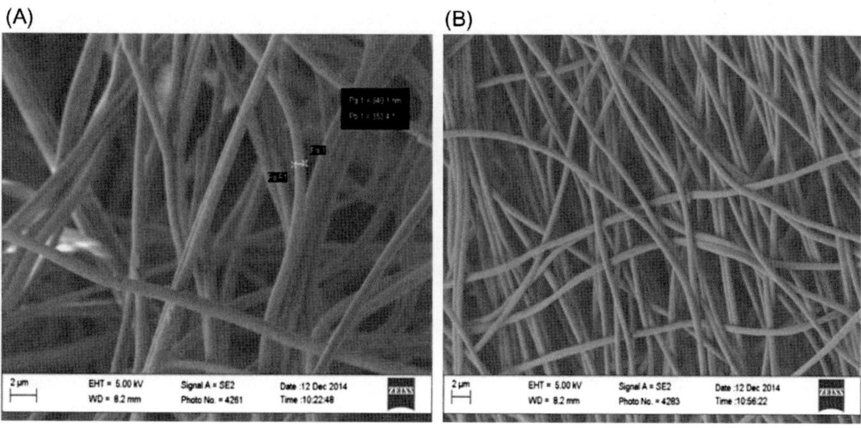

FIGURE 9.4

SEM images of electrospun PAN nanofibers: (A) before the carbonization process, and (B) after the carbonization process [20].

Ding et al. mentioned the global increasing demands on highly sensitive sensors for monitoring, inspection, and medical diagnostics using electrospun nanofibers and nanoscale films [21]. It was reported that excellent specific surface areas, flexibility, and porosity of nanofibers could attract ultrasensitive nanosensors to various industries. The authors summarized that the recent developments in electrospun nanomaterials in sensing approaches (e.g., impedance, resistive, acoustic wave, optical, photoelectric, and amperometric) demonstrate with examples how the systems work, and also discuss the basic fundamentals and optimization designs [21].

Aussawasathien et al. prepared lithium perchlorate-doped polyethylene oxide (PEO) electrospun nanofibers for humidity sensing and camphosulfonic acid (HCSA)-doped polyaniline (PANI)/PS electrospun nanofibers for sensing hydrogen peroxide and glucose [22]. The diameters of the nanofibers were between 400 and 1000 nm. Because of the high surface area and relatively good electrical conductivity values, the sensitivity of the nanofiber sensors was significantly increased compared to the conventional film-type counterparts. SEM studies showed that after some humidity measurements, $LiClO_4$-doped PEO nanofibers had some distortions, but not on the HCSA-doped PANI/PS nanofiber sensors after measurements of H_2O_2 and/or glucose [22].

Ding et al. also stated that electrospun nanofibers have up to two orders of magnitude higher specific surface areas compared to flat thin films, which makes them excellent materials for nanosensing applications [23]. The gas sensors were made of nanofibers using polyelectrolytes, conducting polymer composites, and semiconductors using different sensing techniques (e.g., resistive, photoelectric, acoustic wave, and optical techniques). They concluded that electrospun

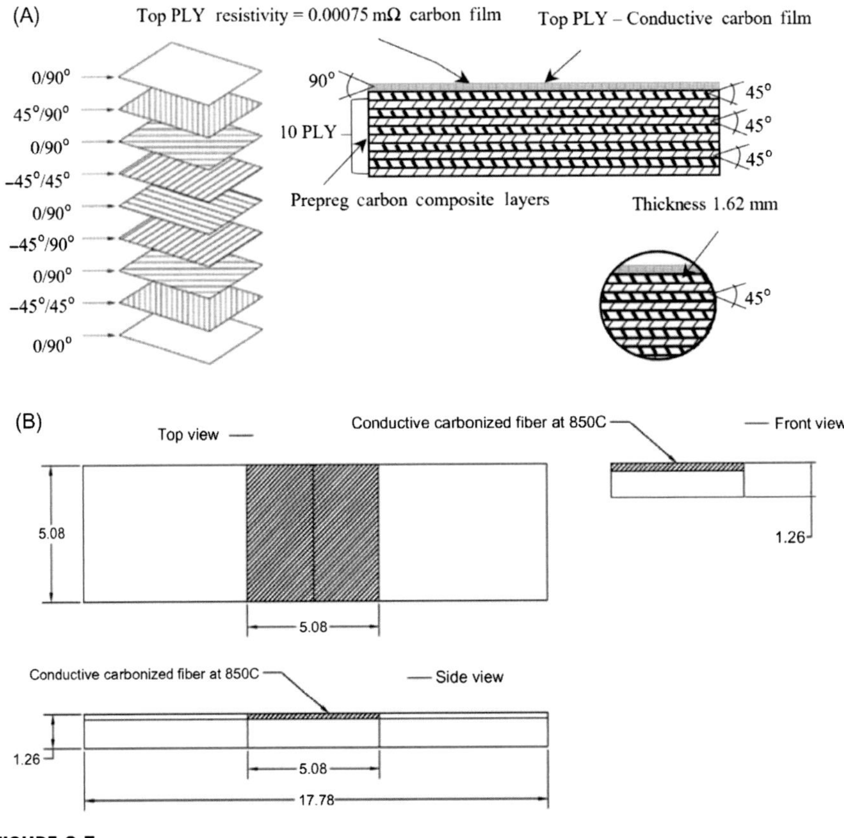

FIGURE 9.5

Schematic images of SHM testing setups: (A) layup fabrication scheme of carbon fiber composite incorporated with carbonized PAN nanofibers, and (B) specimen used in strain sensing [20].

nanofiber-based sensors displayed significantly higher sensitivity and quicker response rates to targeted gases when compared to the flat film sensors [23].

Wang et al. discussed the detection and monitoring of reusable, explosive, and chemical hazardous materials, such as heavy metal ions, gases, and other organic and inorganic contamination using the nanosensors [24]. The authors specified that electrospun nanofiber membranes with relatively high surface area, porosity, interconnected pore structures, and flexibility were ideal nanomaterials for improved performances of nanosensors. These nanofiber sensors further enhanced the sensitivity, selectivity, stability, and response rates of the sensors. They also summarized the recent progress in the field, described the designing and fabrication of various nanosensors, and discussed performances of these nanosensors for the detection of various analytes [24].

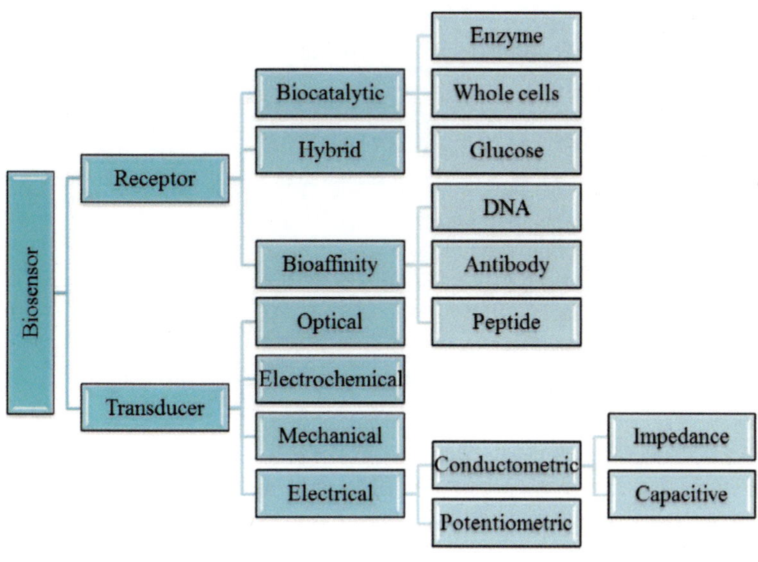

FIGURE 9.6

The classification of various biosensors developed for different biological applications [12].

9.2 NANOFIBERS FOR BIOSENSORS

9.2.1 BIOSENSORS

Similar to the nanosensor, the biosensor is a transducer/receptor that converts a biological quantity or parameter into a measurable signal for quantification. In other words, the biosensor translates one form of energy into another form and detects the information to a measurable quantity. These are specific types of sensors which are capable of sensing different biological elements, molecules, and structures. In general, a biosensor consists of two major parts: a biological sensing element and a transducer for measurement. The role of the biological sensing element is to interact with the analyte, while that of the transducer is to produce a measurable signal proportional to analyte concentration/specifications [1,12]. Fig. 9.6 shows the classification of various biosensors developed for different biological applications [12]. Table 9.2 also provides the detection performance parameters and working principles of numerous biosensors designed for C-reactive protein (CRP) detection [2,13].

9.2.2 NANOFIBERS FOR BIOSENSORS

A number of different nanofibers have been employed in sensor and actuator devices as sensing elements due to the enhanced efficiency and performance of new generations of biosensors [25]. Their unique features, including flexibility,

Table 9.2 The Detection Performance Parameters and Working Principles of Numerous Biosensors Designed for CRP Detection

Signal Transduction	Assay Format and/or Nanostructures Used	LOD	Sample Volume (μL)	Assay Time (h)
Fluorescence	Agarose microbeads	1 ng/mL	–	1.25
SPR	Co-sputtered Au and SiO_2 nanoparticles	1 μg/mL	–	>24
Chemiluminescence	Luminol substrate and HRP enzyme	1.85 μg/mL	75 nL	–
SPR	Aptamer	0.005 ppm	20 μL	–
LSPR	Gold-capped porous alumina	1 pg/mL	15 μL	5
Fiberoptic	Cylindrical microcavity-based waveguide	10 pg/mL	–	<0.5
LSPR	High aspect ratio triangular silver nanoplate	5 ng/mL	–	–
Chemiluminescence	Aptamer conjugated magnetic beads	12.5 ng/mL	5 μL	2.5
Capacitive	Gold interdigitated electrode on nanocrystalline diamond surface	25 ng/mL	20 μL	9
Capacitive	Microelectrode and microarray	2.2 ng/mL	2 μL	3
Dielectrophoresis	Polystyrene microbeads	1 μg/mL	–	–
Capacitance	Nanoporous alimuna	200 pg/mL	–	–
Impedance	Biogenic nanoporous silica	1 pg/mL	10 μL	0.75
Impedance	Electrospun PS fiber coated with PPy	1 pg/mL	200 μL	3
FET	Au/NiCr as channel	3 μg/mL	–	–
Impedance	3D nanogap gold interdigitated microelectrode	13 ng/mL	–	12
FET	Silicon nanowire array	1 fM	–	<0.5
Capacitance	Aptamer	100 pg/mL	–	12
Mass loading	Piezoresistive microcantilever with functionalized gold surface	100 ng/mL	–	4
Mass loading	Microcantilever with gold surface	1 μg/mL	–	1.25
Acoustic	Gold-coated quartz wafer	3 ng/mL	–	<0.5
SH-SAW	Gold-coated quartz substrate	50 ng/mL	–	<0.5

high porosity, high surface area, and tailored mechanical properties make them appealing materials [25,26]. These nanofibers can be fabricated via different techniques to provide controlled sizes, shapes, porosity, morphology, and mechanical strengths. These nanomaterials improve the sensitivity of the biosensors through huge immobilization sites, hence increasing the biomolecular interactions and masses [26]. Also, the substantially high porosity and hydrophilic nature of nanofibers enables rapid biomolecular penetration, which in turn drastically lowers the detection time and increases the response rate [27]. Surface modifications and functionalizations can be easily conducted on the entire surfaces of nanofibers for different molecular interactions (e.g., antigen—antibody or avidin—biotin) when compared to nanotubes and nanowires. For instance, nanofibers and their mats are simpler to functionalize owing to the oxygen atoms available on their active sites. A number of polymeric fibers, including polyaniline and chitosan, can be functionalized using a variety of chemical agents (surfactants, acids, base, amino acids, DNA, proteins, enzymes, nanoparticles, and other additives) [28—30]. There are many approaches to producing nanofibers for biomedical applications, including electrospinning, interfacial polymerization, chemical vapor deposition, drawing, template synthesis, phase separation, and self-assembly [31]. Among these techniques, electrospinning is an easy way of producing nanofibers with controlled dimensions [12]. These nanofibers can be easily modified through adsorption, covalent bonding, crosslinking, blending, and dispersion of inclusions. The modifications can drastically improve the sensing performance via interactions between the target analyte and the sensing materials.

9.2.3 APPLICATIONS OF ELECTROSPUN NANOFIBERS FOR BIOSENSORS

Recently, functionalized electrospun nanofibers have been widely investigated to integrate them into biosensors as sensing platforms. The effectiveness of functionalized nanofibers in many biosensors improved their applications in every biomedical fields and they have been the main focus of multiple researches conducted so far [26—32]. A brief review of several published research articles using electrospun nanofibers for biosensing applications is given below.

Mao et al. investgated electrospun carbon nanofiber mats with controlled density for the biosensor platform [32]. In order to improve the electrochemical detection levels, the electronic properties of nanofibers were modified and the density of the electronic states was adjusted through the graphitization of the nanofibers. The carbonized nanofibers increased the efficiency and sensitivity of the biosensing applications. Similar studies were also conducted by other researchers on the highly porous electrospun Mn_2O_3-Ag nanofibers for the sensor platform for glucose detection [33]. The prepared biosensor provided a very high sensitivity of $40.6\,\mu A \times mM^{-1} \times cm^{-2}$ and considerably low detection limit of 1.73 µM, which is mainly because of the novel Mn_2O_3-Ag nanofiber properties [33].

In order to enhance the sensitivity and reliability of glucose biosensors, palladium (IV)-copper oxide composite nanofibers were fabricated via an electrospinning method [34]. Test results indicated that palladium (IV)-copper oxide composite nanofiber platforms provided a fast response, high sensitivity, and low limit of detection using amperometric glucose detection systems. A similar approach was also followed by other studies for the detection of a nonenzymatic glucose biosensor using electrospun Co_3O_4 nanofibers [35]. These metal oxide nanofiber sensors also offered high selectivity and sensitivity, fast response time, and a low detection limit.

Wang et al. introduced the electrospun nanofibers of multiwalled carbon nanotubes and poly(acrylonitrile-co-acrylic acid) mixtures [36]. These nanofibers were incorporated with platinum electrodes for the glucose oxidase (or enzyme) immobilization to improve the glucose biosensor detection level. The experimental test results indicated that multiwalled carbon nanotube concentration in nanofibers generally correlated with the enhancement. Also, the increase in the maximum current during the amperometric characterization disturbed the secondary structure of the enzyme [36].

The other study was conducted on *Escherichia coli* O157:H7 for the detection of colony-forming unit (CFU)/mL in liquid using electrospun cellulose nitrate nanofibers [37]. To be able to enhance the selectivity of the biosensor and separate the target analyte from biomolecules in the system, magnetic nanoparticles were attached to the *E. coli* O157:H7 antibody. Because of the extraordinary structures and properties of electrospun fibers incorporated with magnetic nanoparticles, these biosensors had faster and reliable detection rates with relatively linear sensing responses compared to other sensors [37].

Guo et al. studied the novel amperometric hydrogen peroxide biosensor using the electrospun hemoglobin (Hb)−collagen composite [38]. The prepared nanocomposite fibers were transferred onto the electrode surface to produce Hb−collagen microbelt modified electrodes. These biosensors were tested using cyclic voltammetry to prove the potential of the unit for hydrogen peroxide detection. The test studies demonstrated that electrospun Hb−collagen composite nanosensor increased the sensitivity, response time, and stability [38].

Most of the biosensors were incorporated with nanostructured elements (e.g., CNTs, nanowires, nanoparticles, membranes, diatoms, and nanofibers) because of the selectivity, sensitivity, response rate, and lower detection limits. Some of them were linked to the electrospun nanofibers and other related fibers in the forms of polymers, ceramics, composites, and metals. Fig. 9.7 shows the nanomaterials used in biosensors, including CNTs, nanowires, nanoparticles, membranes, diatoms, and nanofibers [2,13].

It was stated that electrospinning has demonstrated a tremendous amount of flexibility in both bio- and nanosensor developments for different applications [39]. Marx et al. reported on electrospun gold nanofiber electrodes for biosensing

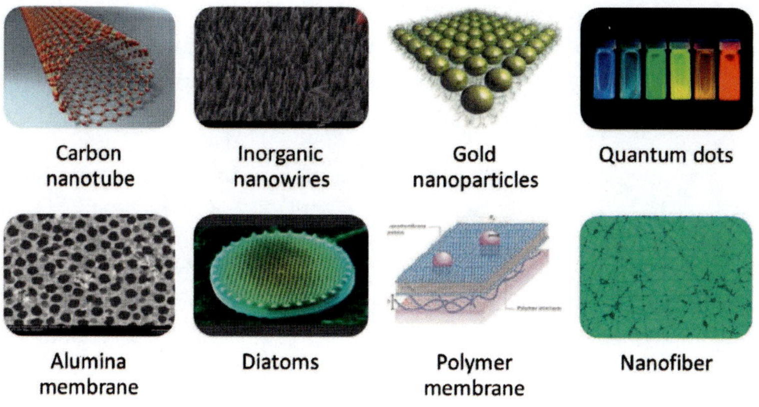

FIGURE 9.7

The nanomaterials used in biosensors, including CNTs, nanowires, nanoparticles, quantum dots, membranes, diatoms, and nanofibers in biosensors.

applications [40]. In this research, surface areas of gold electrodes were significantly improved using the electrospun gold nanofibers. These sensors, used for fructose detection, exhibited high stability (over 20 cycles), and short response time (less than 2.2 s) with very high accuracy rates. The same sensor was modified through the addition of glucose isomerase for glucose detection as the second analyte, hence improving the abilities of sensors for the parallel detections of other substances [12]. Generally, gold is stable and does not corrode in normal conditions, so the reliability, service life, and stability of gold-based sensors were substantially higher compared to the many other metal-based sensors.

Zhang et al. investigated the highly sensitive and stable humidity nanosensors incorporated with LiCl-doped TiO_2 electrospun eanofibers [41]. In this study, polystyrene—poly(styrene-co-maleic anhydride) (PS-PSMA) blends were dissolved in an appropriate solvent and electrospun at various voltages, distances, and pump speeds. The prepared fiber mats were used as sensing elements for protein immobilization. Fluorescence labeling materials and spectroscopy were employed during the tests to measure analyte concentrations in the liquid stock solutions. The sensitivity of the same electrospun nanofiber biosensor was also compared with the 96-microwell plate format, and the test results indicated that nanofiber-based biosensor provided 2500-fold higher sensitivity because of the excellent properties of electrospun nanofibers [41].

Sidek et al. studied the polyvinylpyrrolidone/polyaniline nanofiber composites for H_2 gas detection using the surface acoustic wave method [42]. The prepared sensors showed high stability, sensitivity, reliability, and a low limit detection of about 0.06% H_2 gas concentrations. Many other studies also focused on detection of other gases, including NO_2, CO, NH_3, O_2, H_2S, CO_2, and organic volatile compounds using different electrospun nanofibers, such as TiO_2 [43], ZnO, WO_3,

MoO_3, SnO_2 [44], and In_2O_3 [12] in order to improve the mass transfer rate and lower the detection times. These nanofibers were mainly in organic phases, and then heat-treated to produce inorganic phase nanofibers.

Stafiniak et al. developed a novel electrospun ZnO nanofiber biosensor for different biomedical applications [45]. After the fabrication process, protective films of AlNx were sputter coated onto the surfaces of nanofibers to enhance the stability of the sensors. Similar studies were also conducted on the single ZnO nanofiber synthesized by electrospinning methods for amperometric-based glucose detection [46]. The surface of ZnO nanofiber was coated with gold nanofilms and then the fiber surface was modified for the physical adsorption of glucose oxidase prior to electrochemical measurement. The characterization studies showed that high sensitivity and a low limit of detection as well as long-term stability could be achieved using this approach.

Wang et al. fabricated a facile and highly sensitive colorimetric sensor for the detection of formaldehyde using electrospun nanofiber mats at a low detection limit of 50 ppb [47]. A blend of methyl yellow-impregnated electrospinning with nylon 6 was fabricated for the biosensing element. The prepared biosensor was tested at various volatile organic compounds, and confirmed that it had very high selectivity towards formaldehyde detection when compared to the other biosensors [47].

Kang et al. fabricated electrospun 1-butyl-3 methylimidazolium hexafluorophosphate (BMIMPF6)-nylon 6,6 nanofiber chemiresistors for organic vapor sensing purposes [48]. The prepared nanofibers were transferred onto the microelectrodes to create a sensitive platform for cyclic exposure testing using four different organic vapors including methanol, ethanol, acetone, and tetrahydrofuran as targeted analytes. The test results indicated that these nanofiber sensors had good characteristics, demonstrating the useful applications of the nanofibers as biosensing elements for organic vapor detection at room temperature.

Li et al. ivnestigated the highly sensitive and stable humidity nanosensors using LiCl-doped TiO_2 electrospun nanofibers [49]. The test results indicated that their prepared nanosensors had promising characteristics, including reproducibility, faster response, and high linear detection range. In addition to that, the other sensors made of SnO_2 nanofibers also had confirmed high detection levels of analytes in the air, such as ammonia [50], methanol [51], and ethanol [52] owing to the novel properties of the nanofiber sensors, including wide band gap, high transparency, and chemical-sensing capabilities [53].

Zampetti et al. developed a highly sensitive NO_2 gas sensor based on PEDOT$-$PSS/TiO_2 nanofibers [54]. The detection level was down to 1 ppb depending on the stoichiometric oxidation of NO into NO_2. The electrospun TiO_2 nanofibers were directly placed on the electrodes after a thin film of PEDOT$-$PSS was applied to the sensing elements through the dipping process. These sensitive and functional nanofibers can be used for early diagnosis of human lung diseases (e.g., asthma) because of the changes in NO concentrations during the sickness.

Choi et al. demostorated a new study on electrospun nanofibers of CuO-SnO_2 nanocomposites as semiconductor gas sensors for H_2S detection in the atmosphere [55]. Electrospun CuO (p-type) and SnO_2 (n-type) semiconductor nanofibers improved the creation of p-n junctions. During the tests, two operating temperatures of 150 and 300 °C were chosen, and sensitivity, response time, and recovery time of the nanosensors were analyzed. The test results indicated that the electrospun nanofibers were excellent candidates for H_2S sensor development at lower concentrations.

Kacmaz et al. studied the selective sensing of Fe^{3+} at a pico-molar level with ethyl cellulose-based electrospun nanofibers [56]. Ethyl cellulose nanofibers were incoporated with fluorescent ion-selective chromoionophore as an indicator. The prepared nanosensors provided higher response rate and sensitivity, and lower detection limits. These studies can be updated to analyze other ions and molecules at lower concentrations.

Urrutia et al. investigated the possibility of developing nanofiber mats for evanescent optical fiber sensors [57]. The authors fabricated several nanofibers of poly(acrylic acid) (PAA) with different diameters and densities using an electrospinning method and deposited onto an optical fiber core as sensing elements. The prepared sensors were subjected to the various relative humidity conditions and then different transfer function patterns were studied in detail. The diameter, thickness, and porosity/density had major effects on the transmitted optical power. Also, human breathing test cycles were used to measure the response and recovery times of the biosensors.

9.2.4 POLYANILINE-BASED NANOFIBERS FOR BIOSENSORS

Polyaniline is one of the most studied polymers, which were widely employed as immobilization platforms for nano- and biosensing applications [58]. It has its own intrinsic conductive property, which makes it an appealing material for different industries. Nanostructured polyaniline can enhance biomolecule interactions and diffusion rates by providing highly hydrophilic and porous structures with smaller fiber diameters [59]. Recent studies showed that polyaniline-based nanofibers have been incorporated with different actuators and transducers (especially in electrochemical/electrical transducers) for the sensing and detections of enzymes, drugs, and DNA detections [60–62].

Lin et al. reported about a polyaniline nanofiber humidity sensor produced via an electrospinning method [63]. The test results indicated that these sensors provided high sensitivity, response time, linearity, and repeatability compared to existing similar models. The same group produced similar nanofibers using a blend of polyaniline and poly(vinyl butyral) (PVB), and reported that the surface acoustic wave resonator results were very sensitive to humidity, with fast response rate and good linearity range [64]. Poly(methyl methacrylate) (PMMA)-polyaniline nanofibers with different diameters were produced, and then they were deposited on the gold electrodes for triethylamine vapor detections [65]. The experimental results indicated that these biosensors could detect triethylamine vapors in the environment at the 500 ppm level with linear response correlations. Furthermore, the nanofibers at smaller diameters increased the sensitivity of the sensor because of the high surface-area-to-volume ratio and more available binding sites.

A new chemreistor device was developed using electrospun polyaniline nanofibers doped with palladium nanoparticles [66]. The prepared nanofibers were placed onto the gold electrodes to detect the H_2 level using the electrical characterization method. The test results revealed that about 1.8% resistance change was recorded with a 0.3% hydrogen concentration change. The concentration change had significant reading measurements for hydrogen detection. Different surface acoustic wave sensors were also produced for hydrogen detection [12]. These biosensors were developed using electrospun polyvinylpyrrolidone/polyaniline nanofibers. It was concluded that the nanofibers had a limitation of 0.25% H_2 concentrations.

Pinto et al. studied the electric response rates of isolated electrospun polyaniline nanofibers for the detection of aliphatic alcohols [67]. The electrospun nanofibers were produced using HCSA doped with polyaniline nanofibers and then the sensitivity values of the biosensor were compared with the polyaniline mat-based sensor. The test results confirmed that the produced nanofiber-based biosensor had a faster response rate for large molecules, which may be because of the novel properties of the polyaniline nanofibers.

Chen et al. developed the dielectrophoresis carbon nanotube and conductive polyaniline nanofiber-based NH_3 gas sensors [68]. Polyanilin nanofibers were deposited on CNT electrodes using an electrospinning method. Under different ammonia concentrations, this biosensor was tested to determine the response rate, sensitivity, selectivity, and repeatability levels. The sensor provided a detection level of 1 ppm in 60-s response time with a good linearity range ($>$20 ppm). In another study, nanoflowers were electrodeposited on the surfaces of electrospun polyaniline nanofibers using cyclic voltammetry, and tested for urea detection [69]. The test results indicated that this sensor could provide a wide linear range and good detection limit of 10 μM. The other study was conducted on electrospun polyaniline nanofibers for amperometric cholesterol detection [70]. Using a layer-by-layer self-assembly deposition technique, 10 layers of cholesterol oxidase were deposited on the surface of polyaniline nanofibers. Quartz crystal microbalance technique was employed to analyze deposition efficiency and nanofilm growth rates. After the fifth bilayer of cholesterol deposition on the surfaces, cholesterol levels were accurately detected.

Veisi et al. reported the integration of electrospun polyaniline nanofibers as sensing elements in interdigitated gold microelectrodes for selective and label-free detection of cyclooxygenase-2 (COX-2) enzyme as a biomarker [1,12]. This is a pain biomarker for any inflammation and cancer cell proliferation in the body. The performance of the biosensor was analyzed using electrochemical impedance spectroscopy at different biomarker concentrations. Fig. 9.8 shows the

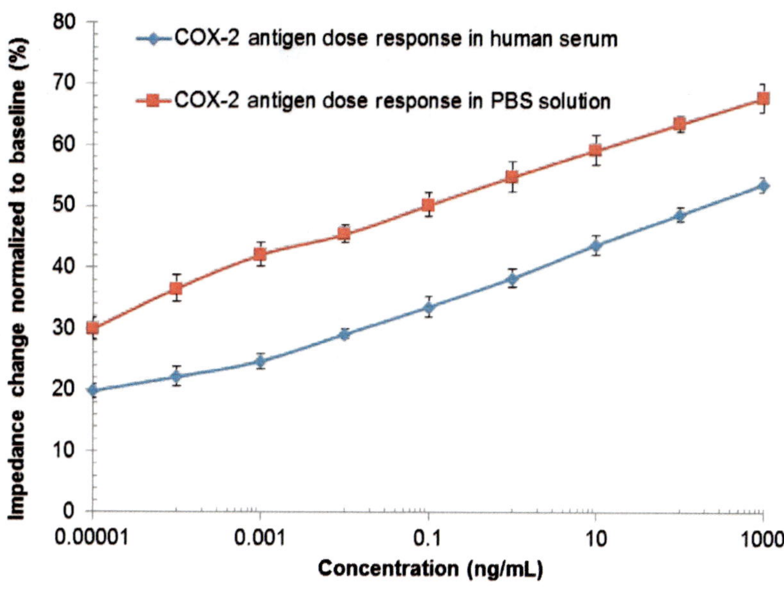

FIGURE 9.8

COX-2 biomarker dose–response of the biosensor integrated with electrospun polyaniline nanofibers at 0.5 mL/h flow rate in buffer and serum solutions.

COX-2 biomarker dose–response of the biosensor integrated with electrospun polyaniline nanofibers at 0.5 mL/h flow rate in buffer and serum solutions [12]. As can be seen, there is a significant improvement in the sensitivity of the electrospun polyaniline nanofiber for robust and rapid detection of the COX-2 biomarker (as low as 0.01 pg/mL and 10 fg/mL in pure and human serum samples, respectively). This improvement was mainly because of the high specific surface area porosity and flexibility of electrospun nanofibers, which can improve the size-matched confinement, transduction, signal strength, and sensitivity of the biosensor [1,12].

9.3 CONCLUSIONS

The first part of this chapter discusses nanomaterials for nanosensors in general, and electrospun nanofibers in particular. Electrospinning parameters (voltage, pump speed, and collector distance), as well as types of polymers, solvents, and additive materials are explained in detail. It is determined that the properties of electrospun nanofibers can be easily changed to increase the sensitivity, selectivity, response rates, and times of the nanosensors. In the second part of the chapter, biosensors and their working principles are evaluated. This part also includes the selective nanofibers produced through electrospinning methods for biosensor applications. It is found that a number of polymeric, composite, metallic, and ceramic nanofibers can be produced and used for the detection of biological entities in liquid and gas forms. In the last part, the polyaniline-based nanofibers and their biosensor applications are analyzed.

REFERENCES

[1] Z. Veisi, M. Ceylan, A. Mahapatro, R. Asmatulu, "An Electrospun Polyaniline Nanofibers as a Novel Platform for Real-time Cox-2 Biomarker Detection," ASME International Mechanical Engineering Congress and Exposition, San Diego, CA, November 15–21, 2013, 7p.

[2] R. Anwar, K. Vattipalli, E. Mayrah, R. Asmatulu, S. Prasad, Highly sensitive conductive polymer nanofibers for application in cardiac biomarker detection, Adv. Sci. Eng. Med. 5 (2013) 633–640.

[3] R. Asmatulu, Nanotechnology Safety, Elsevier, Amsterdam, The Nederland, 2013.

[4] W.S. Khan, R. Asmatulu, M. Ceylan, A. Jabbarnia, Recent progress on conventional and non-conventional electrospinning processes, Fibers Polym. 14 (2013) 1235–1247.

[5] N. Nuraje, W.S. Khan, M. Ceylan, Y. Lie, R. Asmatulu, Superhydrophobic electrospun nanofibers, J. Mater. Chem. A 1 (2013) 1929–1946.

[6] R. Asmatulu, M. Ceylan, N. Nuraje, Study of superhydrophobic electrospun nanocomposite fibers for energy systems, Langmuir 27 (2) (2011) 504–507.

[7] M. Ceylan, R. Asmatulu "Enhancing the Superhydrophobic Behavior of Electrospun Fibers via Graphene Addition and Heat Treatment," ASME International Mechanical Engineering Congress and Exposition, Vancouver, Canada, November 12–18, 2010, 7p.

[8] N. Nuraje, R. Asmatulu, G. Mul, Green Photo-Active Nanomaterials: Sustainable Energy and Environmental Remediation, RSC Publishing, Cambridge, England, 2015.

[9] V.K. Khanna, Nanosensors: Physical, Chemical, and Biological, CRC Press, London, 2016.

[10] K.C. Honeychurch, Nanosensors for chemical and biological applications: sensing with nanotubes, Nanowires and Nanoparticles, Woodhead Publishing, New York, 2014.

[11] M. Ceylan, "Superhydrophobic Behavior of Electrospun Nanofibers with Variable Additives," M.S. Thesis, Wichita State University, December, 2009.

[12] Z. Veisi, "Detection of Cox-2 Enzyme using Highly Sensitive Electrospun Polyaniline Nanofiber-based Biosensor," M.S. Thesis, Wichita State University, May, 2013.

[13] M.R. Anwar, "Electrospun Fiber-based Biosensors for Ultrasensitive Protein Detection," M.S. Thesis, Wichita State University, July, 2011.

[14] I. Alarifi, "Fabrication and Characterization of Electrospun Carbonized Polyacrylonitrile Fibers as Strain Gauges in Composite Aircraft for Structural Health Monitoring Applications," Ph.D. Dissertation, Wichita State University, April 4, 2017.

[15] I.M. Alarifi, A. Alharbi, W.S. Khan, R. Asmatulu, Carbonized electrospun PAN nanofibers as highly sensitive sensors in SHM of composite structures, J. Appl. Polym. Sci. (2015). Available from: https://doi.org/10.1002/app.43235.

[16] I.M. Alarifi, W.S. Khan, A.K.M.S. Rahman, Y. Kostogorova-Beller, R. Asmatulu, Synthesis, analysis and simulation of carbonized electrospun nanofibers infused carbon prepreg composites for improved mechanical and thermal properties, Fibers Polym. 17 (2016) 1449−1455.

[17] I.M. Alarifi, A. Alharbi, W.S. Khan, A.K.M.S. Rahman, R. Asmatulu, Mechanical and thermal properties of carbonized PAN nanofibers cohesively attached to surface of carbon fiber reinforced composites, Macromol. Symp. 365 (2016) 140−150.

[18] I.M. Alarifi, A. Alharbi, O. Alsaiari, R. Asmatulu, Training the engineering students on nanofiber-based SHM systems, Trans. Tech. STEM Educ. 1 (2016) 59−67.

[19] I.M. Alarifi, A. Alharbi, W.S. Khan, A. Swindle, R. Asmatulu, Thermal, electrical and surface properties of electrospun polyacrylonitrile nanofibers for structural health monitoring, Materials 8 (2015) 7017−7031.

[20] I.M. Alarifi, "Fabrication and Characterization of Electrospun Carbonized Polyacrylonitrile Fibers as Strain Gauges in Composite Aircraft for Structural Health Monitoring Applications," Ph.D. Dissertation, Wichita State University, April 4, 2017.

[21] B. Ding, M. Wang, X. Wang, J. Yu, G. Sun, Electrospun nanomaterials for ultrasensitive sensors, Materialstoday 13 (2010) 16−27.

[22] D. Aussawasathien, J.H. Dong, L. Dai, Electrospun polymer nanofiber sensors, Synth. Met. 154 (2005) 37−40.

[23] B. Ding, M. Wang, J. Yu, G. Sun, Gas sensors based on electrospun nanofibers, Sensors 9 (2009) 1609−1624.

[24] X. Wang, Y. Li, B. Ding, Electrospun nanofiber-based sensors, Electrospun Nanofibers Energy Environ. Appl. Part Nanostruct. Sci. Technol. Book Ser. (NST) 9 (2014) 267−297.

[25] Y.A. Dzenis, Spinning continuous fibers for nanotechnology, Science 304 (2004) 1917−1919.

[26] J. Wang, Y. Lin, Functionalized carbon nanotubes and nanofibers for biosensing applications, TrAC Trends Anal. Chem. 27 (2008) 619−626.

[27] Y. Luo, S. Nartker, H. Miller, D. Hochhalter, M. Wiederoder, S. Wiederoder, et al., Surface functionalization of electrospun nanofibers for detecting *E. coli* O157: H7 and BVDV cells in a direct-charge transfer biosensor, Biosens. Bioelectron. 26 (2010) 1612−1617.

[28] R. Jayakumar, M. Prabaharan, S.V. Nair, H. Tamura, Novel chitin and chitosan nanofibers in biomedical applications, Biotechnol. Adv. 28 (2010) 142−150.

[29] Y. Du, X.-L. Luo, J.-J. Xu, H.-Y. Chen, A simple method to fabricate a chitosan-gold nanoparticles film and its application in glucose biosensor, Bioelectrochemistry 70 (2007) 342−347.

[30] X. Chen, Z. Chen, J. Zhu, C. Xu, W. Yan, C. Yao, A novel H_2O_2 amperometric biosensor based on gold nanoparticles/self-doped polyaniline nanofibers, Bioelectrochemistry 82 (2011) 87−94.

[31] C. Feng, K. Khulbe, T. Matsuura, Recent progress in the preparation, characterization, and applications of nanofibers and nanofiber membranes via electrospinning/ interfacial polymerization, J. Appl. Polym. Sci. 115 (2010) 756−776.

[32] X. Mao, F. Simeon, G.C. Rutledge, T.A. Hatton, Electrospun carbon nanofiber webs with controlled density of states for sensor applications, Adv. Mater. (2012). n/a-n/a.

[33] S. Huang, "Glucose Biosensor Using Electrospun Mn_2O_3-Ag Nanofibers," 2011.

[34] W. Wang, Z. Li, W. Zheng, J. Yang, H. Zhang, C. Wang, Electrospun palladium (IV)-doped copper oxide composite nanofibers for non-enzymatic glucose sensors, Electrochem. Commun. 11 (2009) 1811−1814.

[35] Y. Ding, Y. Wang, L. Su, M. Bellagamba, H. Zhang, Y. Lei, Electrospun Co_3O_4 nanofibers for sensitive and selective glucose detection, Biosens. Bioelectron. 26 (2010) 542−548.

[36] Z.-G. Wang, Y. Wang, H. Xu, G. Li, Z.-K. Xu, Carbon nanotube-filled nanofibrous membranes electrospun from poly (acrylonitrile-co-acrylic acid) for glucose biosensor, J. Phys. Chem. C 113 (2009) 2955−2960.

[37] Y. Luo, S. Nartker, M. Wiederoder, H. Miller, D. Hochhalter, L.T. Drzal, et al., Novel biosensor based on electrospun nanofiber and magnetic nanoparticles for the detection of *E. coli* O157: H7, Nanotechnol. IEEE Trans. 11 (2012) 676−681.

[38] F. Guo, X. Xu, Z. Sun, J. Zhang, Z. Meng, W. Zheng, et al., A novel amperometric hydrogen peroxide biosensor based on electrospun Hb−collagen composite, Colloids Surf. B Biointerfaces 86 (2011) 140−145.

[39] B. Guo, S. Zhao, G. Han, L. Zhang, Continuous thin gold films electroless deposited on fibrous mats of polyacrylonitrile and their electrocatalytic activity towards the oxidation of methanol, Electrochim. Acta 53 (2008) 5174−5179.

[40] S. Marx, M.V. Jose, J.D. Andersen, A.J. Russell, Electrospun gold nanofiber electrodes for biosensors, Biosens. Bioelectron. 26 (2011) 2981−2986.

[41] S. Jin Lee, R. Tatavarty, M.B. Gu, Electrospun polystyrene-poly (styrene-co-maleic anhydride) nanofiber as a new aptasensor platform, Biosens. Bioelectron. (2012).

[42] A. Sidek, R. Arsat, X. He, K. Kalantar-Zadeh, W. Wlodarski, "Polyvinylpyrrolidone/ polyaniline composite based 36° YX $LiTaO_3$ Surface Acoustic Wave H 2 gas sensor," in Enabling Science and Nanotechnology (ESciNano), 2012 International Conference on, 2012, pp. 1−2.

[43] I.-D. Kim, A. Rothschild, B.H. Lee, D.Y. Kim, S.M. Jo, H.L. Tuller, Ultrasensitive chemiresistors based on electrospun TiO_2 nanofibers, Nano Lett. 6 (2006). 2009−2013, 2006/09/01.

[44] Q. Qi, T. Zhang, L. Liu, X. Zheng, Synthesis and toluene sensing properties of SnO_2 nanofibers, Sens. a Actuat. B Chem. 137 (2009) 471−475.

[45] A. Stafiniak, B. Boratyński, A. Baranowska-Korczyc, A. Szyszka, M. Ramiączek-Krasowska, J. Prażmowska, et al., A novel electrospun ZnO nanofibers biosensor fabrication, Sens. Actuat. B Chem. (2011).

[46] M. Ahmad, C. Pan, Z. Luo, J. Zhu, A single ZnO nanofiber-based highly sensitive amperometric glucose biosensor, J. Phys. Chem. C 114 (2010) 9308−9313.

[47] X. Wang, Y. Si, J. Wang, B. Ding, J. Yu, S.S. Al-Deyab, A facile and highly sensitive colorimetric sensor for the detection of formaldehyde based on electro-spinning/ netting nano-fiber/nets, Sens. Actuat. B Chem. 163 (2012) 186−193.

[48] E. Kang, M. Kim, J.S. Oh, D.W. Park, S.E. Shim, Electrospun BMIMPF 6/Nylon 6, 6 nanofiber chemiresistors as organic vapour sensors, Macromol. Res. (2012) 1−7.

[49] Z. Li, H. Zhang, W. Zheng, W. Wang, H. Huang, C. Wang, et al., Highly sensitive and stable humidity nanosensors based on LiCl doped TiO_2 electrospun nanofibers, J. Am. Chem. Soc. 130 (2008) 5036−5037.

[50] Q. Qi, T. Zhang, L. Liu, X. Zheng, G. Lu, Improved NH_3, C_2H_5OH, and CH_3COCH_3 sensing properties of SnO_2 nanofibers by adding block copolymer P123, Sens. Actuat. B Chem. 141 (2009) 174−178.

[51] M. Epifani, J. Arbiol, E. Pellicer, N. Sergent, T. Pagnier, J.R. Morante, Synthesis and structural properties of ultra-small oxide (TiO_2, ZrO_2, SnO_2) nanoparticles prepared by decomposition of metal alkoxides, Mater. Chem. Phys. 124 (2010) 809−815.

[52] X. Song, Z. Wang, Y. Liu, C. Wang, L. Li, A highly sensitive ethanol sensor based on mesoporous $ZnO-SnO_2$ nanofibers, Nanotechnology 20 (2009) 075501.

[53] Y. Li, F. Qian, J. Xiang, C.M. Lieber, Nanowire electronic and optoelectronic devices, Mater. Today 9 (2006) 18−27.

[54] E. Zampetti, S. Pantalei, A. Muzyczuk, A. Bearzotti, F. De Cesare, C. Spinella, et al., A high sensitive NO_2 gas sensor based on $PEDOT-PSS/TiO_2$ nanofibres, Sens. Actuat. B Chem. 176 (2013) 390−398.

[55] S.-W. Choi, A. Katoch, J. Zhang, S.S. Kim, Electrospun nanofibers of $CuO-SnO_2$ nanocomposite as semiconductor gas sensors for H_2S detection, Sens. Actuat. B Chem. 176 (2013) 585−591.

[56] S. Kacmaz, K. Ertekin, M. Gocmenturk, A. Suslu, Y. Ergun, E. Celik, Selective sensing of Fe^{3+} at pico-molar level with ethyl cellulose based electrospun nanofibers, React. Funct. Polym. 73 (4) (2013) 674−682.

[57] A. Urrutia, J. Goicoechea, P.J. Rivero, I.R. Matías, F.J. Arregui, Electrospun nanofiber mats for evanescent optical fiber sensors, Sens. Actuat. B Chem. (2012).

[58] J.Y. Shimano, A.G. MacDiarmid, Polyaniline, a dynamic block copolymer: key to attaining its intrinsic conductivity? Synth. Met. 123 (2001) 251−262.

[59] T. Yang, Q. Li, X. Li, X. Wang, M. Du, K. Jiao, Freely switchable impedimetric detection of target gene sequence based on synergistic effect of ERGNO/PANI nanocomposites, Biosens. Bioelectron. 42 (2013) 415−418.

[60] F. Berti, S. Todros, D. Lakshmi, M.J. Whitcombe, I. Chianella, M. Ferroni, et al., Quasi-monodimensional polyaniline nanostructures for enhanced molecularly imprinted polymer-based sensing, Biosens. Bioelectron. 26 (2010) 497−503.

[61] A. Kausaite-Minkstimiene, V. Mazeiko, A. Ramanaviciene, A. Ramanavicius, Enzymatically synthesized polyaniline layer for extension of linear detection region of amperometric glucose biosensor, Biosens. Bioelectron. 26 (2010) 790−797.

[62] E. Spain, R. Kojima, R.B. Kaner, G.G. Wallace, J. O'Grady, K. Lacey, et al., High sensitivity DNA detection using gold nanoparticle functionalised polyaniline nanofibres, Biosens. Bioelectron. 26 (2011) 2613−2618.

[63] Q. Lin, Y. Li, M. Yang, Polyaniline nanofiber humidity sensor prepared by electrospinning, Sensors Actuat. B Chem. 161 (2012) 967−972.

[64] Q. Lin, Y. Li, M. Yang, Highly sensitive and ultrafast response surface acoustic wave humidity sensor based on electrospun polyaniline/poly (vinyl butyral) nanofibers, Anal. Chim. Acta (2012).

[65] S. Ji, Y. Li, M. Yang, Gas sensing properties of a composite composed of electrospun poly (methyl methacrylate) nanofibers and in situ polymerized polyaniline, Sensors Actuat. B Chem. 133 (2008) 644−649.

[66] D. Chen, X. Guo, Z. Wang, P. Wang, Y. Chen, L. Lin, "Polyaniline nanofiber gas sensors by direct-write electrospinning," in Micro Electro Mechanical Systems (MEMS), 2011 IEEE 24th International Conference on, 2011, pp. 1369−1372.

[67] N.J. Pinto, I. Ramos, R. Rojas, P.-C. Wang, A.T. Johnson Jr, Electric response of isolated electrospun polyaniline nanofibers to vapors of aliphatic alcohols, Sensors Actuat. B Chem. 129 (2008) 621−627.

[68] D.-J. Chen, S. Lei, R.-H. Wang, M. Pan, Y.-Q. Chen, Dielectrophoresis carbon nanotube and conductive polyaniline nanofiber NH_3 gas sensor, Chin. J. Anal. Chem. 40 (2012) 145−149.

[69] W. Jia, L. Su, Y. Lei, Pt nanoflower/polyaniline composite nanofibers based urea biosensor, Biosensors Bioelectron. 30 (2011) 158−164.

[70] Y.J. Shin, J. Kameoka, Amperometric cholesterol biosensor using layer-by-layer adsorption technique onto electrospun polyaniline nanofibers, J. Ind. Eng. Chem. (2011).

CHAPTER

Toxicity of nanofibers and recent developments in protections

10

CHAPTER OUTLINE

10.1 NANOMATERIALS

Research and development on nanotechnology and nanomaterials have been increasing significantly. When compared to bulk materials, nanomaterials mostly provide superior physical, chemical, physicochemical, and biological properties [1]. Nanomaterials can be in the form of nanofibers, nanotubes, nanoparticles, nanowires, nanofilms, quantum dots, and buckyballs [2]. Nanomaterials can already be found in more than 1600 different industrial products, including construction, defense, electronics, biomedical, separation, energy, environment, and so on [3–10]. Fig. 10.1 gives a classification of nanomaterials based on the structural characteristic features [11]. Some of the major applications of those nanomaterials are summarized below [12]:

- Biological applications of nanomaterials;
- Molecular electronics and nanoelectronics;
- Nanobots (nanomedicine);
- Catalysis by gold, platinum, and other nanomaterials;
- Band gap engineered quantum devices;
- Nanomechanics and nanorobots;
- Carbon nanotube emitters;
- Photoelectrochemical cells;
- Photonic crystals and plasmon waveguides;

Synthesis and Applications of Electrospun Nanofibers. DOI: https://doi.org/10.1016/B978-0-12-813914-1.00010-9

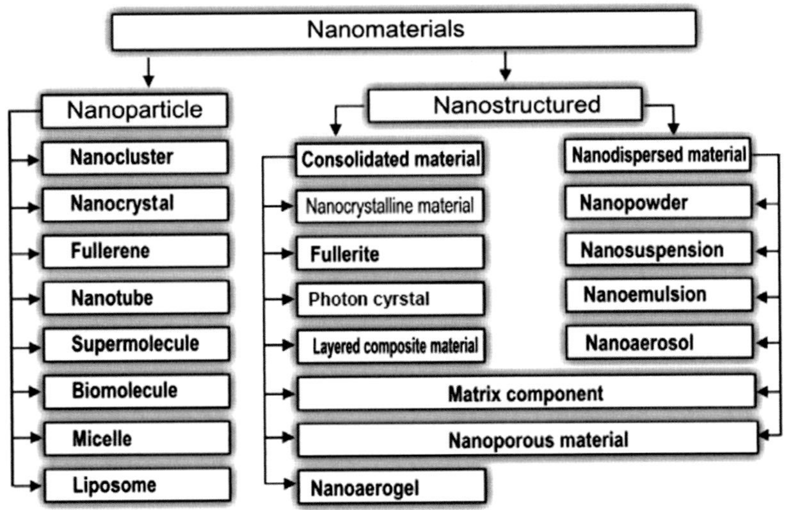

FIGURE 10.1

Classification of nanomaterials based on the structural characteristic features [11].

- Filtration of liquids and fine dust;
- Noise absorption;
- Damping of vibrations;
- Corrosion and prevention;
- Aircraft and spacecraft application;
- Textiles (antibacterial and invisible clothing);
- Oil refineries;
- Food and drinking water;
- Defense (coating, sensing, radar absorption);
- Self-cleaning and deicing;
- Solar cells and fuel cells.

Recent studies have specified that some nanomaterials can be found in air, water, soil, and consequently in human and animal bodies; thus, these nanomaterials have caused some public debate on their toxicological, health, and environmental effects [3–7,12,13]. The experts point out that risks of nanomaterial toxicity can increase through the fabrication, transportation, handling, usage, waste disposal, and recycling of nanomaterials [2]. Even though some progress has been made in the field of toxicity and safe handling of nanomaterials, a lot of new research efforts are needed to standardize how these nanomaterials can affect human life in the long term and how they can be handled properly without concerns during their manufacturing, transportation, storage, and consumption.

It has recently been reported that some nanomaterials with different surface areas, sizes, shapes, surface charges, and energies can interact with cells, tissues,

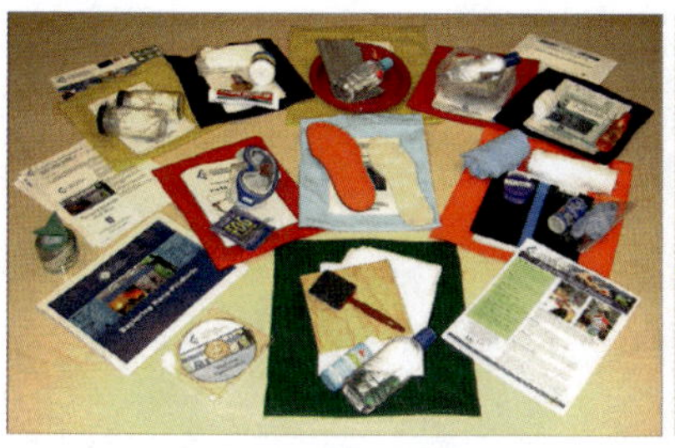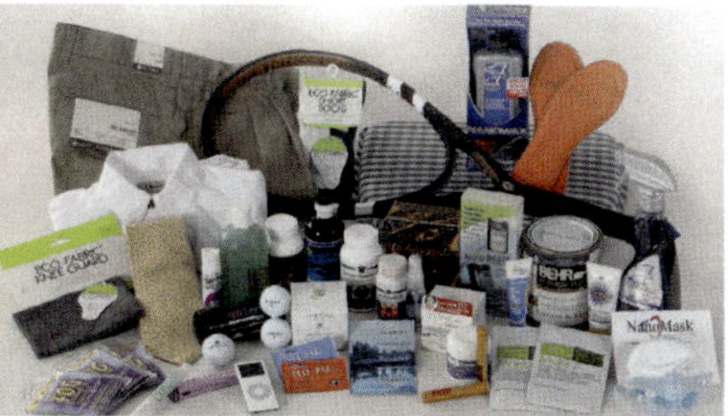

FIGURE 10.2

Commercially used nanomaterials in different products that are used in daily life.

body parts, or organs, and can damage or kill those organs and cells, block blood flow, and also result in a number of serious diseases or permanent organ failures. The diseases related to some of the nanomaterials include bronchitis, asthma, liver and lung cancer, Crohn's disease, Alzheimers, Parkinsons, heart disease, and colon cancer. As a result of nanomaterial exposures, environmental and public health concerns arisen in many industrialized countries [13]. If the mechanisms of nano-material effects on human health and environment are studied and understood, those deadly diseases associated with nanomaterials can be substantially elimi-nated. A number of nanoparticles and their protection methods have been devel-oped and safely used worldwide, but it is always advisable to manufacture nontoxic nanomaterials and devices for costumers. In these studies, nanomaterials in general and nanofibers/nanowires in particular are analyzed and the recent developments in the field are reported. Fig. 10.2 shows the different products incorporated with nanomaterials, mainly in consumer products [2,14].

10.2 NANOFIBERS

Although the length of nanofibers can be several centimeters, diameter ranges of nanofibers are usually between 1 and 100 nm. Some studies can extend the dime-ter of nanofibers up to 500 nm and consider them to be still nanofibers. It was stated that a number of different micro- and nanofibers can be produced via wet spinning, melt spinning, and extrusion molding with minimum initial investment and the shortest possible production time; however, electrospinning is a relatively easier, simpler, and more direct process of fabricating woven and nonwoven poly-meric fibers in both micro- and nanoscales compared to conventional methods. Nanofibers can be manufactured from various natural and synthetic polymers after dissolving them in appropriate solvents. Based on the selected materials and methods, nanomaterials and nanofibers can be in the forms of polymers, ceramics, metals, composites, as well as conductors, insulators, and semiconductors [1,2,12,13].

(A) (B)

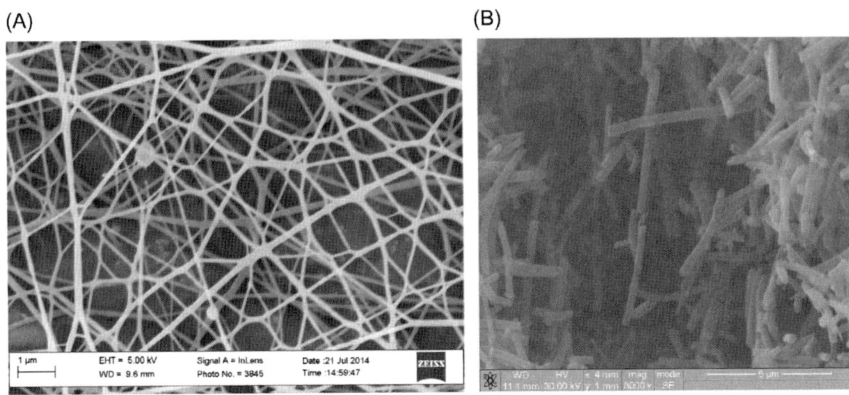

FIGURE 10.3

SEM images showing the electrospun (A) PCL nanofibers for cell culturing and (B) chopped TiO_2 nanofibers for dye-sensitized solar cell (DSSC) purposes.

Electrospinning is currently being investigated as a cost-effective and easy source of manufacturing scaffolds for the production of artificial human tissues and organs. These scaffolds generally have similar requirements to the extracellular matrix in natural tissue. Biodegradable polymers, such as polycaprolactone (PCL), poly(lactic acid) (PLA), poly(glycolic acid), chitosan, and poly(lactic-co-glycolic acid) (PLGA) can be used for this purpose. These fibers are then coated with collagen to promote cell attachment, although collagen has successfully been spun directly onto membranes. Fig. 10.3 shows electrospun nanofibers developed for cell culturing and solar cell (chopped) purposes.

Electrospun nanofibers have a large surface area per unit mass, so these woven and nonwoven fibers can be used in various industrial fields [1−3,12,13]:

- Filtration and separation of micron, submicron, and nanosize organic, inorganic, and biological (bacteria, mold, fungus, and virus) particles;
- HF antenna fabrication for effective signaling;
- Light weight, colorful, and invisible fabric productions;
- Biomedical applications, such as wound dressings in the medical industry, tissue engineering scaffolds and artificial blood vessels, as well as drug delivery, gene therapy, and crop engineering;
- Environmental monitoring for solid, liquid, and gas contamination;
- Energy (solar and fuel cells, supercapacitors, battery membranes, and catalysts);
- Bio- and nanosensors for bacteria, viruses, and other pathogens and hormones, biomarkers and cancer detection, and health scanning and monitoring;
- The other promising areas of electrospun nanofibers include advanced composite fabrications to improve crack resistance, structural health

monitoring (SHM), lightning strike prevention, EMI shielding, and aircraft interior noise reduction.

10.3 SURFACE INTERACTION OF NANOFIBERS

10.3.1 PHYSICAL INTERACTION OF NANOFIBERS

Nanofibers are long and flexible, most of them are produced through an electrospinning process and they are safe to handle; however, after some additional processes (e.g., carbonization, chopping, calcination, or grinding) are applied to those nanofibers or when they become highly brittle, the long fibers can transform into short/sharp fibers and may act like sharp needle nanomaterials. These fine nanomaterials may get into the body through inhalation, ingestion, or contact through the skin, and stay in the body for a long time [1]. Fig. 10.4 shows the

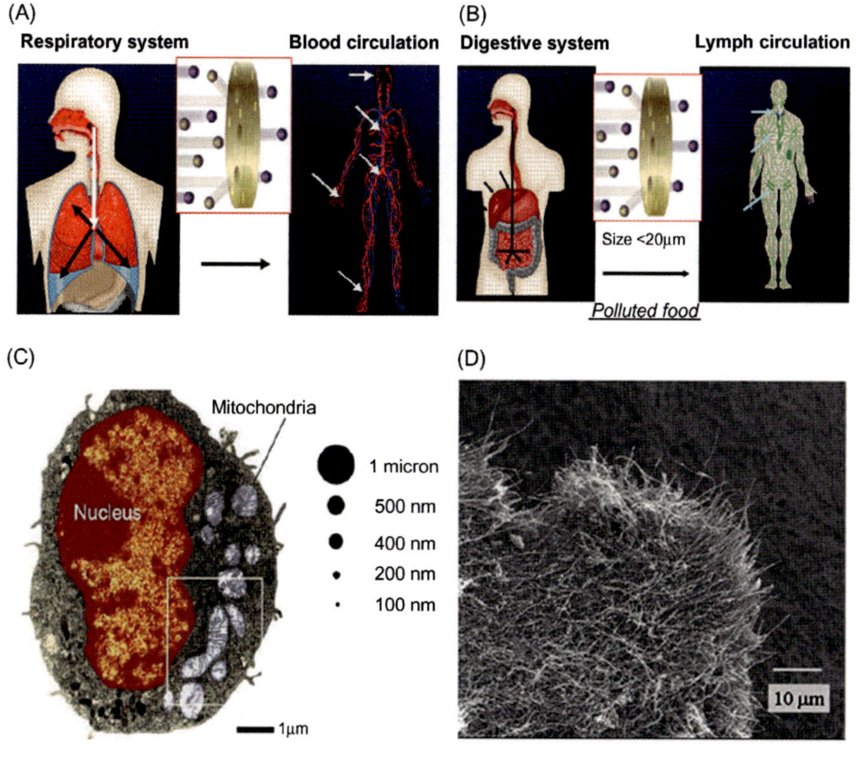

FIGURE 10.4

The entrance of nanoparticles through the (A) respiratory and (B) digestive systems, (C) comparison of a rat macrophage cell size with different particles and (D) nanofibers with sharp needle-like edges.

entrance of nanomaterials into the body, a comparison of rat macrophage cell size with nanoparticles, and nanofibers with sharp needle-like edges [2,6,12,13,15].

Usually, the first defense layer against the outside is skin, which can be exposed to the toxicity of nanomaterials in the forms of liquids and solids. Particle size, surface chemistry, surface potential, and surface area are the major factors in the toxicity of nanomaterials when they are exposed to the body [16]. It is reported that a variety of nanomaterials can interact with the cells, tissues, and organs of the human body, stay inert, and/or react with the system of the body based on the surface properties of the nanomaterials [14,16]. These nanomaterials are many orders of magnitude smaller than human, plant, and animal cells. They can be much smaller than DNA. The lengths of DNA are about 50 nm and the width/diameter is around 2 nm. A human cell's DNA totals about 3 m in length. It is stated that human macrophages are approximately twice the size of rat macrophages, so chopped nanofibers and other nanoparticles can easily penetrate human macrophages (Fig. 10.4B) [12,13].

Recent studies have indicated that the toxicity of nanomaterials is primarily dependent on their surface properties, sizes, shapes, and morphologies [1,2,12,13]:

- Surface chemistry (hydrophobicity, hydrophilicity, aggregation, and clustering);
- Particle size (the smaller the size, the higher the toxicity risk);
- Surface charge (including zeta potential and surface potential—lower is better for clustering/agglomeration);
- Surface area (playing a major role in the interaction of materials with cells);
- Oxidative stress and reactive oxygen species: Oxidative stress is thought to be involved in the developments of many illnesses, such as cancer, Parkinson's, and Alzheimer's diseases, heart failure, fragile X, autism, and many other genetic and chronic diseases.

The surface area of nanomaterials is one of the major health and environmental considerations, because it is exponentially high when the size is near the atomic scale. At this scale, the surface charge and zeta potential of nanomaterials are considerably changed and become the dominant factors. In addition to that, surface energy can also be altered and provide some useful information about the toxicity and its mechanisms for the cell interaction [17−21]. For instance, some inert metals, such as Au, Ag, and Pt, are usually inert at bulk scale and cannot interact with cells, tissues, and organs, and cannot cause any inflammations or rashes in the body. Nevertheless, when those highly inert materials become nanoscale (one of the dimensions is less than 100 nm), the surfaces of these materials is activated and starts interacting with cells, tissues, and organs in the body, resulting in major cell and tissue damage (Fig. 10.4C) [2]. If the nanofibers have higher aspect ratios, the toxicity may be drastically reduced; otherwise, they can also be responsible for the shape-dependent surface reactivity and toxicity to the human body.

The toxicity of nanofibers and other nanoscale materials, devices, and products can be eliminated or substantially minimized via various surface treatment processes, including surface modification, polarization, neutralization, blending, and functionalization. Some studies have indicated that some organs exposed to nanoscale materials have substantially more nanomaterials than other parts of the body because the smaller particles could diffuse into the respiratory system faster than larger particles [16−21]. Nanomaterials penetrated into epithelial and endothelial cells can interfere with cell function, causing other damage to sensitive parts of the body, such as the brain, bone marrow, nervous system, kidneys, liver, lymph nodes, pancreas, spleen, and heart [20−22]. The side effects of these nanostructured materials on the human body can be reduced by using surface modifications, and charge neutrality approaches [22].

Recent studies have stated that the smaller the particle size, the higher the toxicity if they are individually suspended in the system [16]. Hussain et al. reported on the toxicological behavior of nanomaterials and demonstrated that objects less than 100 nm could enhance toxicity in cell-cultured models, as well as animal models [23]. Table 10.1 shows the relationship between nanomaterial size and their possible toxicity to living cells [1,2,12,13,24]. In this table, some of the materials were not nanofibers, but recent studies showed that most of the particles can be in the form of nanofibers by changing the fabrication methods. If nanoparticles are toxic to the human body, most likely their nanofibers can be toxic too because of their needle-like sharp edges and structures. Fig. 10.5 exhibits the general pathways of nanoparticles in the body [25]. As can be seen, a number of different health-related sicknesses can be expected from nanomaterials, some of which include bronchitis, lung and liver cancers, heart disease, asthma, Crohn's disease, Parkinson's, Alzheimer's, and colon cancer.

10.3.2 CHEMICAL INTERACTION OF NANOFIBERS

Chemical structures of chopped nanofibers are also major concerns for the toxicity of nanofibers. If they are not properly functionalized, most of them cannot be stabilized in in vitro conditions. Thus, nanofiber stability will be dependent on some factors, including ionic strength, concentration of the nanomaterials, pH, temperature, thermodynamic feasibility, kinetic facility of electron transfers, and redox reactions in the biological media. The toxicity of nanofibers and other nanomaterials can take place based on three different mechanisms in the body conductions [2]: (1) dissolution process of nanomaterials in biological media, (2) catalyst properties of nanomaterials, and (3) reduction and oxidation (Redox) evolution of the surface [26].

The human body is considered to be highly corrosive due to anions and cations, salt, spicy foods, water, and plasma in the system, so some nanofibers and other nanomaterials can be easily dissolved in the body, depending on the solubility of the particles and surface area where the diffusion and dissolution processes take place faster. Even some of the most inert nanomaterials can be

Table 10.1 Relationship Between Nanomaterial Size and Toxicity

Type of Material	Particle Size (nm)	Surface Area (m²/g)	Charge/Zeta Potential (mV)	Biological Toxicity
TiO_2 short nanofibers	100–200 (diameter)	–	30–35	Toxic solvents during fabrications; acting like nanoparticles and CNTs
MWCNTs	10–20	40–300	–	Cytotoxicity: alveolar macrophages at high dose
SWCNTs	1.4	270	–	Cytotoxicity: alveolar macrophages at low dose; transient inflammatory and cell injury
Alumina	116	13.37	45–50 (pH 5.5–6.5)	Protein (BSA) adsorption with time / IEP shifts with pH and surface area
PEG Quantum (Q) dots	10	–	–	Retention of Q dots in liver, spleen, and bone marrow
Titania	300 nm(rutile)	6	–	Short-term reversible inflammation
	Rods (20–233) (anatase rods)	26.5	–	Short-term reversible inflammation, minor adverse lung tissue reaction
	Spherical (5–6) (anatase spherical powder)	169.4	–	Short-term reversible inflammation minor adverse lung tissue reaction
Quartz	1.5 μm	4	–	High pulmonary toxicity
PTFE	20	–	–	Cell death–15 min exposure
	130	–	–	No ill effects
Emulsifying wax	74.7 ± 53.4 (neutral)	–	–14.1 ± 2.1	No BBB; permeation ability in low conc.
	127.1 ± 70.6 (anionic)	–	–59 ± 2.9	No BBB; permeation ability in low conc.
	97.2 ± 68.9 (cationic)	–	45.2 ± 3.5	Toxic at brain microvasculature endothelium
Ceria	3–5	–	–	Radio protection, nontoxic at low/medium dose
Yttria	50	–	–	Neuroprotection against oxidative stress

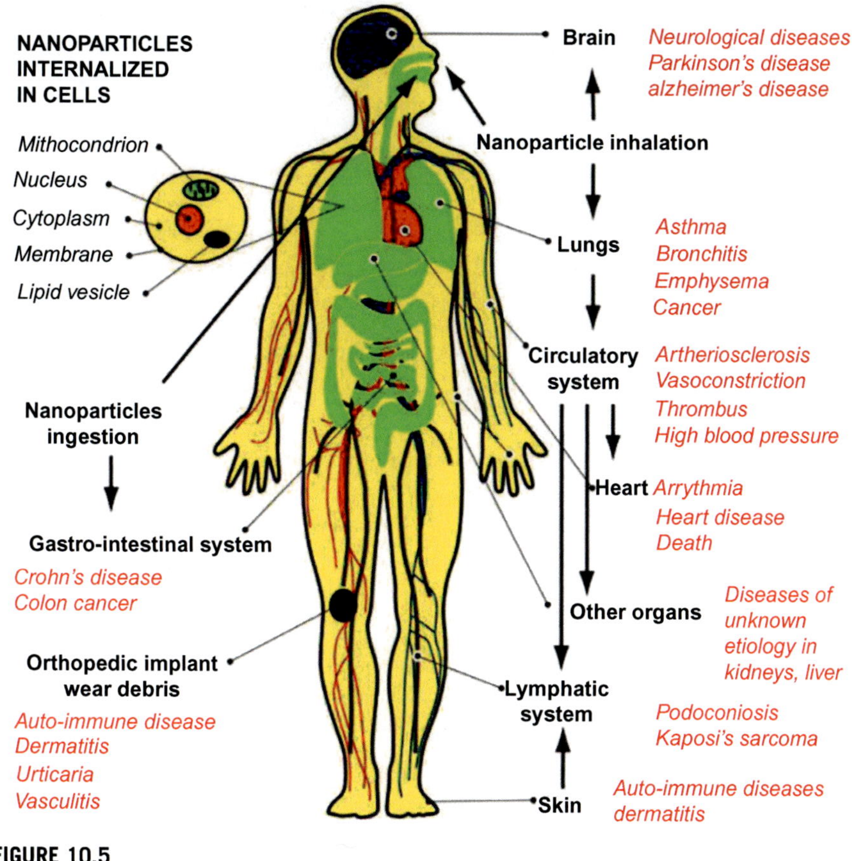

FIGURE 10.5

Diagram showing the several ways of internalization of nanoparticles in a body, and their side effects.

degraded in the body (e.g., silicon, silica, titania, carbon, other metallic, polymeric, and ceramic nanomaterials) over a longer period. Sickness is based on the body reactivity and defense mechanisms of the body against foreign objects. Also, surface chemistry, temperature, ion concentration, pH, chemical structures, porosity/voids, and roughness of the nanofibers can accelerate the solubility rate, and subsequently reduce the toxicity of those nanomaterials [21]. For instance, it was reported that more soluble ZnO nanoparticles could provide a stronger toxicity on mammalian cells when compared to the less soluble TiO_2 nanoparticles with a low solubility rate [26,27]. The dissolution products of ZnO are Zn^{+2} and $Zn(OH)_2$, and at higher concentrations of ions, the body cannot tolerate a larger volume of the reaction products, resulting in severe illnesses. By controlling the surface charges, energy level, pH, and other system and process parameters, one can control the dissolution rates of nanomaterials to minimize the toxicity rates [21].

It is also reported that the major driving forces for catalytic activities are related to surface-to-volume ratios and special binding sites of nanofibers and other nanomaterials [2]. TiO_2 consists of two crystalline structures of anatase and rutile with a size-dependent photocatalytic behavior. For instance, rutile has {001} and {011} faces, while the anatase phase has {011} and {110} faces. Hence, anatase has more efficient active sites for oxidation and reduction processes than the other phase. The size-dependent phase transformations are involved in the toxicity of TiO_2 towards cellular structures. Furthermore, the anatase phase of TiO_2 nanomaterials can cause more oxidative damage to DNA structures [26−28].

During the redox reaction and evolution process, the surfaces of nanofibers can oxidize organic parts (e.g., lipids, DNA, enzyme, peptides, and proteins), and produce reactive oxygen species (ROS) in the form of anions and cations or very small molecules, including oxygen ions, free radicals, and peroxides in the cells and tissues [14]. Any of these species formed during the oxidation or redox reactions can induce an oxidative stress in the body. These chemical reaction processes can also lead to the release of other ions and hormones in the body. As a result, the reaction can take place on the surface of the nanofibers to change the surface properties and toxicity of nanofibers and other nanomaterials [16].

10.4 CARBON-BASED NANOFIBERS

Some of the electrospun nanofibers (e.g., PAN, PVDF, pitch, etc.) can be stabilized at 200−300 °C for about 1 h in ambient conditions and then carbonized at 500−900 °C in argon for an additional 1 h. After the carbonization process, major structural changes take place and nanofibers can become highly brittle [29−33]. Carbon nanofibers (CNFs) produced through chemical vapor deposition are graphitic nanostructures of atomic layers arranged as stacked cones, cups, or plates. CNFs with graphene layers wrapped into perfect cylinders are called carbon nanotubes (CNTs). Additionally, carbon-based nanomaterials (CBNs) in different forms of fibers, fullerenes, single- and multiwalled carbon nanotubes (SWCNTs and MWCNTs), carbon nanoparticles, and nanowhiskers have been produced and used in a number of different industrial applications [21,23].

The size, diameter, and shape of CBNs are key factors in determining the toxicity of those nanomaterials. It is reported that some CBNs can have similar structural properties to asbestos, which increases the primary concerns of some nanofibers. The use of CBNs may lead to mesothelioma, cancer of the lining of the lungs, caused by exposure to these nanomaterials. Particularly, the needle-like fiber shapes have more toxicity than many other nanomaterials to human skin and lungs. CNTs of various size and shapes can have strong chemical stability in normal conditions, and they may not be easily dissolved in the body after inhalation. Depending on the exposure time and concentration, they may damage cells,

DNA, surrounding tissues, and even whole organs. Commonly, SWCNTs can be more toxic than MWCNTs because of the size effects, but other factors are also involved in their toxicity [21]. New studies have indicated that the functionalized CNTs could be less toxic to the human body when compared to nonfunctionalized ones [2]. Thus, surface treatment makes a huge impact on the toxicity of nanofibers. In addition to that, CNTs synthesized by catalytic chemical vapor deposition are not toxic to human umbilical vein endothelial cells compared to other options [34–37].

As was indicated earlier, the surface chemistry and surface interactions of various nanomaterials, such as particle size, surface potential, and surface area are the primary factors for the toxicity of those nanomaterials [7]. Depending on the surface properties, structures, and functionalities, some nanomaterials can easily interact with the human body, react with the tissues and organs, and stay inert and in the body for a long time [7,13,24]. Some of the major properties of nanoparticles and their concerns are provided in Table 10.2 [7,13,24].

The risk control method can be applicable for the planning, chain of assessment, corrective action, and implementation steps that can be repeated to reduce major exposures to employees and participants to unwanted nanomaterials, hence allowing participants to work in acceptable and convincing environmental conditions. The three main categories of risk controls include engineering techniques, administrative measures, and personal protection for those high-risk environments [7,24]. Fig. 10.6 shows the risk control hierarchy applied to the assessment of some of the major nanomaterials [7].

As can be seen, engineering controls, such as modeling, designing, elimination, substitution, confinement, isolation, ventilation, and optimization are more effective than administrative measures, including information, training, personal

Table 10.2 Some of the Major Properties of Nanoparticles and Their Toxicity Concerns

Primary concerns	Secondary concerns
Specific surface area	Solubility/dissolution
Number of particles	Surface shape and geometry
Size and granulometric distribution	Clustering/agglomeration
Concentration	Crystalline structures
Chemical compositions	Surface oxidation
Surface properties	Surface hydrophobicity/hydrophilicity
Zeta potential	Manufacturing techniques
Functional groups	Inertness/reactivity
Oxidative stresses	Biocompatibility/biodegradability
Free radicals	Metal and alloy, ceramic, polymer, composite
Surface coverage	Impurities and defects
Cell viability	Dispersion/settlement

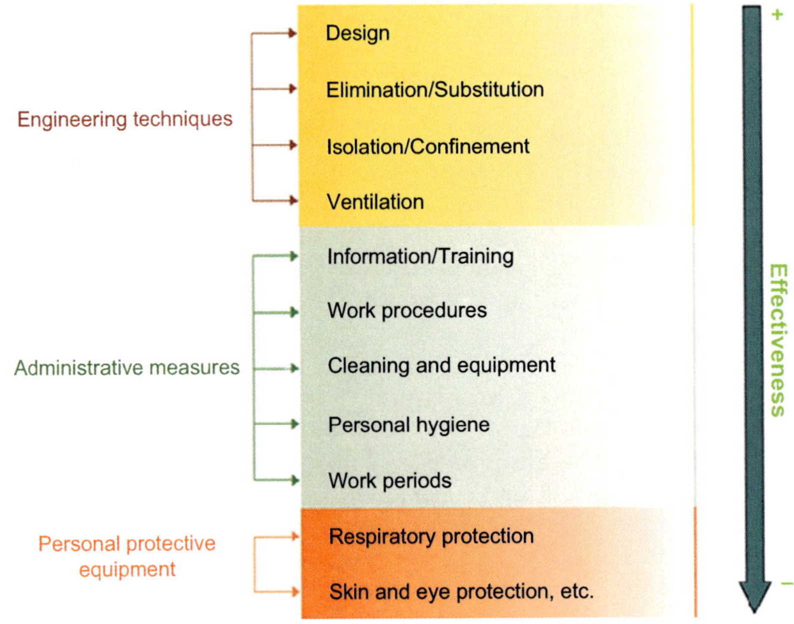

FIGURE 10.6

Major risk control hierarchy applied to the assessment of some of the major nanomaterials [7].

hygiene, work procedures, cleaning and equipment, and work periods [24]. Some of the primarily used personal protective equipment, such as skin and eye protection, and respiratory protection, may be less effective when compared to the other parameters. As long as the engineering techniques and environmental conditions are satisfied, participants will likely decrease the possibility of contact and potential risks from the nanomaterials [7,24]. Based on Fig. 10.6, the effectiveness of those three categories is related to the overall risk control hierarchy, as well.

10.5 PROTECTION METHODS OF NANOFIBERS

It is estimated that, by 2020, approximately 6 million workers will be employed in nanotechnology manufacturing around the world, and 2 million of these are expected to be in the USA. It is assumed that nanotechnology will bring great discoveries and innovations to many countries through developing nanotechnology products, and it will change the regional economies, education, health, and societies. Although nanotechnology is developing faster than any other technologies, handling this technology in a moral and ethical manner will benefit human kind [13].

Nanofibers with different sizes and shapes (especially chopped nanofibers) can enter the body during production, handling, packaging, transportation, maintenance, and cleanup activities [2]. The major routes of entrance can be inhalation, ingestion, dermal exposure, and/or combinations of them. Hazard reduction of chopped nanofibers is done for workers in production and processing facilities, as well as consumers in contact with commercial products [38–43]. Some of the protection methods (Fig. 10.7) during the production and use of nanomaterials are given below [2,12,21,44]:

- Workers, engineers, and scientists who are working on nanofibers, nanomaterials, and devices are recommended to wear a disposable, typically plastic, body covering over their work clothes during high-exposure activities and to wear long gloves pulled over sleeves to minimize wrist exposure. Other recommendations include antistatic shoes to prevent ignition by static charges, and sticky mats at laboratory entrances to prevent accidental nanomaterial transfer.

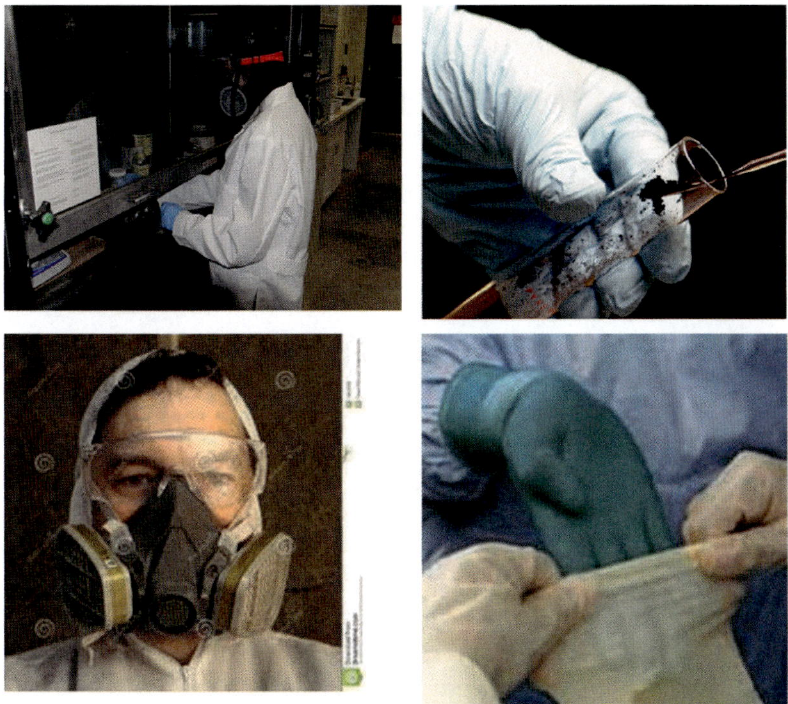

FIGURE 10.7

Photographs showing simple protection methods for potentially hazardous engineered nanomaterials.

- Hazardous effects of nanomaterials need to be reduced during production and processing. The waste nanomaterials should be limited. Outputs are sometimes more hazardous than products or wastes from such activities.
- Employees who inhale nanomaterials are advised to consume milk and unrefined sugar to reduce the initial toxicity level of nanomaterials.
- Nitrile gloves or wrist-length disposable nitrile gloves with extended sleeves must be worn during the handling of nanomaterials (Fig. 10.7) [12,13,24,43]. The gloves need to be changed very often during the operation.
- For eye protection, safety glasses with side shields must be worn during the use of nanomaterials in the forms of solids, liquids, and aerosols.
- Volumes of liquid-based nanomaterials must be limited to the milliliter range (<200 mL) and placed in a sealed container when not in use.
- Total particle masses must be limited to the milligram range (<200 mg) and be manipulated within a high-efficiency particulate air (HEPA)-filtered laboratory exhaust hood over water-soaked absorbent paper to capture any spilled materials.
- Nanomaterial containers must be labeled with all the necessary information.
- Nanomaterials are considered to be hazardous materials, so workers should take all the safety rules necessary in the field.

In addition to these, the National Institute for Occupational Safety and Health (NIOSH) has sequential steps for workers and engineers who are involved in nanotechnology-related research and development processes. These steps, given in Fig. 10.8, will potentially reduce the risk of exposing the employee to nanomaterials [44].

A number of technical research studies have been conducted on the safe handling of nanofibers and other nanomaterials, and more research studies are needed to identify the new classes of nanomaterials and their adverse effects on the human body and environment. Some of them are summarized below, which is also a part of NIOSH agenda [25,44]:

- Overall characterization of nanomaterials and their processes;
- Prudent work practices to minimize and manage exposure to nanomaterials, engineering controls and personal protective equipment (PPE);
- Worker training that addresses potential hazards which may be associated with nanomaterials;
- Evaluation of analytical instruments and methods used to measure exposure to nanomaterials;
- Quantitative and qualitative measurements of exposure to nanomaterials;
- Environmental health and safety (EH&S) helps in protection from the hazards of toxic nanomaterials.

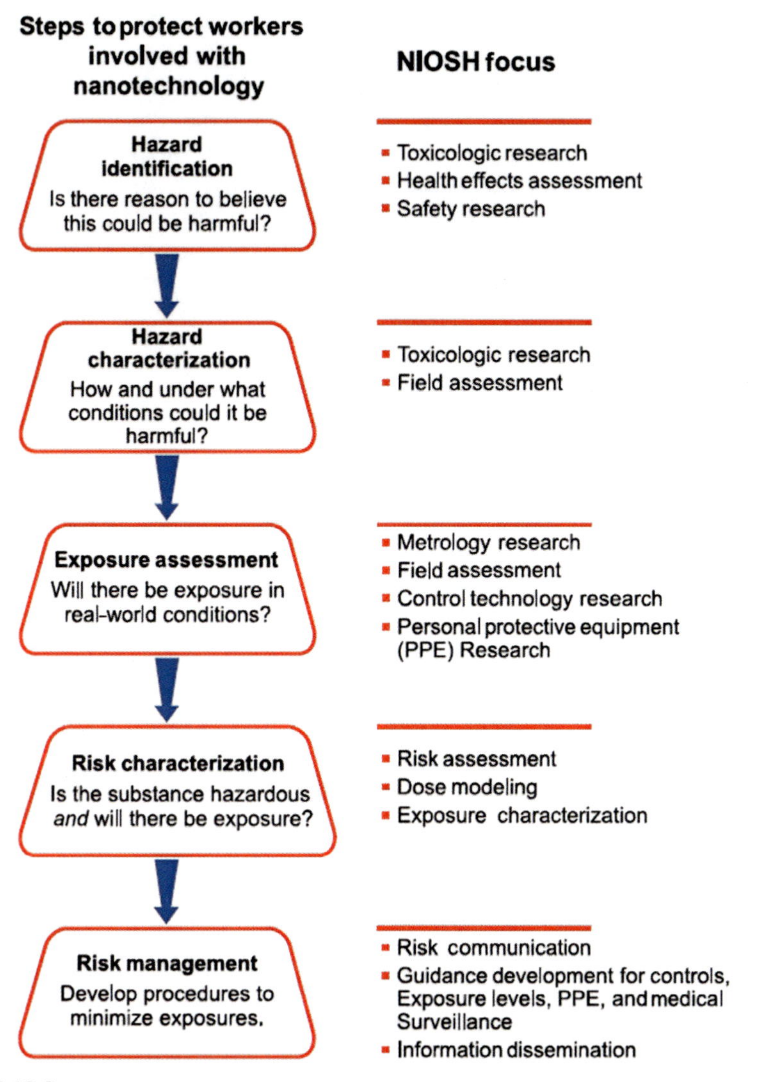

FIGURE 10.8

NIOSH steps for workers involved in nanotechnology [44].

10.6 CONCLUSIONS

Some of the nanomaterials in the forms of nanofibers, nanotubes, nanoparticles, nanofilms, nanocomposites, and nanoflakes are considered to be the future generation of advanced materials for advanced sensors, faster cars and planes, more powerful computers and satellites, and better microchips and batteries owing to their outstanding mechanical, optical, thermal, electrical, and magnetic properties. Although every year hundreds of new nanofibers and other related nanomaterials have been developed and used for different scientific and commercial uses, the toxicity of these nanomaterials has not been completely identified yet. Most studies have mainly focused on the skin, lung, and liver toxicity of these nanomaterials. Their long-term toxicity and examination of chronic exposure must be studied in detail to understand the toxicology mechanisms of nanomaterials. A number of different factors are involved in the toxicity of nanomaterials, so more effort and time are needed to be spent on nanotechnology products.

REFERENCES

[1] R. Asmatulu, E. Asmatulu, A. Yourdkhani, "Importance of Nanosafety in Engineering Education" ASEE Midwest Conference, Lincoln, NB, September, 2009, 8p.

[2] R. Asmatulu, E. Asmatulu, A. Yourdkhani, "Toxicity of Nanomaterials and Recent Developments in the Protection Methods," SAMPE Fall Technical Conference, Wichita, KS, October 19−22, 2009, 12p.

[3] W.S. Khan, R. Asmatulu, The importance of safety for manufacturing nanomaterials, in: D. Fazarro, W. Trybula (Eds.), Nano- Safety: What We Need to Know to Protect Workers, DeGruyter Publisher, 2017, pp. 61−84.

[4] R. Asmatulu, W.S. Khan, Nanotechnology safety in the energy industry, in: R. Asmatulu (Ed.), Nanotechnology Safety, Elsevier, 2013, pp. 127−140.

[5] M. Srikanth, R. Asmatulu, Nanotechnology safety in the construction and infrastructure industries, in: R. Asmatulu (Ed.), Nanotechnology Safety, Elsevier, 2013, pp. 99−114.

[6] H. Haynes, R. Asmatulu, Nanotechnology safety in aerospace industry, in: R. Asmatulu (Ed.), Nanotechnology Safety, Elsevier, New York, 2013, pp. 85−98.

[7] C. Ostiguy, B. Roberge, L. Ménard, C.A. Endo, "Best Practices Guide to Synthetic Nanoparticle Risk Management," Studies and Research Projects REPORT R-599, IRSST − Communications Division, Montréal, Quebec, Canada, 2009.

[8] R. Asmatulu, B. Zhang, E. Asmatulu, Safety and ethics of nanotechnology, in: R. Asmatulu (Ed.), Nanotechnology Safety, Elsevier, 2013, pp. 31−42.

[9] W.S. Khan, R. Asmatulu, Fundamentals of safety, in: R. Asmatulu (Ed.), Nanotechnology Safety, Elsevier, 2013, pp. 17−30.

[10] W.S. Khan, R. Asmatulu, nanotechnology emerging trends, markets and concerns, in: R. Asmatulu (Ed.), Nanotechnology Safety, Elsevier, 2013, pp. 1−16.

[11] A.D. Pogrebnjak, V.M. Beresnew, Structural features of nanocrystalline materials, Nanocoatings Nanosystems Nanotechnologies 12 (2012) 3−14.

[12] R. Asmatulu, August Nanotechnology Safety, Elsevier, Amsterdam, The Nederland, 2013.

[13] N. Nuraje, R. Asmatulu, G. Mul, Green Photo-Active Nanomaterials: Sustainable Energy and Environmental Remediation, RSC Publishing, Cambridge, England, 2015.

[14] http://nano-cemms.illinois.edu/materials/exploring_nano_products_full.html (accessed 23.12.17).

[15] L. Schlagenhaur, F. Nuesch, J. Wang, Review: release of carbon nanotubes from polymer nanocomposites, Fibers 2 (2) (2014) 108−127.

[16] A.S. Karakoti, L.L. Hench, S. Seal, The potential toxicity of nanomaterials—the role of surfaces, JOM J. Miner. Met. Mater. Soc. 58 (2006) 77−82.

[17] T.M. Osman, Environmental, health, and safety considerations for producing nanomaterials, JOM J. Miner. Met. Mater. Soc. 60 (2008) 14−17.

[18] N. O'Brien, E. Cummins, Nanomaterials: Risks and Benefits, Book Chapter: Development of a Three-Level Risk Assessment Strategy for Nanomaterials, Springer, Netherlands, 2008, pp. 161−178.

[19] S. Singh, H.N. Sing, Nanotechnology and health safety-toxicity and risk assesments of nanostructured materials on human health, J. Nanosci. Nanotechnol. 7 (2017) 3048−3070.

[20] G. Oberdörster, E. Oberdörster, J. Oberdörster, Nanotoxicology: an emerging discipline evolving from studies of ultrafine particles, Environ. Health Perspect. 113 (2005) 823−839.

[21] C. Kumar, Nanomaterials - Toxicity, Health and Environmental Issues, 2006, Wiley-VCH. ISBN-10 3-527-31385-0.

[22] http://www.ianano.org/Presentation-ICNT2005/Gatti-S-ICNT2005.pdf (accessed 1209.17).

[23] S.M. Hussain, L.K. Braydich-Stolle, A.M. Schrand, R.C. Murdock, K.O. Yu, D.M. Mattie, et al., Toxicity evaluation for safe use of nanomaterials: recent achievements and technical challenges, Adv. Nanomater. 21 (2009) 1549−1559.

[24] R. Asmatulu, P. Nguyen, E. Asmatulu, Nanotechnology Safety in Automotive Industry, in: R. Asmatulu (Ed.), Nanotechnology Safety, Elsevier, 2013, pp. 57−72.

[25] R. Asmatulu, Advanced Biomaterials Text Notes, Wichita State University, 2016.

[26] M. Auffan, J. Rose, M.R. Wiesner, J.Y. Bottero, Chemical stability of metallic nanoparticles: A parameter controlling their potential cellular toxicity in vitro, Environ. Pollut. 157 (2009) 1127−1133.

[27] L. Wang, D.K. Nagesha, S. Selvarasah, M.R. Dokmeci, R.L. Carrier, Toxicity of CdSe nanoparticles in Caco-2 cell cultures, J. Nanobiotechnol. 6 (2008) 11.

[28] Y. Lu, J. Liu, Catalyst-functionalized nanomaterials, nanomedicine and nanobiotechnology, Wiley Interdiscipl. Rev. Nanomed. Nanobiotechnol. 1 (2009) 35−46.

[29] I.M. Alarifi, A. Alharbi, W.S. Khan, A.K.M.S. Rahman, R. Asmatulu, Mechanical and thermal properties of carbonized PAN nanofibers cohesively attached to surface of carbon fiber reinforced composites, Macromol. Symp. 365 (2016) 140−150.

[30] I. Alarifi, A. Alharbi, W.S. Khan, R. Asmatulu, Carbonized electrospun PAN nanofibers as highly sensitive sensors in SHM of composite structures, J. Appl. Polym. Sci. (2015). Available from: https://doi.org/10.1002/app.43235.

[31] I. Alarifi, A. Alharbi, O. Alsaiari, R. Asmatulu, Training the engineering students on nanofiber-based SHM systems, Trans. Tech. STEM Educ. 1 (2016) 59−67.

[32] I.M. Alarifi, W.S. Khan, A.K.M.S. Rahman, Y. Kostogorova-Beller, R. Asmatulu, Synthesis, analysis and simulation of carbonized electrospun nanofibers infused carbon prepreg composites for improved mechanical and thermal properties, Fibers Polym. 17 (2016) 1449−1455.

[33] I. Alarifi, A. Alharbi, W.S. Khan, A. Swindle, R. Asmatulu, Thermal, electrical and surface properties of electrospun polyacrylonitrile nanofibers for structural health monitoring, Materials 8 (2015) 7017−7031.

[34] N.A. Monteiro-Riviere, A.O. Inman, Challenges for assessing carbon nanomaterial toxicity to the skin, Carbon 44 (2006) 1070−1078.

[35] C. Grabinski, S. Hussain, K. Lafdi, L. Braydich-Stolle, J. Schlager, Effect of particle dimension on biocompatibility of carbon nanomaterials, Carbon 45 (2007) 2828−2835.

[36] R.H. Hurt, M. Monthioux, A. Kane, Toxicology of carbon nanomaterials: Status, trends, and perspectives on the special issue, Carbon 44 (2006) 1028−1033.

[37] R. Brayner, The toxicological impact of nanoparticles, Nanotoday 3 (2008) 48−55.

[38] L. Reijnders, Hazard reduction in nanotechnology, J. Ind. Ecol. 12 (2008) 297−306.

[39] http://thepumphandle.wordpress.com/2008/05/12/safety-in-the-nano-workplace/ (accessed 12.12.17).

[40] http://www.er.doe.gov/bes/DOE_NSRC_Approach_to_Nanomaterial_ESH.pdf (accessed 15.12.17).

[41] D.E. Fazarro, W. Trybula, J. Tate, C. Hanks, Nano-Safety: What We Need to Know to Protect Workers, De Gruyter, Boston, MA, 2017.

[42] http://www.azonano.com/Details.asp?ArticleID = 1073 (accessed 24.12.17).

[43] A.D. Maynard, Safe handling of nanotechnology, Nature 444 (2006) 267−269.

[44] Approaches to Safe Nanotechnology: Managing the Health and Safety Concerns Associated with Engineered Nanomaterials, DHHS (NIOSH) Publication Number 2009-125.

Electrospun nanofibers for tissue engineering

11

CHAPTER OUTLINE

11.1 ELECTROSPUN NANOFIBERS FOR WOUND HEALING

11.1.1 GENERAL BACKGROUND

Wound healing is a complex physiological and dynamic process, which may be associated with disruption of function and structure. The main mechanism of healing is repair where damaged tissue is replaced by connective tissue. Wound healing is a process of regenerating dermal and epidermal tissues [1]. When an individual is wounded, biochemical actions occur in a closely orchestrated cascade to rehabilitate the damage. These events can be classified into inflammatory, proliferative, and remodeling (maturation) phases. The inflammatory phase is generally considered the body's natural response to injury. When the person is initially wounded, the blood vessels in the wound bed contract and a clot is formed around the location. Throughout the proliferation process, the wound is rebuilt by itself with new granulation tissues, which are composed of extracellular matrix and collagen in the wound area, and then a new network of blood vessels are developed for faster healing. This process is termed "angiogenesis." Maturation is the final phase for the healing process that occurs once the wound has closed by itself. There is conversion of granulation tissue to fibrous connective tissue and decreased parallelism of collagen to the plane of the wound. Healing is the body's response to wounds in an attempt to restore normal

Synthesis and Applications of Electrospun Nanofibers. DOI: https://doi.org/10.1016/B978-0-12-813914-1.00011-0

structure and functions [1]. Wound dressing means to protect the wound, discharge extra body fluids from the wound area, decontaminate the exogenous microorganism, refine the appearance, and expedite the healing process. A wound dressing material should be a blockade to a wound area, but be permeable to blood fluid, moisture, and oxygen. Therefore, for a deep dermal injury, the adhesion and integration of an artificial layer, containing 3D tissue scaffold with well-cultured dermal fibroblasts, will substantially help the re-epithelialization [1].

The applications of nanotechnology in healthcare have been under investigation in recent times. Nanotechnology has been gaining momentum in healthcare as nanoparticles, nanosponges, nanogels, nanorods, nanowires, nanoplatelets, nanospheres, and nanofibers are being used in the medical industry [2]. The application of nanotechnology in medicine is called nanomedicine. There are many treatment methods that take a longer time and they are also costly. With nanotechnology, much quicker and less costly treatment can be possible. Among the various nanomaterials, nanofibers are the materials of choice due to their outstanding properties, such as small size, flexibility, high surface area to volume ratio, porosity, preferred orientations, and their ability to permeate into the cell membrane. Nanofibers can act as natural extracellular matrix (ECM) and can deliver biomolecules, drugs, hormones, DNA, enzymes, peptides, and genes in a sustained fashion to the targeted site. Nanofibers are now being considered as the most effective nanomaterials that can be exploited in applications, such as healthcare, energy, catalysis, sensors, bioengineering, and environmental.

Electrospun polymeric nanofiber membranes are very good materials for wound dressing applications, due to their excellent properties, such as their highly porous structure, which is permeable to moisture and oxygen, and well-connected pore structures. These structures are extremely important for discharging body fluids from the wound and eliminating infections from external sources. The small pore size and high surface area not only impede exogenous microorganism invasion, but also facilitate the control of fluid discharge. Additionally, electrospinning presents a simple way to incorporate drugs into nanofibers for any medical treatment and antibacterial purpose [1]. Therefore, nanofibers have been extensively used in wound dressing for burns, bruises, surgery, cuts, punctures, and so on. A biodegradable polymer in fibrous form is placed onto the wound. It generally lets the wound heal by the growth of new skin cells. This type of wound dressing has a pore size in nanometers, which restricts the penetration of bacteria and microorganism. This fibrous membrane also eliminates the formation of scar tissue, which is generally formed in traditional treatment. Fig. 11.1 shows some of the applications of electrospun nanofibers in wound dressing.

A study on polyurethane electrospun nanomembrane exhibited that the membranes were effective in discharging the fluid from the wound, without fluid accumulation and no wound desiccation around the wound [3]. Additionally, the electrospun membrane showed controlled water loss through evaporation, high

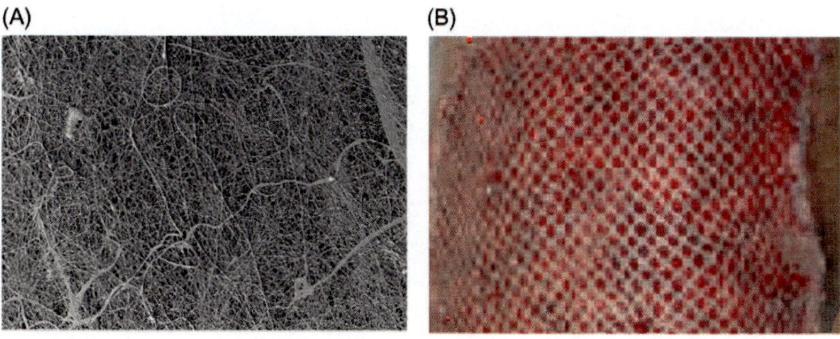

FIGURE 11.1

Images showing electrospun nanofibers (A) in wound dressing application (B).

oxygen permeability, and good fluid discharge, besides impeding the invasion of exogenous microorganisms. A histological study showed that the rate of epithelialization was enhanced and dermis became organized, when the wound was completely covered with membranes [3]. A study of electrospun collagen nanofibers showed that the early-stage healing of an open wound was much faster than using normal cotton gauze [4]. In the first week, the surface of the wound using cotton gauze was covered by fibrinous tissue debris, below which infiltration of polymorphonuclear leukocytes and proliferation of fibroblasts were formed. The surface tissue debris in the collagen nanofiber had disappeared, and proliferation of young capillaries and fibroblasts was found during the healing process. The other stages were similar for both materials [4]. A new study showed that in vivo wound healing of diabetic ulcers was investigated by means of clock copolymer (PCL-PEG) and PCL. After the modification of nanofibers with recombinant human epidermal growth factor (nhEGF), the expression of keratinocyte-specific genes and EGF-receptor was increased [5].

In abdominal surgeries, postsurgery tissue adhesion is an unsolved problem. It causes problems in patients, such as bowel obstruction, female infertility, and chronic debilitating pain [6,7]. An electrospun nanofiber membrane consisting of an antibiotic agent was employed as a barrier to impede postsurgery abdominal adhesion. According to recent developments, the nanomembrane eliminated postsurgery abdominal adhesion substantially and thereby improved the healing process [8]. Some researchers incorporated iodine complex and biocides, such as silver, in the electrospun nanofibers in order to decontaminate bacterial invasion [9−12]. It was determined that polyvinylpyrrolidone (PVP)−iodine complex (PVP-iodine) steadily discharged active iodine. Since iodine possesses microbicidal activity, so PVP nanofibers containing iodine can have external antibacterial, antimycotic, and antiviral applications. Silver ions are reported to have a biocidal effect on many bacteria including *Escherichia coli* and *Staphylococcus aureus* [13].

11.1.2 NANOFIBERS WITH ANTIBACTERIAL ACTIVITY

Generally, the polymers used in wound dressing possess nontoxicity, biocompatibility, and biodegradability features. Some of these polymers include polycaprolactone (PCL), polylactide (PLA), poly(lactic-co-glycolic acid) (PLGA), poly(vinyl alcohol) (PVA), collagen, and chitosan [14]. The common antibacterial materials used in fabricating antibacterial nanofibers include antibiotics, triclosan, chlorhexidine, quaternary ammonium compounds (QACs), biguanides, silver nanoparticles, and metal oxide nanoparticles [14]. Hong et al. [9] fabricated Ag/PVA nanofiber membrane for wound healing. The membrane showed excellent antimicrobial ability and faster release with good effectiveness [9]. Zhang et al. [15] prepared electrospun mats by dispersing Ag nanoparticles in a PVA polymeric solution. These nanofibrous mats displayed excellent antibacterial activities against both Gram-positive *Staphylococcus aureus* (*S. aureus*) and Gram-negative *Escherichia coli* (*E. coli*) microorganisms. Baran [16] fabricated and characterized nanofibrous hybrid yarns of polyvinyl alcohol (PVA) and poly-L-lactide acid (PLLA) that has strong antibacterial property. The prepared nanofibrous materials consisted of 0, 5, 10, 20, and 30 wt.% of silver nanoparticles. His results showed that the bactericidal efficiencies of all samples were 99.99%. Gentamicin-loaded nanofibers and other materials have also been investigated for a number of biomedical applications [17–19]. Fig. 11.2 shows the antibacterial test results of PCL nanofibers with different concentrations of gentamicin [20].

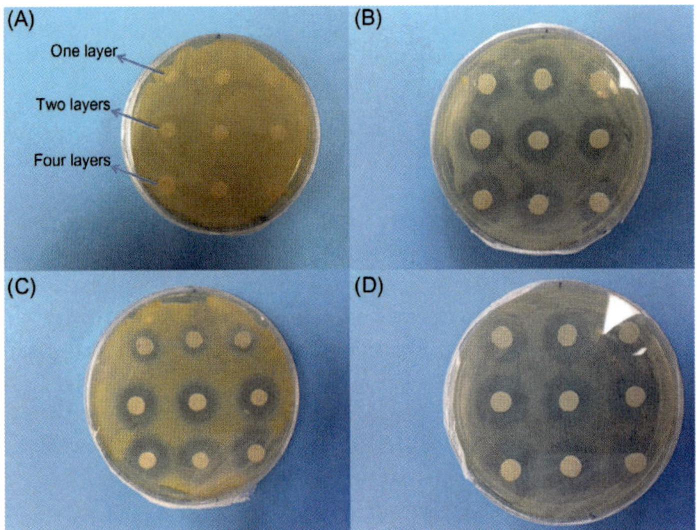

FIGURE 11.2

Antibacterial test results of PCL nanofibers with different concentrations of gentamicin: (A) 0 wt.%, (B) 2.5 wt.%, (C) 5 wt.%, and (D) 10 wt.%, after 7 days of in vitro tests. Top row samples have one layer, middle row samples have two layers, and bottom row samples have four layers of PCL nanofibers [20].

Nguyen et al. [21] demonstrated the fabrication of silver nanoparticles loaded in poly(vinyl alcohol) (PVA) nanowire mats by conjugation of the electrospinning method and the microwave-assisted process. The antibacterial activity of PVA loaded with silver nanoparticles at different irradiation times was tested on Gram-positive bacteria, *S. aureus,* Gram-negative bacteria, and *E. coli.* The test results showed that PVA loaded with silver nanoparticles was very effective in preventing the growth of bacteria. Another work demonstrated the preparation of new poly(3-hydroxybutyrate) (PHB)/poly(ethylene glycol) (PEG)-based fibrous materials containing natural phenolic compound caffeic acid (CA) [22]. The desired materials were obtained by coating with polyelectrolyte complex of alginate (Alg) and *N,N,N*-trimethylchitosan (TMCh). Microbiological assessment of the mats and MTT cell viability studies exhibited that, in contrast to the bare mats, the CA-containing nanofibrous mats were effective in suppressing the growth of *S. aureus* and *E. coli.* Çalamak et al. [23] used antibacterial polyethylenimine (PEI) blended with silk fibroin (SF) to fabricate mats by electrospinning. Furthermore, in order to test the antibacterial activity of silk fibroin (SF), the authors functionalized the surfaces with a sulfate group. The test results showed that PEI/fibroin bionanotextiles displayed strong antibacterial activities against Gram-positive *S. aureus* and Gram-negative *Pseudomonas aeruginosa.* Besides using antibacterial materials, antimicrobial nanofibers can be fabricated by directly using antimicrobial polymers, such as polyurethanes, which contain quaternary ammonium groups. The prepared polyurethane nanofibers showed strong antimicrobial activities against *S. aureus* and *E. coli* [24].

11.2 ELECTROSPUN NANOFIBERS FOR TISSUE ENGINEERING

Tissue engineering involves designing tissue scaffolds for temporary and spatially organized structures to simulate the cellular microenvironment. Tissue engineering is the application of scientific principles and life science to design, modify, construct, and grow living tissues by employing biomaterials. Tissue engineering is a multidisciplinary area of research and innovations, which employs the scientific principles and knowledge of life sciences to develop biological substitutes that can maintain, restore, and enhance the tissue function of an organ without causing any harmful effects to the body. The field of tissue engineering was developed to address problems associated with the replacement of tissues lost owing to trauma or disease. Tissue replacement must address issues such as rejection, chronic inflammation, and organ donor shortage.

Thousands of patients lose their lives every year waiting for organ transplantation [25]. The objectives behind tissue engineering are to avoid such problems by either repairing or regenerating damaged or diseased tissue or the reconstruction of tissue function ex vivo. The designing and evaluation of the biologically active materials are an essential component of this work. Tissue engineering is relatively new and emerging interdisciplinary field that involves the knowledge of bioengineering, life sciences, and clinical sciences towards solving crucial medical problems of tissue loss and failure of organs. This field also involves the use of biochemistry and biomechanics to develop biological substitutes for damaged organs. Tissue engineering aims at designing three-dimensional structures, generally called scaffolds, and activating them in special devices called bioreactors in order to obtain the biological equivalent of a defective organ or tissue. The tissue engineering strategy comprises the isolation of healthy cells from a patient followed by their expansion in vitro for further development [26]. These expanded cells are then seeded onto a three-dimensional biodegradable scaffold that is used to provide support and at the same time act as a reservoir for bioactive molecules. Biomaterials can interact with biological systems and therefore play an important part in tissue engineering by serving as three-dimensional frameworks, which are called scaffolds, matrices, or constructs. The scaffold mimics the architecture of tissue at the nanoscale. The scaffold is fabricated from a biodegradable material; thus, it degrades with time and gets replaced by newly grown tissue from seeded cells to produce new tissue [26].

Generally, scaffolds must be fabricated from a biocompatible material that does not have any side effects on the body. Biodegradable polymers have extensive applications in the biomedical field, especially in tissue engineering. When biodegradable polymer is used as a therapeutic agent, it must possess some criteria, such as: (1) nontoxic; (2) the time taken by the polymer to degrade is equal to the time required for therapy; (3) without any harmful side effects; (4) well sterilized; and (5) admissible shelf-life. In the last 20 years, a fluctuating array of biomaterials have been proposed by many researchers as prominent scaffolds for cell growth yet very few have reached the desired clinical efficiency. Biomaterials, whether naturally occurring or synthetic, need to be biocompatible. One of the most important features of tissue engineering is the designing of polymeric scaffolds with mechanical and biological properties matching the native extracellular matrix (ECM) in order to harmonize or blend cellular behavior [27,28].

Tissue engineering scaffolds work as temporary ECMs until repair or regeneration of new tissues occurs. A scaffold renders a 3D network for cell attachment and develops in vitro [27]. Afterwards, the cell/scaffold construct can be implanted into a site, where the defect tissues are present for repairing tissues and the regeneration process. There are some features that must be considered before recreating new tissues. The first and most important aspect is that the scaffold should be biocompatible, which means that scaffold should integrate well with the host tissue without triggering a major immune response [27]. The scaffold should have high porosity and high surface area to enable cell attachment and growth. Additionally, the porous structure of the scaffold will allow for angiogenesis upon implantation in a defect site. Also, since the scaffold works as a nonpermanent support for the cells to adhere to and proliferate, it should imitate native ECM functionally. Finally, a scaffold should be biodegradable in order to avoid the need for a second surgery to remove the implant [27].

Electrospinning has been recognized as the process of choice for fabricating fibrous tissue and has been excessively used in tissue engineering to generate nanofibrous scaffolds from natural or synthetic biodegradable polymers to simulate the cellular microenvironment [29]. Electrospinning can produce fibers at nanoscale and it usually presents many options to tailor the chemical, physical, and biodegradable properties of a material for particular applications and the cellular environment [26,29]. Electrospinning offers an attractive process for fabricating the polymeric biocompatible materials into nanofibers in a very short time with minimum investment. This process also presents the options for control over thickness and composition of the fibrous material, along with porosity, flexibility, and orientation. The high surface area and high porosity of nanofibers are ideal for cell interactions, which makes electrospun nanofibers an ideal candidate in tissue engineering applications [26]. Electrospinning offers a simple and rapid method to fabricate nanofibrous scaffolds for different tissue engineering. Electrospun scaffolds incorporated with biological polymers, composites, or ceramic precursors can provide bioactive cues [29,30]. Nanofibrous scaffolds possess high porosity and surface area to volume ratios and present a variety of topographical features to improve cellular adhesion and proliferation [31,32]. The applications of electrospun scaffolds as cell delivery vehicles has been increased significantly recently due to the physical similarities between the electrospun nanofiber scaffolds and the extracellular matrix (ECM) generally found in native tissues [31]. The favorable architecture of nanofibrous scaffolds offers an ideal substrate to accurately assess cellular behavior in vitro. Additionally, electrospinning presents the ability to control many processing parameters in tailoring the scaffold for specific biomedical applications. Fig. 11.3 shows SEM micrographs of nanofiber PCL mats developed for bone scaffolding using different percentages of gentamicin and hydroxyapatite for bone scaffolding [18].

11.2.1 APPLICATIONS OF SYNTHETIC POLYMERS IN SCAFFOLDS

There has been a rapid surge in the use of electrospun fibers to create nanofiber scaffolds for tissue engineering. Mostly, the electrospun scaffolds are composed of biocompatible and biodegradable synthetic polymers with good mechanical and surface properties. The synthetic biodegradable polymers used in scaffolds can include poly(lactic acid) (PLA), poly(glycolic acid) (PGA), poly lactic-co-glicolide (PLGA) copolymers, collagen, polycaprolactone (PCL), and a blend of them [33,34]. Literature reviews show that PLGA electrospun fibrous mats have been used extensively in tissue engineering scaffolds. PLGA electrospun fiber mats possess high porosity ($>90\%$) and high surface area to volume ratio, as well as numerous focal adhesion sites on the fiber surface due to nano-sized diameters, which is ideal for cellular attachment [35−37]. Electrospun nanofibers can support a large variety of cell types. It has been found that human umbilical vein endothelial cells attach well and displayed high proliferation when seeded onto 50:50 poly (L-lactic acid-co-ε-caprolactone) (PLCL) fibers with diameters in

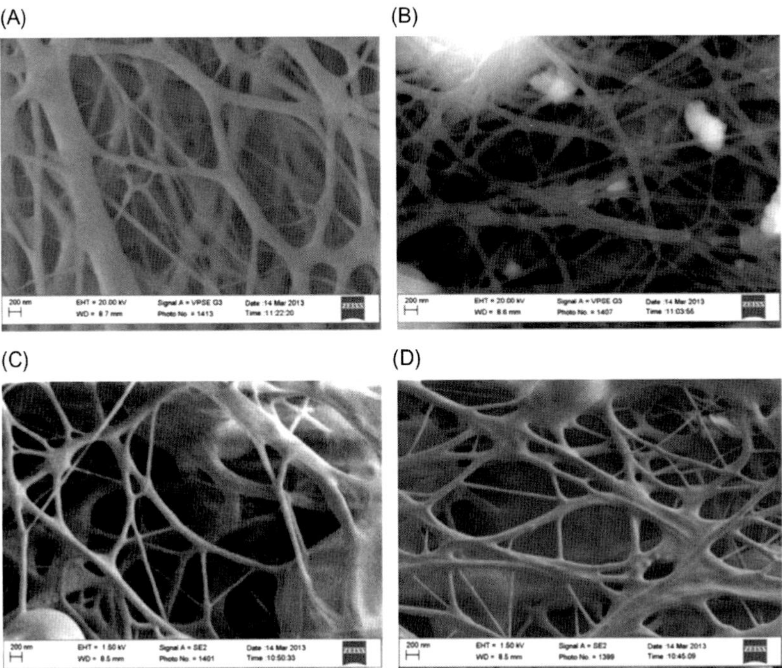

FIGURE 11.3

SEM micrographs of nanofiber PCL mats developed for bone scaffolding: (A) 0% HA and 0% gentamicin, (B) 20% HA and 0% gentamicin, (C) 0% HA and 10% gentamicin, and (D) 20% HA and 10% gentamicin [18].

the nano-range [27]. NIH 3T3 fibroblasts and rat kidney cells seeded onto electrospun polyamide nanofibers arranged their action cytoskeleton to a more in vivo-like morphology [38]. Similarly, breast epithelial cells on the same polymeric surface experienced morphogenesis to form multicellular spheroids. Fetal bovine chondrocytes seeded on electrospun nanofiber of poly(ε-caprolactone) (PLC) scaffolds. The results have shown that fetal bovine chondrocytes were able to maintain the chondrocytic phenotype during 3 weeks of culture and upregulating collagen type 11B expression, which means a mature chondrocyte phenotype [36]. These studies showed that nanofiber scaffolds are cytocompatible and they can be used to stimulate and vitalize cell proliferation and phenotypic behavior [36].

Many synthetic polymers possess physicochemical and mechanical properties similar to those of biological tissues. Synthetic polymers constitute a sizeable group of biodegradable polymers and these polymers can be synthesized under controlled conditions [39]. Generally, they exhibit predictable and reproducible mechanical properties, such as elastic modulus, tensile strength, degradation rate, and percentage elongation at breaking [39]. As discussed earlier, PLA, PGA,

PCL, and PLGA copolymers are the most common synthetic polymers that are used in tissue engineering. Polyhydroxyalkanoates (PHAs) are microbial polyesters and they are now being excessively considered for tissue engineering [39]. A new study demonstrated that smooth muscle cells and endothelial cells adhered to and proliferated in electrospun poly(L-lactide-co-ε-caprolactone) [P (LLA-CL)] after 7 days [40]. Another study also showed that chondrocytes proliferated even after 21 days due to seeding in poly(ε-caprolactone) (PCL) nanofibrous scaffolds [36]. Likewise, poly(ethylene-co-vinyl alcohol), having hydrophilic surface characteristics, was demonstrated to support smooth muscle cells and fibroblasts in fibrous scaffold form. Moreover, mesenchymal stem cells, harvested from neonatal rats, were successfully seeded onto polycaprolactone (PCL) scaffolds and shown to mineralize and produce type 1 collagen after nearly 4 weeks in culture [41]. Kumbar et al. used electrospun PLGA as a wound dressing for skin wounds and studied the effect of fiber size on fibroblasts [42,43]. Fig. 11.4 shows electrospun PCL nanofibers, and PCL nanofibers incorporated with hydroxyapatite nanoparticles and graphene nanoflakes for bone scaffolding applications [44]. Graphene nanoflakes were chosen to increase the mechanical strength and biological properties of produced bone after the biomineralization process.

11.2.2 APPLICATIONS OF NATURAL POLYMERS IN SCAFFOLDS

Besides using synthetic polymers as primary components of electrospun scaffolds, some researchers used naturally occurring polymers, due to their enhanced biocompatibility and biofunctionality. Collagen is frequently used as a scaffold for cells [27]. Additionally, adding collagen, alginate, hyaluronic acid, and starch

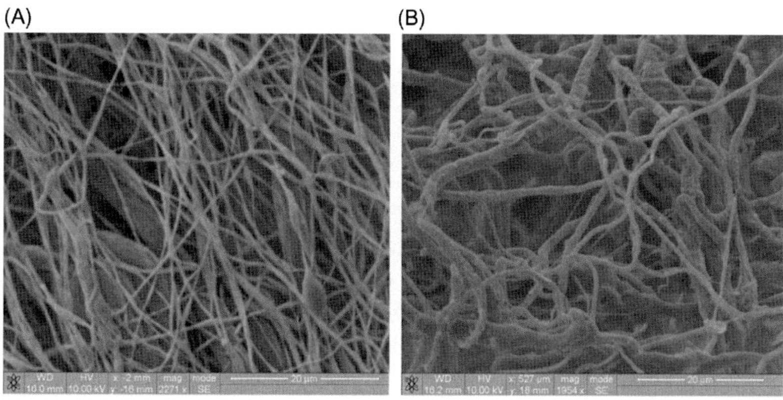

FIGURE 11.4

SEM images showing (A) electrospun PCL nanofibers, and (B) PCL nanofibers incorporated with hydroxyapatite nanoparticles and graphene nanoflakes for bone scaffolding applications.

into synthetic polymers can improve the cytocompatibility of a scaffold [45−47]. Many researchers have used various electrospun biological substances, such as collagen, silk, fibrinogen, cellulose, and chitosan for biomedical applications. The electrospinning of a biological substance is rather difficult, since an appropriate solvent that does not compromise the integrity of biological substances has to be used [48]. Natural polymers present the advantage of being similar to macromolecular substances present in human body. Some of the natural polymers used in biomedical applications include collagen, gelatin, chitosan, silk, wheat protein, and hyaluronic acid [26].

Collagen has been investigated by many researchers as a natural biomaterial for biomedical applications. Matthews et al. prepared collagen nanofibers via electrospinning and demonstrated that collagen nanofibers have good compatibility with a number of cell types, such as myoblasts and chondrocytes [26,49]. Additionally, crosslinking in collagen type 11 scaffolds offers good mechanical properties, therefore these scaffolds are good options for cell growth [50]. Huang et al. [51] blended electrospun collagen (type1) nanofibers with poly(ethylene oxide) (PEO) and their results demonstrated that the mechanical properties were enhanced due to intermolecular interactions between collagen and PEO. These results are very encouraging for using collagen in scaffolds.

Geng et al. [37] used electrospun chitosan, a natural polymer, to make nanofibers for biomedical applications. They prepared chitosan/PEO nanofibers employing electrospinning, and found that these fibers possessed structural integrity in water. They observed enhanced attachment of human osteoblasts and chondrocytes onto the nanofibers and in addition these fibers also showed critical cytocompatibility [52]. Therefore, chitosan nanofibers are selected for scaffold applications in tissue engineering. Hyaluronic acid is a natural compound of extracellular matrix of tissue. Um et al. [53] prepared hyaluronic acid nanofibers by combining electrospinning with an electro-blowing process. The development of consistent nanofibers via electrospinning was difficult, so the authors also used a hot blast of air during the electrospinning process.

Another natural biomaterial that has been studied extensively for biomedical applications is gelatin. Zhang et al. [54] prepared gelatin/PCL composite nanofibrous scaffolds via an electrospinning method, and determined that the mechanical strength and wettability of composite fibrous scaffolds were better than gelation and PCL alone. Hence, these composite nanofibers can be used for biomedical applications. Silk fibroin is another natural biomaterial for scaffolds [55,56]. Jin et al. [55] and Min et al. [56] demonstrated the in vitro cytocompatibility of silk nanofibers with keratinocytes and fibroblasts. Their studies associated with cytocompatibility, fiber diameter, and porosity showed that silk fibroin is an ideal material for scaffold applications. Li et al. [57] prepared electrospun protein fibers for scaffolds in tissue engineering and other biomedical applications. They prepared human tropoelastin for electrospinning. Their results showed that tropoelastin nanofibers seeded with human embryonic palatal mesenchymal cells supported cell adhesion and proliferation successfully [57]. In another study,

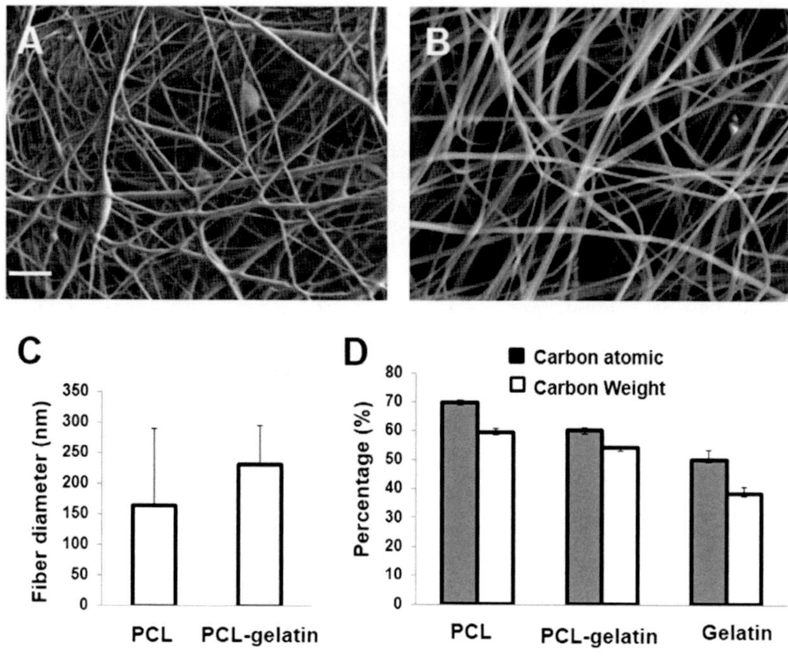

FIGURE 11.5

Characterization of electrospun fibers showing (A) the SEM image of electrospun PCL fibers. (B) SEM image of PCL-gelatin fibers. (C) Quantification of fiber diameter. (D) Carbon atomic and carbon weight ratio in nanofibers (scale bar: 2 μm) [59].

Woerdeman et al. [58] used wheat gluten to prepare electrospun nanofibers for tissue engineering. Hence, a wide variety of naturally occurring polymers have been explored for the synthesis of nanofibers via electrospinning as scaffolds for tissue engineering. Fig. 11.5 shows the characteristics of electrospun fibers: (A) SEM image of electrospun PCL fibers; (B) SEM image of PCL-gelatin fibers; (C) quantification of fiber diameter; and (D) carbon atomic and carbon weight ratio in nanofibers [59]. This study was mainly focused on nanofibers support for oligodendrocyte precursor cell growth and function as a neuron-free model for myelination.

11.3 ELECTROSPUN NANOFIBERS FOR BONE TISSUE ENGINEERING

Bone tissue engineering is a highly interdisciplinary field and involves biomaterials, cells, and growth factors to produce biosynthetic bone grafts with well-organized mineralization for regeneration of damaged or fractured bones. Of the many processes available for scaffold synthesis, electrospinning is the process of choice as it can fabricate nanostructures with minimum investment and in a very short period of time. The electrospun nanofibers possess outstanding properties, such as high surface area, flexibility, porosity, permeability, and controlled morphology similar to that of extracellular matrix [60]. The design of scaffolds for bone tissue engineering is related to the properties of bone tissue, such as hardness, pore size, strength, porosity, and three-dimensional structure of bone tissues [26]. It is generally accepted that for bone tissue engineering, a highly porous microstructure with interconnecting pores and high surface area is essential for tissue in-growth. For bone generation, a scaffold with pore size in the range of 100−130 micrometers and porosity of more than 90% is preferred for better cell/tissue in-growth and enhanced bone generation [61−63]. Polycaprolactone (PCL), a biodegradable polyester, has been studied extensively for bone tissue scaffolds. It has been determined that mesenchymal stem cells (MSCs) penetrated into the PCL matrix associated with large extracellular matrix after 1 week of seeding, and mineralization and type I collagen occurred at 4 weeks [1].

Seyedjafari et al. [64] seeded hydroxyapatite-coated and uncoated electrospun PLLA nanofibers with human cord blood-derived stem cells and then implanted the scaffolds subcutaneously into mice. After 10 weeks, they observed that scaffolds without hydroxyapatite did not show any calcium deposit and were surrounded by a granulomatous inflammatory response, while scaffolds with hydroxyapatite showed noticeable mineralization with inflammatory response. Moreover, trabeculi (tissue element in the form of a small beam, strut, or rod) and bone marrow were seen on the newly formed ectopic bone [64]. Shin et al. [65] conducted an in vivo experiment by implanting a mesenchymal stem cells (MSC)-cultured polycaprolactone (PCL) construct in 4 weeks in the omenta of rats [65]. Their study demonstrated that after 4 weeks, the constructs retained the same size and shape and had a bone-like appearance [1]. Yoshimoto et al. [66] reported the development of PCL scaffolds via electrospinning for bone tissue engineering. They derived MSCs from the bone marrow of neonatal rats and seeded onto nanofibrous scaffold. Their results showed that the MSCs migrated inside the scaffolds and generated abundant extracellular matrix in the fibrous scaffolds [66]. These studies demonstrate that PCL-based nanofiber structures are promising candidates for bone scaffolds.

Ramay et al. [67] reported the use of HA with β-tricalcium phosphate (β-TCP) to synthesize biodegradable nano-composite porous scaffolds. B-TCP/HA scaffolds prepared from HA nanofibers with β-TCP as a matrix were used to fabricate porous scaffolds by employing a process that merged the gel-casting method with the polymer sponge method. The in vitro experimental results indicated that incorporation of HA nanofibers as a second component in β-TCP drastically increased the mechanical strength of the porous composite scaffolds [68]. This study developed nano-composites with HA nanofibers as a potential scaffolding system for bone tissue engineering. Some researchers reported nanofiber fabrication from a blend of gelatin/PCL and PCL composite with calcium carbonate nanoparticles and hydroxyapatite (HAp) nanoparticles for bone scaffolds

[69–73]. The incorporation of gelation (50%) to PCL enhanced the mechanical properties and wettability of composites, thereby resulting in enhanced cell attachment and growth on the scaffold surface.

Cai et al. [74] prepared electrospun PLLA scaffolds incorporated with some biological compounds (e.g., collagenous) for guided bone generation membrane. They used rabbit tibia defect model to implant. The implants contained porous collagen membrane, a PLLA electrospun membrane, or a bilayer combining the two. After around 3 weeks, the defects treated with the bilayer group were 91% filled with new bone tissue compared to 64% for PLLA nanomembrane alone, and approximately 32% for collagen membrane alone [74]. After 6 weeks, the new bone formed on the PLLA nanofiber membrane was estimated to contain a large percentage of cortical bone, and 86% and 77% for bilayer and nanofiber membrane alone, respectively, compared to only 46% for collagen membrane [74]. Woo et al. [75] compared the in vivo bone formation potential of nanofibrous scaffolds versus solid-walled scaffolds. They created a calvarial defect in rats and implanted nanofibrous and a solid-walled scaffold, as well. Their analysis showed that new bone was formed throughout the nanofibrous implants, while only the periphery of the solid-walled scaffold sustained bone ingrowth [75]. This study clearly indicates that nanofibrous membrane facilitates the growth mechanism.

Jin et al. [76] and Kim et al. [77] investigated silk fibroin nanofibers for bone tissue engineering and other related biomedical studies. They found that silk fibroin has excellent biocompatibility with enhanced bone generation. When silk fibroin nanofibers contained bone morphogenetic protein-2 (BMP-2) or HAp nanaoparticles, BMP-2 facilitated higher calcium deposition and increased transcript levels of bond-specific markers, and the presence of HAp nanoparticles in silk fibroin enhanced bone formation [78]. The existence of both materials gave excellent improvement to calcium deposition. There are many other polymers that have been investigated for bone scaffolds, such as polyurethanes, polyphosphazenes, poly(propylene carbonate), poly(3-hydroxybutyrate), poly(3-hydroxybutyrate-co-3-hydroxyvalerate), PLGA, and poly(ethylene oxide terephthalate)-poly(butylene terephthalate).

11.4 ELECTROSPUN NANOFIBERS FOR DENTAL GROWTH

Since structures of electrospun nanofibers behave like natural extracellular matrix, this make them ideal materials in the field of tissue engineering due to ease of fabrication, tailorability of pore size, scaffold shape, and fiber thickness and alignment. For these reasons, polymeric nanofibers have been used in dentistry and their nanostructure and flexibility facilitate cell homing, thereby improving the dental regeneration process [79]. In dental tissue engineering, a wide variety of implanted materials have been introduced in the past and some exceptional results have been revealed. In this connection, bioceramics have shown good biocompatibility with dental tissues and teeth, since their physiochemical and biological properties, such as osteoconductivity and bioactivity, are close to dental tissues [79,80]. However, these bioceramics are very brittle and nonflexible, and therefore their applications in dentistry are limited [81]. In order to overcome this obstacle, highly flexibly, biodegradable, and biocompatible polymers have been proposed by many researchers to facilitate dental generation [82]. These polymers include poly(ethylene glycol) (PEG) and a series of polyesters, such as poly(lactic acid) (PLA), poly(glycolic acid) (PGA), poly(D,L-lactide-co-glycolide) (PLGA), polycaprolactone (PCL), and their copolymers [83].

Dental trauma and caries (tooth decay) generally results in the loss of pulp—dentine complex (mineralized layer and tissue below enamel). There are various forms of pulp-generative dental materials available, such as calcium hydroxide, ferric sulfate, and mineral trioxide [84]. However, electrospun nanofibrous scaffolds have been found to be excellent materials for regeneration using dental pulp stem cells (DPSCs), which are a well-established cell source for the formation of dentine—pulp complex [79]. Odontogenic differentiation of DPSCs from human on PLA electrospun nanofibrous scaffolds was indicated by increased alkaline phosphatase (ALP) activity, mineralization, and dentine marker gene expression [85]. Electrospun nanofibrous scaffolds also facilitate adhesion of DPSCs.

Mineralized PCL electrospun nanofibrous scaffolds have demonstrated odontogenic differentiation and growth of human DPSCs through collagen (type 1) and the integrin-mediated signaling pathway. However, these scaffolds lack mechanical and biofunctional features for clinical practice [86]. Therefore, in order to enhance mechanical and biological properties of nanofibers, nanoparticles are added before electrospinning. These nanoparticles include bioactive glass nanoparticles, magnetic nanoparticles, and hydroxyapatite nanoparticles [79]. Bottino et al. [87] reported the preparation of electrospun scaffolds containing bioactive nanoparticles and incorporated with polydioxanone, in which antibiotics (metronidazole and ciprofloxacin) were also loaded. Their study revealed that these scaffolds delivered antibiotics more efficiently than pastes. Kim et al. [88] prepared biocompatible and biodegradable polyvinyl alcohol and hydroxyapatite (HAp) nanocomposite fibers that mimic mineralized hard tissue for dentine regeneration. Bae et al. [89] prepared collagen-based nanocomposites containing nano-bioactive glass (Col/nBG) as a scaffold for detine—pulp regeneration and investigated the effects of the matrix on the proliferation of human dental pulp cells (HDPCs) and their differentiating into odontoblastic lineage. Their study revealed that the introduction of nBG significantly improved bone mineral-like apatite formation in the simulated body fluid, indicating excellent acellular bone bioactivity. The hDPCs cultured on the Col/nBG nanocomposite indicated active growth behavior during 14 days.

Yun et al. [90] reported the preparation of magnetic nanocomposite scaffolds from magnetite nanoparticles and polycaprolactone (PCL) and studied the effects of the scaffolds on the adhesion, migration, growth, and odontogenic differentiation of HDPCs. Their study demonstrated that magnetic nanocomposite scaffolds can provide excellent matrix conditions for HDPCs in their migration, adhesion, and odontogenic differentiation, this behavior is useful for scaffold-based dentine—pulp tissue engineering.

Periodontal diseases generally cause periodontal tissue destruction and finally result in loss of teeth [91]. Regeneration of destructed periodontal tissue has been a challenging issue in dental tissue engineering. Thus, periodontal tissue engineering is very important for the repair of defects in periodontal tissues such as alveolar bone, periodontal ligament (PDL), and cementum [79]. Generally, nonresorbable materials (expanded polytetrafluoroethylene) have been employed as guide tissue regeneration (GTR) membrane, but they had the disadvantage of requiring a secondary surgery to remove the membrane [79]. Biodegradable polymers, synthetic or natural, such as PLGA, PLA, and PCL, have been investigated but they suffer from biological functionality and physical properties.

Electrospinning has recently emerged to address these issues by increasing the functionality of these membranes and thereby leading to expecting periodontal regeneration [79]. Hence, biodegradable nanofibrous guide tissue regeneration (GTR) membrane via electrospinning has improved the biological aspects and functionality, such as high porosity to attach cells and fiber orientation, and alignment of collagen fibers in periodontal ligament (PDL) regeneration [91,92]. Electrospun collagen membranes were first used in GTR applications due to their favorable biological properties, such as differentiating potential into osteoblast-like cells [49]. The only problem with collagen was their sources. Since most of the collagen is generally originated from animals, that caused some ethical issues and infection. Thus, as an alternative, synthetic biodegradable polymer membranes have been suggested for PDL regeneration. Recent research investigations have demonstrated good cell attachments and proliferation of PCL cells and tissue formation on electrospun PLGA [93].

Park et al. [94] prepared hydroxyapatite-coated biopolymer nanofibrous membrane by mineralizing the electrospun polycaprolactone nanofibers. They studied the effects of hydroxyapatite-coated biopolymer nanofibrous membranes on the proliferation and differentiation of human periodontal ligament fibroblasts, which is essential for periodontal tissue generation. Hydroxyapatite-coated nanofibrous membrane displayed notable mineral formation on day 14. According to this study, it is obvious that hydroxyapatite-coated nanofibrous membrane has positive effects on the proliferation and differentiation of human periodontal ligament fibroblasts and this membrane can be a competent candidate material for periodontal tissue regeneration [94]. Suganya et al. [95] added bioactive agents such as aloe vera (AV) and silk fibroin (SF) with 4% hydroxyapatite (HA) in the poly (lactic acid-co- caprolactone) (PLACL) to fabricate PLACLAV-HA (4%) nanofibrous scaffolds via electrospinning, which mimic natural bone constitution. They cultured human mesenchymal stem cells on this scaffold. Their results showed a significant increment in cell proliferation, osteogenic differentiation, osteocalcin expression, and mineral deposition [95]. Electrospinning has spurred new prospects to the field of periodontal tissue regeneration; however, more research work is still needed to endorse the applications of electrospun nanofibrous scaffolds in clinical practice. Fig. 11.6 shows images of the PCL nanofibers after protonation processes and an SEM image of the electrospun PCL nanofibers developed for tooth cavity regeneration [96]. The bead formations (spherical particles on the surfaces of the nanofibers) were clearly seen in the SEM images.

Overall, nanotechnology is based on the synthesis of engineered nanomaterials, and opens a new era of technological innovations, which is impacting on all sectors of human life: electronics and computers, aviation, energy, food and agriculture industry, cosmetics, paints and, of course, healthcare. The ever-lasting potential impact on healthcare of nanotechnology is prodigious and permeates, and we many expect a new era aptly designated as "nano-medicine." The applications of nanotechnology in healthcare have been under extensive investigation in recent times, since many nanotechnology products, such as nanoparticles, nanosponges, nanogels, nanorods, nanowires, nanoplatelets, nanospheres, and nanofibers, are being extensively used in the medical industry. There are various treatments available, but these treatments are costly and take a long time. With nanotechnology, swift and less-costly treatment is possible. Among the various nanomaterials available, polymeric nanofibers have been aptly termed "materials of choice" due to their remarkable properties and their ability to permeate into the cell membrane. Nanofibers can acts as natural extracellular matrix (ECM) and deliver biomolecules, drugs, etc., in a sustained fashion to the targeted site.

(A)   (B)

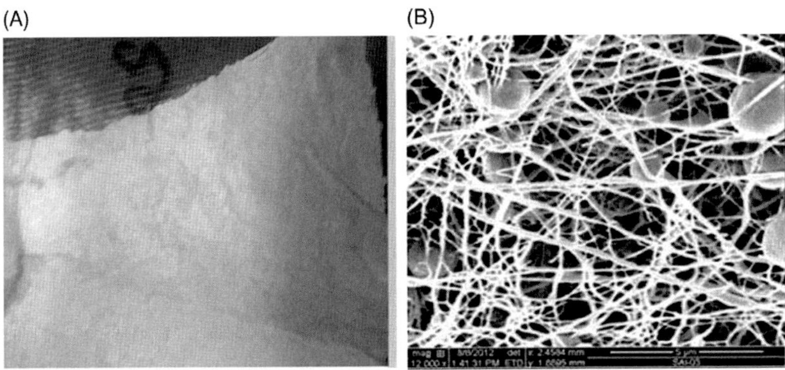

FIGURE 11.6

Images showing (A) the PCL nanofibers after protonation processes and (B) SEM image of the electrospun PCL nanofibers for tooth cavity regeneration. The bead formations (spherical particles) were clearly seen in the SEM image [96].

11.5 CONCLUSIONS

Nanotechnology, in general, and nanofibers, in particular, have been gaining much attention in the tissue engineering field. Mimicking the 3-D architecture of ECM is one of the major challenges of tissue engineering. Amongst all the approaches used to prepare ECM synthetically, the approach using nanofibers has shown the most encouraging results. Electrospun nanofibers have provided porous scaffolds for tissue engineering with high surface area. These properties have been shown to have a substantial effect on cell adhesion, proliferation, and differentiation. Therefore, nanofibrous matrices are currently being explored as scaffolds for musculoskeletal tissue engineering, such as bone, ligament, and skeletal muscle, neural tissue, skin tissue engineering, vascular tissue engineering, and controlled delivery of drugs, proteins, and DNA, etc. Electrospun nanofiber membranes are excellent materials for wound dressing, since their small pore size and high surface area not only impede exogenous microorganism invasion, but also help to control fluid discharge. In the field of dental tissue engineering, various dental materials have been used to create a conducive microenvironment for dental generation. The electrospun nanofibrous scaffolds are most suitable for dental applications due to ease of fabrication, control over scaffold size, flexibility, and fiber alignment. Electrospun nanofibers have provided mechanical properties and biologically favorable functionalities to facilitate biological aspects in dental applications.

REFERENCES

[1] F. Jian, N.I.U. HaiTao, L. Tong, W. XunGai, Applications of electrospun nanofibers, Chin. Sci. Bull. 53 (15) (2008) 2265–2286.

[2] S. Babitha, L. Rachita, K. Karthikeyan, E. Shoba, I. Janani, B. Poornima, et al., Electrospun protein nanofibers in healthcare: a review, Int. J. Pharmaceut. 523 (2017) 52–90.

[3] M.S. Khil, D.I. Cha, H.Y. Kim, I.S. Kim, N. Bhattarai, Electrospun nanofibrous polyurethane membrane as wound dressing, J. Biomed. Mater. Res. Part B Appl. Biomater. 67 (2) (2003) 675–679.

[4] K.S. Rho, L. Jeong, G. Lee, B.M. Seo, Y.J. Park, S.D. Hong, et al., Electrospinning of Collagen Nanofibers: Effects on the behavior of normal human Keratinocytes and early stage wound healing, Biomaterials 27 (8) (2006) 1452–1461.

[5] J.S. Choi, K.W. Leong, H.S. Yoo, In vivo wound healing of diabetic ulcers using electrospun nanofibers immobilized with human epidermal growth factor (EGF), Biomaterials 29 (5) (2008) 587–596.

[6] A.H. DeCherney, G.S. DiZerega, Clinical problem of intraperitoneal postsurgical adhesion formation following general surgery and the use of adhesion prevention barriers, Surg. Clin. N. Am. 77 (3) (1997) 671–688.

[7] D. Menzies, Peritoneal adhesions. Incidence, cause, and prevention, Surg. Ann. 24 (1992) 26–45.

[8] N. Bolgen, I. Vargel, P. Korjusuz, Y.Z. Menceloglu, E. Piskin, In vivo performance of antibiotic embedded electrospun PCL membranes for prevention of abdominal adhesions, J Biomed. Mater. Res. Part B Appl. Biomater. 81 (2) (2007) 530−543.

[9] K.H. Hong, Preparation and properties of electrospun poly (vinyl alcohol)/silver fiber web as wound dressings, Polym. Eng. Sci. 47 (1) (2007) 43−49.

[10] J. Jia, Y.Y. Duan, S.H. Wang, S.F. Zhang, Z.Y. wang, Preparation and characterization of antibacterial silver-containing nanofibers for wound dressing applications, J. US-China Med. Sci. 4 (2) (2007) 52−54.

[11] N.L. Lala, R. Ramaseshan, L. Bojun, S. Sundarrajun, R.S. Barhate, L.Y. Jun, et al., Fabrication of Nanofibers with antimicrobial functionality used as filters: protection against bacterial contaminants, Biotechnol. Bioeng. 97 (6) (2006) 1357−1365.

[12] M. Ignatova, N. Manolova, I. Rashkov, Electrospinning of poly (vinyl pyrrolidone)−iodine complex and poly (ethylene oxide)/poly (vinyl pyrrolidone)−iodine complex − a prospective route to antimicrobial wound dressing materials, Eur. Polym. J. 43 (5) (2007) 1609−1623.

[13] J.A. Spadaro, T.J. Barranco, S.E. Chapin, R.O. Becker, Antibacterial effects of silver electrodes with weak direct current, Antimicrob. Chemother. 6 (5) (1975) 637−642.

[14] M. Liu, X.P. Duan, Y.M. Li, D.P. Yang, Y.Z. Long, Electrospun nanofibers for wound healing, Mater. Sci. Eng. C 76 (1) (2012) 1413−1423.

[15] Z. Zhang, Y. Wu, Z. Wang, X. Zou, Y. Zhao, L. Sun, Fabrication of silver nanoparticles embedded into polyvinyl alcohol (Ag/PVA) composite nanofibrous films through electrospinning for antibacterial and surface-enhanced Raman scattering (SERS) activities, Mater. Sci. Eng. C 69 (1) (2016) 462−469.

[16] H. Baran, Antibacterial continuous nanofibrous hybrid yarn through in situ synthesis of silver nanoparticles: Preparation and characterization, Mater. Sci. Eng. C 43 (2014) 50−57.

[17] M.S. Seraz, M.N. Uddin, S.Y. Yang, R. Asmatulu, Investigating electrochemical behavior of antibacterial polyelectrolyte-coated magnesium alloys for biomedical applications, J. Environ. Sci. Eng. Technol. 5 (2017) 23−33.

[18] R. Asmatulu, S. Patrick, M. Ceylan, I. Ahmed, S.Y. Yang, N. Nuraje, Antibacterial polycaprolactone/natural hydroxyapatite nanocomposite fibers for bone scaffoldings, J. Bionanosci. 9 (2015) 1−7.

[19] A. Usta, L. Saeednia, S. Pyarasani, R. Asmatulu, "Antibacterial Hydrogels Reinforced by Electrospun PVC Nanofibers for Biomedical Applications," SAMPE Fall Technical Conference, Wichita, KS, October 21−24, 2013, 12p.

[20] M. Ceylan, S.Y. Yang, R. Asmatulu, Effects of gentamicin-loaded PCL nanofibers on growth of gram positive and gram negative bacteria, Int. J. Appl. Microbiol. Biotechnol. Res. 5 (2017) 40−51.

[21] T.H. Nguyen, K.H. Lee, B.T. Lee, Fabrication of Ag nanoparticles dispersed in PVA nanowire mats by microwave irradiation and electrospinning, Mater. Sci. Eng. C 30 (7) (2010) 944−950.

[22] M.G. Ignatova, N.E. Manolova, I.B. Rashkov, N.D. Markova, R.A. Toshkova, A.K. Georgieva, et al., Poly (3-hydroxybutyrate)/caffeic acid electrospun fibrous materials coated with polyelectrolyte complex and their antibacterial activity and in vitro antitumor effect against HeLa cells, Mater. Sci. Eng. C 65 (2016) 379−392.

[23] S. Çalamak, C. Erdoğdu, M. Özalp, K. Ulubayram, Silk fibroin based antibacterial bionanotextiles as wound dressing materials, Mater. Sci. Eng. C 43 (2014) 11−20.

[24] E.H. Jeong, J. Yang, J.H. Youk, Preparation of polyurethane cationomer nanofiber mats for use in antimicrobial nanofilter applications, Mater. Lett. 61 (18) (2007) 3991−3994.

[25] R.A.J.R. Felicity, O.C.O. Richard, Bone tissue engineering: hope vs hype, Biochem. Biophys. Res. Commun. 292 (1) (2002) 1−7.

[26] R. Vasita, D.S. Katti, Nanofibers and their applications in tissue engineering, Int. J. Nanomed. 1 (1) (2006) 15−30.

[27] Q.P. Pham, U. Sharma, A.G. Mikos, Electrospinning of polymeric nanofibers for tissue engineering applications: a review, Tissue Eng. 12 (5) (2006) 1197−1211.

[28] R. Langer, J. Vacanti, Tissue Eng. 260 (5110) (1993) 92−926.

[29] D.R. Nisbert, J.S. Forsythe, W. Shen, D.I. Finkelstein, M.K. Horne, A review of the cellular response on electrospun nanofibers for tissue engineering, J. Biomater. Appl. 24 (1) (2009) 7−29.

[30] Z. Ma, M. Kotaki, R. Inai, S. Ramakrishna, Potential of nanofiber matrix as tissue-engineering scaffolds, Tissue Eng. 11 (1−2) (2005) 101−109.

[31] B.J. Chiu, K.Y. Luu, D. Hang, B. Hsiao, B. Chu, M. Hadjuargyrou, Electrospun nanofibrous scaffolds for biomedical applications, J. Biomed. Nanotechnol. 1 (2) (2005) 115−132.

[32] D.R. Nisbet, S. Pattanawong, N.E. Ritchie, W. Shen, D.I. Finkelstein, M.K. Horne, et al., Interaction of embryonic cortical neurons on nanofibrous scaffolds for neural tissue engineering, J. Neural Eng. 4 (2) (2007) 35−41.

[33] K. Kim, M. Yu, X. Zong, J. Chiu, D. Fang, Y.S. Seo, et al., Control of degradation rate and hydrophilicity in electrospun non-woven poly (D, L -lactide) nanofiber scaffolds for biomedical applications, Biomaterials 24 (27) (2003) 4977−4985.

[34] X. Zong, K. Kim, D. Fang, S. Ran, B.S. Hsiao, B. Chu, Structure and process relationship of electrospun bioabsorbable nanofiber membranes, Polymer 43 (2004) 4403−4412.

[35] W.J. Li, C.T. Laurencin, E.J. Caterson, R.S. Tuan, F.K. Ko, Electrospun nanofibrous structure: a novel scaffold for tissue engineering, J. Biomed. Mater. Res. 60 (4) (2002) 613−621.

[36] W.J. Li, K.G. Danielson, P.G. Alexander, R.S. Tuan, Biological response of chondrocytes cultured in three-dimensional nanofibrous poly (epsilon-caprolactone) scaffolds, J. Biomed. Mater. Res. A 67 (4) (2003) 1105−1114.

[37] X. Geng, O.H. Kwon, J.H. Jang, Electrospinning of chitosan dissolved in concentrated acetic acid solution, Biomaterials 26 (27) (2005) 5427−5432.

[38] M. Schindler, I. Ahmed, J. Kamal, A.N. E-Kamal, T.H. Grafe, H.Y. Chung, et al., A synthetic nanofibrillar matrix promotes in vivo-like organization and morphogenesis for cells in culture, Biomaterials 26 (28) (2005) 5624−5631.

[39] B. Dhandayuthapani, Y. Yoshida, T. Maekawa, D.S. Kumar, Polymeric scaffolds in tissue engineering application: a review, Int. J. Polym. Sci. 2011 (2011) 19. Article ID 290602.

[40] X.M. Mo, C.Y. Xu, M. Kotaki, S. Ramakrishna., Electrospun P (LLA-CL) nanofiber: a biomimetic extracellular matrix for smooth muscle cell and endothelial cell proliferation, Biomaterials 25 (10) (2004) 1883−1890.

[41] M. Shin, H. Yoshimoto, J.P. Vacanti, In vivo bone tissue engineering using mesenchymal stem cells on a novel electrospun nanofibrous scaffold, Tissue Eng 10 (1-2) (2004) 33−41.

[42] M.R. Ladd, T.K. Hill, J.J. Yoo, S.J. Lee, Nanotechnology and nanomaterials, electrospun nanofibers, in: Tony Lin (Ed.), Tissue Engineering, Intech, 2011.

[43] S.G. Kumbar, S.P. Nukavarapu, R. James, L.S. Niar, C.T. Laurenin, Electrospun poly (lactic acid-co-glycolic acid) scaffolds for skin tissue engineering, Biomaterials 20 (30) (2008) 4100–4107.

[44] A. Jabbarnia, V.R. Patlolla, H.E. Misak, R. Asmatulu, "Electrospun Fibers Incorporated with Hydroxyapatite Nanoparticles and Graphene Nanoflakes for Bone Scaffolding," ASME International Mechanical Engineering Congress and Exposition, Houston, TX, November 9–15, 2012, 5p.

[45] J.S. Wayne, C.L. McDowell, K.J. Shields, R.S. Tuan, In vivo response of polylactic acid-alginate scaffolds and bone marrow-derived cells for cartilage tissue engineering, Tissue Eng. 11 (5–6) (2005) 953–963.

[46] H.S. Yoo, E.A. Lee, J.J. Yoon, T.G. Park, Hyaluronic acid modified biodegradable scaffolds for cartilage tissue engineering, Biomaterials 26 (14) (2005) 1925–1933.

[47] M.P. Pavlov, J.F. Mano, N.M. Neves, R.L. Reis, Fibers and 3d mesh scaffolds from biodegradable starch-based blends: production and characterization, Macromol. Biosci. 4 (8) (2004) 776–784.

[48] Y. Zhang, H. Ouyang, C.T. Lim, S. Ramakrishna, Z.M. Huang, Electrospinning of gelatin fibers and gelatin/PCL composite fibrous scaffolds, J. Biomed. Mater. Res. B Appl. Biomater. 72 (1) (2005) 156–165.

[49] J.A. Matthews, G.E. Wnek, D.G. Simpson, G.L. Bowlin, Electrospinning of collagen nanofibers, Biomacromolecules 3 (2) (2002) 232–238.

[50] K.J. Shields, M.J. Beckman, G.L. Bowlin, J.S. Wayne, Mechanical properties and cellular proliferation of electrospun collagen type II, Tissue Eng. 10 (9–10) (2004) 1510–1517.

[51] L. Huang, K. Nagapudi, R.P. Apkarian, E.L. Chaikof, Engineered collagen-PEO nanofibers and fabrics, J. Biomater. Sci. Polym. Ed. 12 (9) (2001) 979–993.

[52] N. Bhattarai, D. Edmondson, O. Veiseh, F.A. Matsen, M. Zhang, Electrospun chitosan-based nanofibers and their cellular compatibility, Biomaterials 26 (31) (2005) 6176–6184.

[53] I.C. Um, D. Fang, B.S. Hsiao, A. Okamoto, B. Chu, Electrospinning and Electro blowing of hyaluronic acid, Biomacromolecules 5 (4) (2004) 1428–1436.

[54] Y. Zhang, H. Ouyang, C.T. Lim, S. Ramakrishna, Z.M. Huang, Electrospinning of gelatin fibers and gelatin/PCL composite fibrous sacffolds, J. Biomed. Mater. Res. Part B Appl. Biomater. 72 (1) (2005) 156–165.

[55] H.J. Jin, S.V. Fridrikh, G.C. Rutledge, D.L. Kaplan, Electrospinning Bombyx more silk with poly (ethylene oxide), Biomacromolecules 3 (6) (2002) 1233–1239.

[56] B.M. Min, G. Lee, S.H. Kim, Y.S. Nam, T.S. Lee, W.H. Park, Electrospinning of silk fibroin nanofibers and its effect on the adhesion and spreading of normal human keratinocytes and fibroblasts in vitro, Biomaterials 25 (7-8) (2004) 1289–1297.

[57] M. Li, M.J. Mondrinos, M.R. Gandhi, F.K. Ko, A.S. weiss, P.I. Lelkes, Electrospun protein fibers as matrices for tissue engineering, Biomaterials 26 (30) (2005) 5999–6008.

[58] D.L. Woeddeman, P. Ye, S. Shenoy, R.S. Parnas, G.E. Wnek, O. Trofimova, Electrospun fibers from wheat protein: investigation of the interplay between molecular structure and the fluid dynamics of the electrospinning process, Biomacromolecules 6 (2) (2005) 707–712.

[59] Y. Li, M. Ceylan, B. Sherstha, H. Wang, Q.R. Lu, R. Asmatulu, et al., Nanofibers support oligodendrocyte precursor cell growth and function as a neuron-free model for myelination study, Biomacromolecules 15 (2014) 319–326.

[60] J.M. Holzwarth, P.X. Ma, Biomimetic nanofibrous scaffolds for bone tissue engineering, Biomaterials 32 (36) (2011) 9622–9629.

[61] S.P. Bruder, A.L. Caplan, Bone Generation through cellular Engineering, in: R.P. Lanza, R. Langer, J. Vacanti (Eds.), Principles of tissue Engineering, Academic Press, New York, USA, 2000, pp. 683–696.

[62] D.W. Hutmacher, Scaffolds in tissue engineering bone and cartilage, Biomaterials 21 (24) (2000). 2529-1543.

[63] Y. Hu, D.W. Grainger, S.R. Winn, H. Jo, Fabrication of poly (alpha-hydroxy acid) foam scaffolds using multiple solvent systems, J. Biomed. Res. Mater. 59 (3) (2002) 563–572.

[64] E. Seyedjafari, M. Soleimani, N. Ghaemi, I. Shabani, Nanohydroxyapatite-coated electrospun poly (L-lactide) nanofibers enhance osteogenic differentiation of stem cells and induce ectopic bone formation, Biomacromolecules 11 (11) (2010) 3118–3125.

[65] M. Shin, H. Yoshimoto, J.P. Vacanti, *In vivo* bone tissue engineering using mesenchymal stem cells on a novel electrospun nanofibrous scaffold, Tissue Eng. 10 (1-20) (2004) 33–41.

[66] H. Yoshimoto, Y.M. Shin, H. Terai, J.P. Vacanti, A biodegradable nanofiber scaffold by electrospinning and its potential for bone tissue engineering, Biomaterials 24 (12) (2003) 2077–2082.

[67] H.R. Ramay, M. Zhang, Biphasic calcium phosphate nanocomposite porous scaffolds for load-bearing bone tissue engineering, Biomaterials 25 (21) (2004) 5171–5180.

[68] H.R. Ramay, M. Zhang, Preparation of porous hydroxyapatite scaffolds by combination of the gel-casting and polymer sponge methods, Biomaterials 24 (19) (2003) 3293–3302.

[69] Y. Zhang, H. Ouyang, C.T. Lim, S. Ramakrishna, Z.M. Huang, Electrospinning of gelatin fibers and gelatin/PCL composite fibrous scaffolds, J. Biomed. Mater. Res. Part B Appl. Biomater. 72 (1) (2005) 156–165.

[70] K. Fujihara, M. Kotaki, S. Ramakrishna, Guided bone regeneration membrane made of polycaprolactone/calcium carbonate composite nanofibers, Biomaterials 26 (19) (2005) 4139–4147.

[71] P. Wutticharoenmongkol, S. Pataharaporn, N. Sanchavanakit, P. Pavasant, P. Supaphol, Novel bone scaffolds of electrospun polycaprolactone fiber filled with nanoparticles, J. Nanosci. Nanotechnol. 6 (2) (2006) 514–522.

[72] P. Wutticharoenmongkol, S. Pataharaporn, N. Sanchavanakit, P. Pavasant, P. Supaphol, Preparation and characterization of novel bone scaffolds based on electrospun polycaprolactone fibers filled with nanoparticles, Macromol. Biosci. 6 (1) (2006) 70–77.

[73] P. Wutticharoenmongkol, P. Pavasant, P. Supaphol, Osteoblastic phenotype expression of MC3T3-E1 cultured on electrospun polycaprolactone fiber mats filled with hydroxyapatite nanoparticles, Biomacromolecules 8 (8) (2007) 2602–2610.

[74] Y.Z. Cai, L.L. Wang, H.X. Cai, Y.Y. Qi, X.H. Zou, H.W. Ouyang, Electrospun nanofibrous matrix improves the regeneration of dense cortical bone, J. Biomed. Mater. Res. Part A 95A (1) (2010) 49–57.

[75] K.W. Woo, V.J. Chen, H.M. Jung, T.I. Kim, H.I. Shin, J.H. Baek, et al., Comparative evaluation of nanofibrous scaffolding for bone regeneration in critical-size calvarial defects, Tissue Eng. Part A 15 (8) (2009) 2155–2162.

[76] H.J. Jin, J. Chen, V. Karageorgious, G.H. Altman, D.L. Kaplan, Human bone marrow stromal cell responses on electrospun silk fibroin mats, Biomaterials 25 (6) (2004) 1039–1047.

[77] K.H. Kim, L. Jeong, H.N. Park, S.Y. Shin, W.H. Park, S.C. Lee, et al., Biological efficacy of silk fibroin nanofiber membranes for guided bone regeneration, J. Biotechnol. 120 (3) (2005) 327–339.

[78] X. Xu, X. Zhuang, X. Chen, X. Wang, L. yang, X. Jing, Preparation of core-sheath composite nanofibers by emulsion electrospinning, Macromol. Rapid Commun. 27 (19) (2006) 1637–1642.

[79] S.J. Seo, H.W. Kim, J.H. Lee, Electrospun nanofibers applications in dentistry, J. Nanomater. 2016 (2016) 7. Article ID 5931946.

[80] T. Thamaraiselvi, S. Rajeswari, Biological evaluation of bioceramic materials—a review, Trends Biomater. Artif. Organs 18 (1) (2004) 9–17.

[81] Y. Zhou, C. Wu, Y. Xiao, Silicate-based bioceramics for periodontal regeneration, J. Mater. Chem. B 2 (25) (2014) 3907–3910.

[82] Y. Ikada, Biodegradable polymers as scaffolds for tissue engineering, Handbook of Biodegradable Polymers: Isolation, Synthesis, Characterization and Applications (2013) 341–362.

[83] R. Nayak, R. Padhye, I.L. Kyratzis, Y.B. Truong, L. Arnold, Recent advances in nanofiber fabrication techniques, Textile Res. J. 82 (2) (2012) 129–147.

[84] N. Salako, B. Joseph, P. Ritwik, J. Salonen, P. John, T.A. Junaid, Comparison of bioactive glass, mineral trioxide aggregate, ferric sulfate, and formocresol as pulpotomy agents in rat molar, Dental Traumatol. 19 (6) (2003) 314–320.

[85] J. Wang, H. Ma, X. Jin, J. Hu, X. Liu, L. Ni, et al., The effect of scaffold architecture on odontogenic differentiation of human dental pulp stem cells, Biomaterials 32 (31) (2011) 7822–7830.

[86] J.J. Kim, W.J. Bae, J.M. Kim, E.J. Lee, H.W. Kim, E.C. Kim, Mineralized polycaprolactone nanofibrous matrix for ontogenesis of human dental pulp cells, J. Biomater. Appl. 28 (7) (2014) 1069–1078.

[87] M.C. Bottino, K. Kamocki, G.H. Yassen, J.A. Platt, M.M. Vail, Y. Ehrlich, et al., Bioactive nanofibrous scaffolds for regenerative endodontics, J. Dental Res. 92 (11) (2013) 963–969.

[88] G.M. Kim, A.S. Asran, G.H. Michler, P. Simon, J.S. Kim, Electrospun PVA/HAp nanocomposite nanofibers: biomimetics of mineralized hard tissues at a lower level of complexity, Bioinspir. Biomimet. 3 (4) (2008) 1–12. Article ID 046003.

[89] W.J. Bae, K.S. Min, J.J. Kim, J.-J. Kim, H.W. Kim, E.C. Kim, Odontogenic responses of human dental pulp cells to collagen/nanobioactive glass nanocomposites, Dental Mater. 28 (12) (2012) 1271–1279.

[90] H.M. Yun, E.S. Lee, M.J. Kim, J.J. Kim, J.H. Lee, H.H. Lee, et al., Magnetic nanocomposite scaffold-induced stimulation of migration and odontogenesis of human dental pulp cells through integrin signaling pathways, PLOS One 10 (9) (2015) e0138614.

[91] S. Shang, F. Yang, X. Cheng, X.F. Walboomers, J.A. Jansen, The effect of electrospun fiber alignment on the behavior of rat periodontal ligament cells, Eur. Cells Mater. 19 (2010) 180–192.

[92] W. Jiang, L. Li, D. Zhang, S. Huang, Z. Jing, Y. Wu., et al., Incorporation of aligned PCL-PEG nanofibers into porous chitosan scaffolds improved the orientation of collagen fibers in regenerated periodontium, Acta Biomater. 25 (2015) 240−252.

[93] B. Inanç, Y.E. Arslan, S. Seker, A.E. Elçin, Y.M. Elçin, Periodontal ligament cellular structures engineered with electrospun poly(DL-lactide-co-glycolide) nanofibrous membrane scaffolds, J. Biomed. Mater. Res. Part A 90 (1) (2009) 186−195.

[94] S.H. Park, T.I. Kim, Y. Ku, C.P. Chung, S.B. Han, J.H. Yu, et al., Effect of hydroxyapatite-coated nanofibrous membrane on the responses of human periodontal ligament fibroblast, J. Ceram. Soc. Japan 116 (1349) (2008) 31−35.

[95] S. Suganya, J. Venugopal, S. Ramakrishna, B.S. Lakshmi, V.R.G. Dev, Aloe Vera/silk fibroin/hydroxyapatite incorporated electrospun nanofibrous scaffold for enhanced osteogenesis, J. Biomater. Tissue Eng. 4 (1) (2014) 9−19.

[96] S.M. Hughes, A. Pham, K.H. Nguyen, R. Asmatulu, Training undergraduate engineering students on biodegradable PCL nanofibers through electrospinning process, Trans. Tech. STEM Educ. 1 (2016) 19−25.

FURTHER READING

K. El- Refaie, J.M. Layman, J.R. Watkins, G.L. Bowlin, J.A. Matthew, D.G. Simpson, et al., Electrospinning of poly (ethylene-co-vinyl alcohol) fibers, Biomaterials 24 (6) (2003) 907−913.

Electrospun nanofibers for photonics and electronics applications

12

12.1 ELECTROSPUN NANOFIBERS FOR PHOTONIC AND LIGHT-EMITTING SOURCES

Polymer fibers with diameter comparable to or less than the wavelength of light have attracted attention as organic optical fibers for manipulating light at the nanoscale. Various techniques for fabricating polymer nanofibers have been developed in recent years. Among these techniques, electrospinning is a versatile, cost-effective, fast, and easy technique for fabricating polymeric nanofibers from a broad range of polymer materials with excellent surface qualities and controlled morphology, which are highly required for low-loss optical wave-guiding. Researchers and scientists have developed electrospinning to realize nanofibers from optically inert polymers that generally display good processing and visco-elastic behavior [1−3]. Electrospun nanofibers and nanotubes have recently been utilized in the fabrication of optical sensing devices, photonic, light sources, waveguide, sensors and light detectors, and optoelectronic applications.

Electrospun light-emitting nanofibers have spurred increased attention on their potential uses in photonic applications, as miniaturized light source optical sensors, waveguides, and detectors [1]. Polymer nanofibers can function as subwavelength optical waveguides, display some outstanding features, such as optical confinement, very minute allowable bending radius, a high fraction of evanescent

Synthesis and Applications of Electrospun Nanofibers. DOI: https://doi.org/10.1016/B978-0-12-813914-1.00012-2

fields, high molecular diffusion, and small footprint [2]. Polymer nanofibers are hospitable to a variety of functional dopants, such as dye molecules, quantum dots, and metallic nanoparticles [2]. Fluorescent electrospun nanofibers can be apprehended by embedding emissive systems, such as quantum dots, dyes, and bio-chromophores in optically inert polymers and by employing light-emitting conjugate polymers. Polymer nanofibers possess a perm-selectivity nature with respect to gas molecules, flexibility of surface functionalities, high surface area, and biocompatibility, thereby presenting high possibilities in applications, such as lasers, optical sensing, light-emitting devices, photonics, biophotonics, and photo-detectors [2].

The special features of an electrospinning process present an ideal solution for obtaining fluorescent fibers with improved properties, such as enhanced photolu-minescence quantum yield and radiative rates, polarized emission, and self-waveguiding of the emitted light. These properties, coupled with high surface area to volume ratio and flexibility, make these nanoengineered materials an ideal nanomaterial for high-performance optical sensing [1]. Electrospinning can pro-duce various functional nanofibers since this method is cost-effective and suitable for flexible morphology tuning and high-yield continuous production. Optically active electrospun nanofibers can be synthesized by employing transpar-ent polymers incorporated with low-molar-mass active molecules or inorganic quantum dots and semiconducting polymers [3]. Fig. 12.1 shows a basic sche-matic view of the electrospinning setup used to produce various nanofibers for photonics and electronics materials and devices. The electrospun nanofibers embedded with luminescent or active compounds exhibited enhanced optical properties, polarization properties, waveguiding, and lasing properties than their bulk counterpart [3].

In the electrospinning process, the fiber morphology and orientation can be controlled by adjusting the process parameters. However, for photonic applica-tions, the requirement for fiber synthesis is very stringent, since the propagation of light in a fiber with nanosize (wavelength size) can be adversely affected by a nonuniform shape and/or composition of fibers [1]. The electrospun nanofibers have diameters from 3 nm to several micrometers. The application of polymeric nanofibers spurred many researchers to demonstrate the feasibility of fabricating hybrid light emitter/polymer nanofiber composites. The organic chromophores and semiconductor nanoparticles can be used as an internal light source. The flex-ibility of an electrospinning process allows changing of the diameters of fibers, and the type of embedded light emitters [4]. Besides nanofibers, the polymer beads that are formed during electrospinning also possess photonic properties sim-ilar to those found in dielectric microspheres [4]. Some researchers studied the electrospun beaded structures of poly(methyl methacrylate) and poly(ethylene oxide), and found that the chromophores were differently distributed over the bead structure, thereby indicating different configurations of the beads. The sig-nificant effect of the fiber and beads confinement on the emission properties of the embedded dyes was determined by studying the fluorescence lifetime. It was

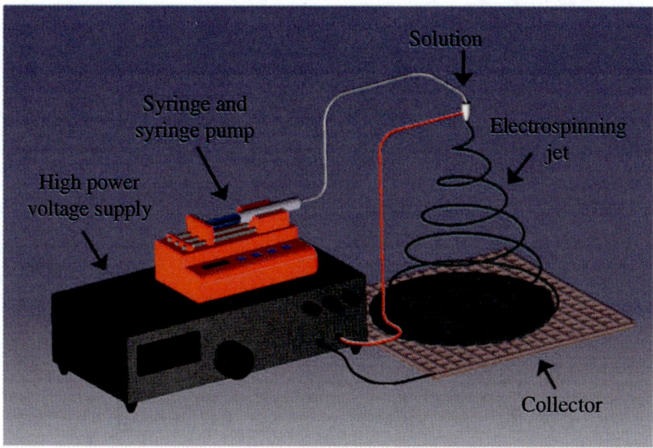

FIGURE 12.1

A basic schematic view of the electrospinning process used to produce various nanofibers for photonics and electronics materials and devices.

found that the fluorescence lifetime of the embedded dyes depends on the materials utilized in the process [3].

12.2 NANOFIBER-BASED LIGHT-EMITTING SYSTEMS

The electrospinning process has been recently applied to light-emitting conjugate polymers for the realization of optoelectric and light-emitting devices. Electrospun fluorescent nanofibers can be recognized by two approaches: the first comprises the use of optically inert polymers, which can be doped with a luminescent system, such as organic quantum dots, organic chromophores, and bio-chromophores, and the second is based on the conjugated polymers, which are intrinsically luminescent prior to the electrospinning process [3]. The rheological properties, solubility, and optical properties of these polymers vary broadly, which needs specific approaches for fabricating electrospun nanofibers with good morphology and fluorescence capabilities [3].

12.2.1 QUANTUM DOT- AND DYE-DOPED ELECTROSPUN NANOFIBERS

One of the effective approaches to obtain luminescent electrospun nanofibers uses transparent thermoplastic polymers or polymeric photoresists as matrix material, and inorganic quantum dots or nanowires as fluorescent constitutes [4–6]. In this approach, optically inert polymers can be doped by small fluorescent molecules or by inorganic fluorescent quantum dots or nanowires. The photoluminescence properties can easily be tailored by using inorganic nanoparticles, which feature emission in a broad range, generally from visible to infrared. These nanoparticles belong to group II—VI elements, such as CdS, CdSe, and CdTe, and have diameters in the nanoscale range and electronic bandgaps and photoluminescence wavelengths tunable depending upon their size [7]. In these inorganic nanoparticles, quantum confinement can be realized for particle size approaching Bohr radius, which is approximately in the nanosize range (3 nm for CdS) [8]. A blueshift of the emission can be achieved by reducing the particle size below the Bohr radius. As mentioned earlier, the fluorescent composite nanofibers can be fabricated by two approaches: (1) incorporating quantum dots and nanowires that can be synthesized ex situ to produce a light-sensitive electrospinning solution, and (2) incorporating suitable molecular precursors in the polymeric matrix, which permits the in situ synthesis of quantum dots in electrospun fibers by means of precursor decomposition via thermal treatment, gas reaction, or optical and electron beam exposure [9—11].

In the first approach, the use of the colloidal quantum dots may alter the rheological properties of the electrospinning solution, thereby leading to clogging of the capillary tube and subsequent failure of the electrospinning process. This problem can be addressed by in situ synthesis, since the properties of polymeric solution are not much affected by the presence of molecular precursors [9,12]. Apart from quantum dots, organic luminescent chromophores can also be embedded in the polymer matrices for fabricating light-emitting electron nanofibers. Fluorescent organic molecules have emissions from ultraviolet to near-infrared region and can be incorporated in polymers, such as poly(methyl methacrylate) (PMMA), polystyrene (PS), poly(vinyl pyrrolidone), (PVP) and poly(ethylene oxide) (PEO) [13,14]. By employing this approach, a wide variety of luminescent fibers can be produced, which can be used in optical sensing and photonic applications, including light-emitting diodes and other sources [15]. Some researchers used laser dyes in PMMA solution with chloroform solvent to prepared nanofibrous mats. The test results showed a bright and uniform emission of light from the surface of fibers.

12.2.2 NANOFIBERS EMBEDDED WITH BIO-CHROMOPHORES

Among the different systems that can be embedded in electrospun nanofibers for optical applications, bioluminescent chromophores receive extensive attention, due to their various industrial applications. The green fluorescent protein (GFP) from the jellyfish *Aequorea victoria* is among the most studied luminescent biomolecules [16,17]. Most of the fluorescent proteins contain fluorophores distinct from the amino acid sequence, whereas the chromophore of the GFP is generally generated by means of a reaction employing three amino acid residues. Additionally, GFP shows a relationship between its fluorescence features and conformation changes, a characteristic which opens up the chances of using GFP in a wide variety of sensing and actuating applications [18]. Tomczak et al. [19]

fabricated luminescent PMMA and poly(ethylene oxide) (PEO) fibers with beaded structures having different spatial distributions of the embedded chromophores. The chromophores in PMMA were distributed evenly over the entire volume of beaded structures. However, the chromophores in PEO were located at the edges due to the fact that the crystallinity was disturbed during the electrospinning process. Moreover, their polarization capability exhibited that the chromophores in PEO were uniformly oriented.

The charges accumulated during electrospinning may alter the functionality of the bio-macromolecules and of biological species, such as proteins, DNA, enzymes, viruses, and bacteria [20]. In PEO electrospun nanofibers, the fluorescence may not be present due to the denaturation of GPF in PEO matrix. However, in coelectrospinning the charges are present on the outer surface and inner solution and systems are not charged, thereby allowing to preserve the properties of the species embedded in the core. Therefore, in coelectrospinning the functionality of proteins and its fluorescence are better protected.

12.2.3 NANOFIBERS FABRICATED BY CONJUGATED POLYMERS

Conjugated polymers are a class of organic macromolecules that are characterized by a backbone chain of alternating double- and single-bonds. Their overlapping p-orbitals create a system of delocalized π-electrons, which can result in outstanding optical and electronic properties. They are semiconductors or conductors and interact with light. Almost all conjugate polymers possess conjugated π electrons, that is, electrons that are delocalized rather than being part of one valence bond. The excitation energies of conjugated π electrons are generally in the visible spectra and, therefore, they are optically active. A thin film of poly(phenylene vinylene) (PPhV) covered with indium-tin oxide alloy (ITO) and calcium electrodes is placed on a glass substrate and then connected to an external source of current. As soon as the voltage surpasses 2 V, the PPhV film begins to emit light, with an intensity proportional to the current. The effect is known as electroluminescence. A setup with such a performance is called a light-emitting diode, abbreviated as LED. Conjugate polymers can be blended with optically transparent matrices, such as PMMA and polystyrene, which can be easily electrospun. When PMMA and poly(9,9-dioctylfluoreny-2,7-diyl) (PFO) are blended and electrospun, a blue-emitting conjugate polymer forms [21]. The blue-emitting electrospun nanofibers with diameters as low as 180 nm have been fabricated with a solution of a conjugate polymer dissolved in an appropriate solvent and the addition of some organic salt [22]. The addition of organic salt does not alter the emission properties, but it helps in increasing the electrospinnability of the polymer solution. Fig. 12.2 shows the fluorescence microscopy images of poly(phenylene vinylene) (PPV) nanofibers embedded with CdS quantum dots [23].

Moran-Mirabal et al. [24] prepared electrospun light-emitting nanofibers of ruthenium (11) tris(bipyridine) ($[Ru(bpy)_3]^{2+}$ $(PF_6^-)_2$) and a PEO mixture. These functional nanofibers were then laid down on Au interdigitated electrodes having

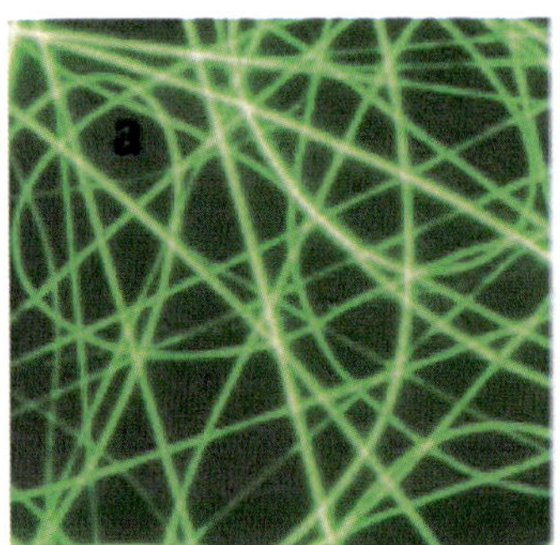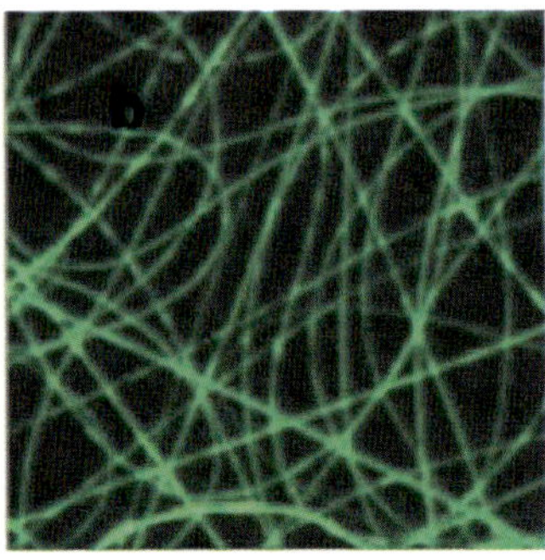

FIGURE 12.2

Fluorescence microscopy images of poly(phenylene vinylene) (PPV) nanofibers embedded with CdS quantum dots [23].

a gap of approximately 500 nm, and emitted confined light of around 240×325 nm^2 between Au interdigitated electrodes was detectable with a CCD camera at a voltage of 3.2 V. Yang et al. [25] used the coaxial electrospinning process to fabricate core-shell fiber OLEDs (organic light-emitting diodes) from an ionic transition-meta complex (iTMC). First, a galinstan liquid metal core and a ([Ru (bpy)$_3$]$^{2+}$(PF$_6^-$)$_2$)/PEO blend shell were coaxially electrospun to fabricate the cathode and an electroluminescent layer, respectively. Then, an ITO anode layer was deposited by means of evaporation and then the electroluminescence of the device was turned on at 4.2 V.

Vohra et al. [26] fabricated electrospun F8BT/PEO blend nanofibers and applied them as an active layer in OLDEs. The authors used an annealing process, which transformed an active layer into a ribbon-like structure and separated PEO from nanofibers. They found that the luminance was 2300 cd m^{-2} at 6 V. This study is significantly important in nanofiber photonics, since this work demonstrated the possibility of using nanofibers as active layers in OLDE structures for the first time. Wang et al. [27] prepared electrospun nanofibers from the ternary blends of poly(9,9-dioctylfluorenyl-2,7-diyl) (PFO)/poly(2,3-dibutoxy-1,4-phenylene vinylene) (DB-PPV)/PMMA using a single-capillary spinneret and found different PFO and DB-PPV phase-separated structures in the electrospun nanofibers by two different solvents: ellipsoidal DB-PPV (10−40 nm) and fiber-like PFO (20−40 nm) in the PMMA using chloroform, and fiber-like DB-PPV (10−20 nm) and fiber-like PFO (20−30 nm) using chlorobenzene. These different PFO and DB-PPV structures displayed various energy transfer/emission colors in electrospun nanofibers. Additionally, aligned luminescence PFO/DB-PPV/PMMA blend nanofibers prepared from chlorobenzene solvent showed higher polarized emission when compared to the nonwoven, and the emission colors changed from blue to greenish-blue to green as the DB-PPV composition increased. Some researchers have electrospun ruthenium (II) tris(bipyridine) with polyethylene oxide. The electrospun fibers showed light emission at lower voltage, which can be seen on a CCD camera.

12.2.4 LUMINESCENT NANOFIBER ARRAYS

One of the most common issues with electrospun nanofiber applications is the precise control on the fiber morphology and positioning so that the connection of the active fiber element to excitation sources and detectors becomes easy. This can be done by electrospun nanofiber arrays. These arrays can be manufactured by employing a patterned collector composed of two metallic strips. The instabilities that an electrified jet experiences during electrospinning is the main drawback for not obtaining the ordered array of nanofibers. Near-field electrospinning is a new approach for precise positioning of individual nanofibers and for realizing an ordered array of nanofibers [28−30]. In conventional electrospinning, the applied voltage is generally increased to cause greater stretching of the polymer jet and thereby reduced the fiber diameter into the nano range. The key strategy for producing nanosized fibers via near-field electrospinning is to reduce the size of the polymer jet from the cone by reducing the applied electrostatic potential below threshold value. The near-field electrospinning starts with the application of a subcritical electric voltage to distort polymer meniscus without electrospinning, and then mechanical drawing is applied using a tungsten probe with a 1 micrometer diameter to poke inside the meniscus. The probe is then quickly pulled away from the meniscus in order to activate the electrospinning process under subcritical electrostatic potential. This process exploits the stable region of the electrospun jet and the collector screen is placed a few millimeters away from the needle, before the onset of instability. This allows precise control of fiber morphology and uniform deposition of individual nanofibers. The near-field electrospinning can be used to produce arrays of ordered nanofibers by moving the substrate perpendicular to the jet axis. This process has been applied to many polymeric solutions, such as polyethylene oxide, polyvinylidene fluoride, polycaprolactone, SnO_2, TiO_2, and light-emitting conjugate polymers [31−35]. Several studies have been conducted on aligned arrays of light-emitting nanofibers based on conjugate polymers by an electrospinning process, and exploited the polarized emission for the photo-excitation of chromophores in prototype microfluidic channels. Literature survey revealed that there are numerous reports on the synthesis of metallic oxide fibers via an electrospinning process, followed by heat treatment in order to obtain pure metallic nanofibers. These nanofibers were subjected to photoluminescence testing, showed different emission at low voltages. These nanofibers can be used as light emitting devices in nanoscale optoelectronic applications.

12.3 ELECTRICALLY CONDUCTIVE ELECTROSPUN NANOFIBERS

Since the discovery of polyacetylene (CH)x, which is commonly referred to as the prototype conducting polymer, the development of the conducting polymers field has been stimulated at an accelerated rate and a wide variety of other conducting polymers and derivatives have been discovered. The study of conducting polymers has been referred to as synthetic metals and it has attracted both industries and academia. Conducting polymers, such as poly-(p-phenylene), poly (phenylenevinylene), polypyrrole, polythiophene, poly(heteroaromatic vinylenes), and polyaniline have been a focus of attention by many researchers worldwide. Of the conducting polymers, polyaniline (PANi) is unique due to its ability to fabricate tunable conductive forms in bulk with a very low cost [36]. The highly conducting doped form of PANi can be reached by two processes: proton acid doping and oxidative doping [36]. PANi has a drawback of yielding carcinogenic products upon degradation due to the presence of benzidine moieties, which limits its applications [36]. However, polypyrrole and polythiophene, as well as poly(p-phenylene vinylene), are environmentally friendly systems but they are insoluble and infusible. In order to make them soluble, various substituted derivatives of these polymers have been developed that carry alkyl, alkoxy, and other substituents along their backbones.

Researchers at the Bayer AG research laboratories developed a new conducting polymer. This is a polythiophene derivative, poly(3,4-ethylenedioxythiophene), commonly abbreviated as PEDT or PEDOT, which was originally developed to provide a soluble conducting polymer that lacked the presence of undesired α,β- and β,β-coupling within the backbone [36]. PEDT possesses very high electrical conductivity of around $300\ \mathrm{S\ cm}^{-1}$ and it is transparent in its oxidized state [36]. The solubility issue was overcome by using a water-soluble polyelectrolyte, poly(styrene-sulfonic acid) (PSS), as a dopant to provide charge during polymerization of PEDT/PSS. The intrinsically conducting polymers are organic polymers that possess electrical, electronic, magnetic, and optical properties of a metal and maintain their mechanical properties. Conjugate organic polymers that are either electrical insulators or semiconductors can have their conductivities increased extensively by doping with conducting polymers. Intrinsic conducting polymers have a wide range of applications, such as lightweight batteries, electrochromic displays, electrodes for integrated circuits, etc.

An outstanding feature of an electrospinning process is the exceptionally expeditious formation of nanofibers on a millisecond scale [37]. The other important features of electrospinning are enormous fiber elongation rate on the order of $1000\ \mathrm{s}^{-1}$ and cross-sectional area reduction on the order of 10^5 to 10^6, which have been shown to affect the orientation and structure of a fibrous surface [37]. Optimal properties of nanofibers can be obtained by controlling process parameters. Polymers are typically used in the electrical and electronic industry as an insulator due to their high resistivities. Some typical properties of polymers are flexibility, elasticity, stability, moldability, and ease of handling.

Electrically conductive polymers exhibit the chemical and physical properties of organic polymers and the electrical characteristics of metals. The fabrication of electrospun nanofibers employing conductive polymers has been demonstrated in the design and construction of nanoelectrical and nanoelectronic devices in the electronic industry [38,39]. The focus in the electrospinning of conductive polymers is PANi and blending with other polymers. MacDiarmid et al. [39] fabricated highly conducting sulfuric acid-doped polyaniline fibers with an average diameter of 139 nm and determined the conductivity of a single fiber as 0.1 S cm^{-1}. Electrospinning of conductive nanofibers employing poly(3,4-ethylenedioxythiophene), PEDOT, a commercial polythiophene derivative (Baytron Ptype from Bayer), has also been reported in literature reviews [39]. In this study, poly (3,4-ethylenedioxythiophene)/poly(styrene sulfonate) (PEDOT/PSS) dispersions were employed and polyacrylonitrile was used as a carrier. Norris et al. [38] prepared conducting electrospun nanofibers with PANi doped with camphorsulfonic acid and blended with polyethylene oxide. They found that the conductivity of fibers was slightly low compared to a cast film, due to the high porosity of the fibers. Chronakis et al. [37] prepared electrospun nanofibers using polypyrrole (PPy) polymer and polypyrrole (PPy) doped with polyethylene oxide as a carrier to improve electrospinnability. The electrical conductivity of pure polypyrrole (PPy) was about three orders of magnitude higher than that of ppy/polyethylene oxide. The electrical conductivity of pure $[(ppy3)^+(DEHS)^-]_x$ nanofiber was around 2.7×10^{-2} S cm^{-1}, which is about four orders of magnitude higher than that of PPy/polyethylene oxide [37].

Polypyrrole (PPy) is one of the most studied intrinsic conductive polymers. The alternate single and double bonds in the backbone in the molecular chain represent the main features of the intrinsic conductive bond [40]. Although PPy is a conductive polymer, its conductivity can further be increased by dopant. Dopant generally leads to the formation of counterions by an oxidation and reduction process [40]. Ju et al. [41] prepared electrospun fibers from a solution of PPy/sulfonated-poly(styrene-ethylene-butylenes-styrene) in chloroform or dimethylformamide (DMF). The electrospun fibers had an average diameter of approximately 300 nm and electrical conductivity of 0.52 S cm^{-1}. Cetiner et al. [42] prepared electrospun nanofibers with the solution PPy/poly(acrylonitrile-co-vinyl acetate) in DMF and obtained fibers with diameters between 200 and 400 nm and an electrical conductivity of 10^{-7} S cm^{-1}. Yanilmaz et al. [43] prepared electrospun fibers from the solution of PPy/polyurethane in DMF with an average diameter of around 2 microns and electrical conductivity 1.4×10^{-6} S cm^{-1}. Tavakkola et al. [40] prepared polypyrrole/poly(vinyl pyrrolidone) nanofibers via the electrospinning of pyrrole solution. The electrospun pyrrole—PVP nanofibers were collected in a solution of $FeCl_3 \cdot 6H_2O$ (2% w/v) and dopant in ethanol and oxidized in situ (polymerized to PPy in a one-step method). In another method, the electrospun pyrrole-PVP-dopant nanofibers were collected on an aluminum foil and then detached from it. To polymerize pyrrole, the nanofibrous mats were immersed in $FeCl_3 \cdot 6H_2O$ (2%w/v) solution. P-toluene sulfonic acid (PTSA) and anthraquinone-2-sulfonic acid sodium salt (AQSA) were employed as dopant. Their study showed the highest electrical conductivity of 5.22×10^{-1} S cm^{-1} for the AQSA doped PPy-PVP nanofibers with an average diameter of around 830 nm.

Alarifi et al. carbonized electrospun PAN nanofibers and utilized them as highly sensitive sensors in a structural health monitoring device (SHM) of composite structures for aircraft and wind turbines [44—49]. In this study, electrospun PAN nanofibers were stabilized at 270 °C for 60 min and then carbonized at 650—950 °C for an additional 60 min before SHM applications. The authors also studied the synthesis, analysis, and simulation of carbonized PAN nanofibers for improved thermal and electrical conductivities of the carbon fiber composites [44—49]. It was stated that carbonizing PAN nanofibers is a great way of enhancing the thermal and electrical properties of electrospun nanofibers. Electrically conducting nanofibers have a number of different benefits for various industries as outlined below [44—49]:

- Protect composite surfaces against lightning strikes;
- Provide electromagnetic shielding of electronic circuits and composite aircraft;
- Use as antistatic coating materials to prevent electrical discharge exposure on photographic emulsions;
- Utilize as hole-injecting electrodes for electronic, full-color video and optical devices;
- Display electroluminescent for mobile telephones and other portable devices;
- Employ to produce field-effect transistors;
- Absorb microwaves for stealth communications and other information technologies in both rural and remote locations;
- Enhance the sensing capabilities of structural health monitoring devices on composite structures, bridges, trains, tunnels, ships, and other buildings and infrastructures.

12.4 THERMALLY CONDUCTIVE ELECTROSPUN NANOFIBERS

Polymer nanofibers with high thermal conductivities and good thermal stabilities have a long range of applications in thermal management, heat exchangers, and energy storage. Generally, polymers have low thermal conductivities, which limits their applications in many areas. Bulk polymers have thermal conductivities as low as 0.1 W/m/K. Scientists have made all kinds of efforts to enhance the thermal conductivities of polymers, but with limited success. The low thermal conductivities of polymers are due to their structures, which have numerous molecular chains coiled up in a disorderly manner, thereby limiting heat transfer in a steady way. The other reason for low thermal conductivities in polymers is that the amorphous structure of molecular chains reduces the mean free path of heat-conducting phonons. One way of increasing the thermal conductivity of polymers is to draw polymer fibers that will enhance the crystallinity and chain alignment and thereby facilitate the movement of heat-conducting phonons. Electrospinning can prepare well-aligned polymer fiber arrays with ordered molecular chains in each fiber, thus providing necessary conditions of phonon transfer in a steady way. Electrospinning provides molecular orientation and crystallinity, a condition for enhancing thermal conductivity.

The commercially available polymer microfibers have a much higher thermal conductivity of approximately 20 W/m/K [50]. Surprisingly, recent molecular dynamic simulations have predicted thermal conductivity as high as 350 W/m/K for polyethylene, which has been supported by experimental observations of mechanically drawn polyethylene nanofibers, exhibiting a thermal conductivity of approximately 100 W/m/K. Wang et al. [51] determined the thermal conductivities of several commercially available high-modulus polymer fibers. Their results indicated that liquid crystalline p-phenylene benzobisoxazole (PBO) fibers have the highest thermal conductivity of approximately 20 W/m/K—higher than commercially available polyethylene. Singh et al. [52] demonstrated that pure polythiophene nanofibers can have a thermal conductivity up to ~ 4.4 W/m/K (more than 20 times higher than the bulk polymer value) while remaining amorphous. This study clearly indicates that the thermal conductivity of a polymeric material depends on its molecular orientation along the fiber axis.

Zhong et al. [53] showed that the thermal conductivity of single Nylon-11 electrospun fibers could be as high as 1.6 W/m/K, nearly one order of magnitude higher than the typical Nylon-11; the bulk value is around 0.2 W/m/K. However, all the electrospun nanofibers in the above-mentioned results were fabricated at a relative low spinning voltage (6−7 kV). Canetta et al. [54] reported the thermal conductivity of individual polystyrene nanowire samples between 6.6 and 14.4 W/m/K. These nanowires were electrospun at 7−10 kV. This increase in thermal conductivity can be attributed to molecular chain orientation and crystallinity effect during the electrospinning. Ma et al. [50] determined a clear trend of enhanced thermal conductivity as the spinning voltage increases. They found the thermal conductivity of electrospun polyethylene nanofibers as 0.8 W/m/K at 9 kV, which is about two times more than the bulk polyethylene. They measured the highest thermal conductivity of polyethylene electrospun nanofibers as 9.3 W/m/K at a voltage of 45 kV.

During the electrospinning process, the polymer chains are subjected to extremely high elongation due to the electrostatic field, which gives rise to molecular orientation and better crystallinity. Olubayode et al. [55] determined that there exists an optimum voltage to obtain the highest degree of crystallinity in electrospun poly(L-lactic acid) (PLLA). At electrospinning voltages higher or lower than the optimum voltage, the degree of crystallinity would certainly drop to a lower level. This is due to the fact that a higher voltage exerts higher stretching forces and thereby facilitates crystalline structure formations.

12.5 ELECTROSPUN MAGNETIC NANOFIBERS

The force of attraction or repulsion caused by the movement of polarized particles is known as magnetism. Magnetism is characterized by combined electromagnetic force. It refers to physical phenomena arising from the force caused by magnetic objects that produce fields to repel or attract other objects. The magnetic study of different materials is a major topic due to its wide industrial importance. With the advancement in technology and engineering, the need for specific magnetic materials will rise. All materials can be categorized as either paramagnetic, diamagnetic, ferromagnetic, antiferromagnetic, or ferromagnetic. Before the mid-1930s, the only permanent magnets were special steels. During the 1930s, some improvements in the magnetic strengths of these materials were made. The major developments in this field came in the 1940s and 1950s with the introduction of aluminum-nickel-cobalt alloys in the 1960s, the development of rare earth/cobalt alloys provided some advancement in this field. The real technological advancement came in the 1980s with the development of neodymium-based magnets. Today, permanent magnets have a magnetic strength 100 times greater than those available in the 1930s [56−58].

There has been a recent interest in the fabrication of magnetic nanofibers. The ability to produce superlattices of nanomaterials with uniform material properties will have important consequences on technology and basic comprehension of magnetism. Recently, magnetic nanofibers, nanotubes, and nanowires have been receiving extensive attention due to their remarkable properties and compelling applications in areas, such as photonics, high-density magnetic recording, magnetic filters, magnetic sensors, and information technology [56]. Compared to other fabrication processes, electrospinning is a simple and effective process for producing continuous nanofibers. Electrospinning also has the ability to produce nanofibers with uniform and desired material properties.

Shao et al. [56] employed an electrospinning process to fabricate fine iron fibers. They prepared a solution for electrospinning by using polyvinyl alcohol (PVA) and ferric nitrate (Fe $(NO_3)3 \cdot 9H_2O$), and after electrospinning, burnt out the organic components from the composite and deoxidized the fibers. Their results showed that the PVA/Fe $(NO_3)_3$ composite nanofibers have a uniform surface and the average diameter of fibers was about 350 nm. However, after heat treatment, a continuous structure with a relative rough surface with average diameter of approximately 180 nm was observed. Copper ferrite ($CuFe_2O_4$) has been widely used in the electronic industry owing to its magnetic and semiconducting properties. Pohan et al. [57] prepared copper ferrite nanofibers using the electrospinning process, which contained poly(vinyl pyrrolidone) (PVP) and Cu and Fe nitrite as alternative metals. The as-spun $CuFe_2O_4$/PVP was calcinated to obtain pure metallic nanofibers. Their results showed that the polycrystalline copper ferrite nanofibers had diameters between 60 and 600 nm. The produced nanofibers consisted of packed particles or crystallites of around 60 nm, having specific ferromagnetic properties of 17.73, 20.52, and 23.98 emu g^{-1} at 10 kOe for samples calcinated at 500, 600, and 700°C, respectively.

Thus far, electrospinning the ferrite nanofibers of $CoFe_2O_4$, $NiFe_2O_4$, $MnFe_2O_4$, and $BiY_2Fe_5O_{12}$ has also been reported [58,59]. The electrospinning process has been exploited to fabricate many metallic nanofibers after heat treatment to decompose carrier polymer. Electrospinning has also been exploited to synthesize Co, Cu, Fe, and Ni nanofibers by calcination of the metal precursor/polymer nanofibers in a hydrogen atmosphere [58]. However, this approach has some drawbacks in the case of nickel, since nickel nanofibers will be polluted by hydrogen, which affects the magnetic properties of nanofibers [58]. In order to overcome these issues, Barakat et al. [58] synthesized pure nickel fibers by using a sol-gel containing nickel acetate tetrahydrate and poly(vinyl alcohol) followed by argon atmosphere calcination. Their results showed that the saturation magnetizations at 5°K and 300K were 29.22 and 26.75 emu g^{-1}, respectively. Chen et al. [59] synthesized α-Fe$_2$O$_3$ nanofibers and nanotubes via an electrospinning process using PVP as a carrier polymer. The polymer was then decomposed through heat treatment in order to obtain pure α-Fe$_2$O$_3$ nanofibers and nanotubes. Their study showed ferromagnetic-like behavior for both α-Fe$_2$O$_3$ nanofibers and nanotubes and this behavior was found to be more pronounced in the case of nanotubes, due to a possible interaction among spins on the surface. The saturation magnetization was found to be 0.6 and 1.5 emu g^{-1} for nanofibers and nanotubes, respectively.

Flexible and superparamagnetic electrospun nanofibers were prepared using elastomeric polyurethane containing ferrite nanoparticles (14 nm) of MnZnFe−Ni [60−62]. The flexible mats were characterized in terms of fiber morphology and magnetic properties. Field emission scanning electron microscopy (FESEM) indicated that the diameter of these nanocomposite fibers was 300−500 nm. The induced specific magnetic saturation and the relative permeability were found to increase linearly with increasing weight percent loading of the magnetic nanoparticles in the polymeric structure. A specific magnetic saturation of 1.7−6.3 emu g^{-1} at ambient conditions indicated superparamagnetic behavior for these electrospun nanocomposite samples. Fig. 12.3 shows the specific magnetic saturation vs. magnetic field of electrospun nanocomposite fibers [60,61].

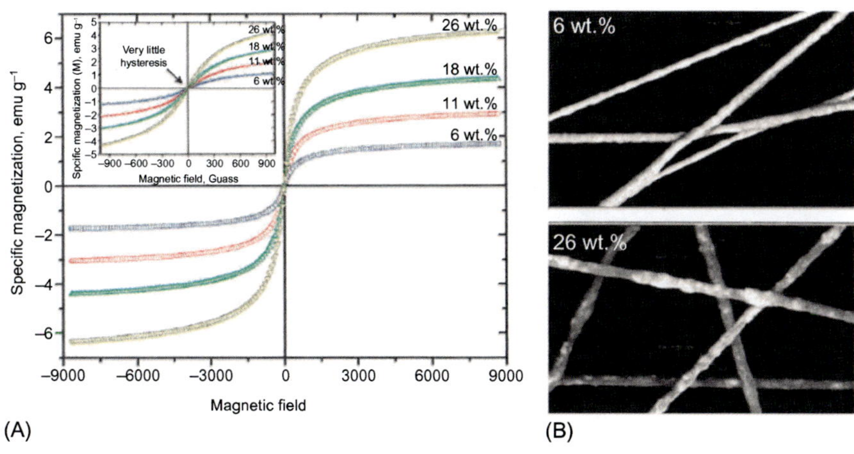

FIGURE 12.3

Specific magnetic saturation values of electrospun magnetic nanocomposite fibers at different wt.% loadings (A) and SEM images of the resulting nanofibers (B).

Overall, electrospun nanofibers and nanotubes have recently been utilized in the fabrication of optical sensing devices, photonics, light sources, waveguides, sensors and light detectors, and optoelectronic applications due to their excellent properties. Polymer nanofibers exhibit some outstanding characteristics, such as optical confinement, very minute allowable bending radius, a high fraction of evanescent fields, high molecular diffusion, and a small footprint. Fluorescent electrospun nanofibers can be realized by embedding emissive systems such as quantum dots, dyes, and bio-chromophores in optically inert polymers and by employing light-emitting conjugate polymers. The outstanding properties of electrospinning present an ideal solution for obtaining fluorescent fibers with better properties, such as enhanced photoluminescence quantum yield and radiative rates, polarized emission, and self-waveguiding of the emitted light. These properties coupled with the high surface area to volume ratio and flexibility make these nanomaterials ideal for high-performance optical sensing.

12.6 CONCLUSIONS

Electrospun nanofibers and nanotubes have recently been utilized in the fabrication of optical sensing devices, photonic, light sources, waveguide, sensors and light detectors and optoelectronic applications due to their excellent properties. Polymer nanofibers exhibit some outstanding characteristics, such as optical confinement, very minute allowable bending radius, a high fraction of evanescent fields, high molecular diffusion and a small footprint. Fluorescent electrospun nanofibers can be realized by embedding emissive systems such as quantum dots, dyes and bio-chromophores in optically inert polymers and by employing light-emitting conjugate polymers. The outstanding properties of electrospinning present an ideal solution for obtaining fluorescent fibers with better properties such as enhanced photoluminescence quantum yield and radiative rates, polarized emission and self-waveguiding of the emitted light. These properties coupled with high surface area to volume ratio and flexibility make these nanomaterials ideal for high performance optical sensing.

After the discovery of polyacetylene $(CH)x$, the interest in conductive polymers has been spurred due to the widespread applications of conductive polymers in many industries. Conducting polymers, such as poly(p-phenylene), poly(phenylenevinylene), polypyrrole, polythiophene, poly(heteroaromatic vinylenes), and polyaniline have been a focus of attention by many researchers. The conductive polymers have been used in applications, such as lightweight batteries, electrochromic displays, electrodes for integrated circuits, etc. Generally, polymers have low thermal conductivities, which limits their applications in many areas. The low thermal conductivities of polymers are due to their structure, which has numerous molecular chains coiled up in a disorderly manner, thereby limiting the mean free path of heat-conducting phonons. Electrospinning can prepare well-aligned polymer fiber arrays with ordered molecular chains in each fiber, thus providing necessary conditions of phonon transfer in a steady way. Electrospinning provides molecular orientation and crystallinity, a condition for enhancing thermal conductivity.

Recently, magnetic nanofibers, nanotubes, and nanowires have been under extensive focus by many researchers because of their remarkable properties and compelling applications in areas, such as photonics, high-density magnetic recording, magnetic filters, magnetic sensors, and information technology. Metallic and metal oxide nanofibers and nanotubes have been synthesized now, by employing an electrospinning process and utilizing a carrier polymer, which can be decomposed by subsequent heat treatment. To date, synthesizing the ferrite nanofibers of $MnFe_2O_4$, $NiFe_2O_4$, $CoFe_2O_4$, and $BiY_2Fe_5O_{12}$ have also been stated in the literature. This chapter has highlighted the synthesis of pure Ni, $CuFe_2O_4$, and α-Fe_2O_4 magnetic nanofibers for various industrial applications.

REFERENCES

[1] A. Camposeo, M. Moffa, L. Persano, Electrospun fluorescent nanofibers and their application in optical sensing, in: A. Macagnano, E. Zampetti, E. Kny (Eds.), Electrospinning for High Performance Sensors. NanoScience and Technology, Springer, Cham, 2015.

[2] P. Wang, Y. Wang, L. Tong, Functionalized polymer nanofibers: a versatile platform for manipulating light at the nanoscale, Light Sci. Appl. 2 (2013). Available from: https://doi.org/10.1038/lsa.2013.58.

[3] A. Camposeo, L. Persano, D. Pisignano, Light-emitting electrospun nanofibers for nanophotonics and optoelectronics, Macromol. Mater. Eng. 298 (5) (2013) 487−503.

[4] H. Liu, J.B. Edel, L.M. Bellan, H.G. Craighead, Electrospun polymer nanofibers as subwavelength optical waveguides incorporating quantum dots, Small 2 (2006) 495−499.

[5] X. Lu, C. Wang, Y. Wei, One-dimensional composite nanomaterials: synthesis by electrospinning and their applications, Small 5 (2009) 2349−2370.

[6] S. Wei, J. Sampathi, Z. Guo, N. Anumandla, D. Rutman, A. Kucknoor, et al., Nanoporous poly (methyl methacrylate)-quantum dots nanocomposite fibers toward biomedical applications, Polymer 52 (2011) 5817−5829.

[7] Y. Shirasaki, G.J. Supran, M.G. Bawendi, V. Bulovic, Emergence of colloidal quantum-dot light-emitting technologies, Nat. Photonics 7 (2013) 13−23.

[8] T. Vossmeyer, L. Katsikas, M. Giersig, I.G. Popovic, K. Diesner, A. Chemseddine, et al., CdS nanoclusters: synthesis, characterization, size dependent oscillator strength, temperature shift of the excitonic transition energy, and reversible absorbance shift, J. Phys. Chem 98 (1994) 7665−7673.

[9] F. Di Benedetto, A. Camposeo, L. Persano, A.M. Laera, E. Piscopiello, R. Cingolani, et al., Light-emitting nanocomposite CdS-polymer electrospun fibers via in situ nanoparticle generation, Nanoscale 3 (2011) 4234−4239.

[10] X. Lu, Y. Zhao, C. Wang, Fabrication of PbS nanoparticles in polymer-fiber matrices by electrospinning, Adv. Mater 17 (2005) 2485−2488.

[11] L. Persano, A. Camposeo, F. Di Benedetto, R. Stabile, A.M. Laera, E. Piscopiello, et al., CdS-polymer nanocomposites and light-emitting fibers by in situ electron-beam synthesis and lithography, Adv. Mater 24 (2012) 5320–5326.

[12] L. Persano, S. Molle, S. Girardo, A.A.R. Neves, A. Camposeo, R. Stabile, et al., Soft nanopatterning on light-emitting inorganic–organic composites, Adv. Funct. Mater 18 (2008) 2692–2698.

[13] A. Camposeo, F. Di Benedetto, R. Stabile, R. Cingolani, D. Pisignano, Electrospun dyedoped polymer nanofibers emitting in the near infrared, Appl. Phys. Lett 90 (4) (2007) 143115.

[14] Y. Ner, J.G. Grote, J.A. Stuart, G.A. Sotzing, Enhanced fluorescence in electrospun dyedoped DNA nanofibers, Soft Matter 4 (2008) 1448–1453.

[15] S. Pagliara, A. Camposeo, S. Polini, R. Cingolani, D. Pisignano, Electrospun light-emitting nanofibers as excitation source in microfluidic devices, Lab Chip 9 (2009) 2851–2856.

[16] J.S. Sparks, R.C. Schelly, W.L. Smith, M.P. Davis, D. Tchernov, V.A. Pieribone, et al., The covert world of fish biofluorescence: a phylogenetically widespread and phenotypically variable phenomenon, PLoS One 9 (1) (2014) e83259.

[17] R.Y. Tsien, The green fluorescent protein, Annu. Rev. Biochem 67 (1998) 509–544.

[18] R. Bizzarri, M. Serresi, S. Luin, F. Beltram, Green fluorescent protein based pH indicators for in vivo use: a review, Anal. Bioanal. Chem 393 (2009) 1107–1122.

[19] N. Tomczak, N.F. Van Hulst, G.J. Vansco, Beaded electrospun fibers for photonic applications, Macromolecules 38 (2005) 7863–7866.

[20] A.L. Yarin, E. Zussman, J.H. Wendorff, A. Greiner, Material encapsulation and transport in core–shell micro/nanofibers, polymer and carbon nanotubes and micro/nanochannels, J. Mater. Chem 17 (2012) 2585–2599.

[21] C.-C. Kuo, C.-H. Lin, W.-C. Chen, Morphology and photophysical properties of light-emitting electrospun nanofibers prepared from poly (fluorene) derivative/PMMA blends, Macromolecules 40 (2007) 6959–6966.

[22] V. Fasano, A. Polini, G. Morello, M. Moffa, A. Camposeo, D. Pisignano, Bright light emission and waveguiding in conjugated polymer nanofibers electrospun from organic salt added solutions, Macromolecules 46 (2013) 5935–5942.

[23] S. Bhattacharyya, A. Patra, Interactions of π-conjugated polymers with inorganic nanocrystals, J. Photochem. Photobiol. C Photochem. Rev. 20 (2014) 51–70.

[24] J.M. Moran-Mirabal, J.D. Slinker, J.A. DeFranco, S.S. Verbridge, R. Ilic, S. Flores-Torres, et al., Electrospun light-emitting nanofibers, Nano Lett. 7 (2) (2007) 458–463.

[25] H. Yang, C.R. Lightner, L. Dong, Light-emitting coaxial nanofibers, ACS Nano 691 (2012) 622–628.

[26] V. Vohra, U. Giovanella, R. Tubino, H. Murata, C. Botta, Electroluminescence from conjugated polymer electrospun nanofibers in solution processable organic light-emitting diodes, ACS Nano 5 (7) (2011) 5572–5578.

[27] C.T. Wang, C.C. Kuo, H.C. Chen, W.C. Chen, Non-woven and aligned electrospun multicomponent luminescent polymer nanofibers: effects of aggregated morphology on the photophysical properties, Nanotechnology 20 (37) (2009) 10.

[28] D. Sun, C. Chang, S. Li, L. Lin, Near-field electrospinning, Nano Lett. 6 (2006) 839–842.

[29] G.S. Bisht, G. Canton, A. Mirsepassi, L. Kulinsky, S. Oh, D. Dunn-Rankin, et al., Controlled continuous patterning of polymeric nanofibers on three-dimensional substrates using low-voltage near-field electrospinning, Nano Lett. 11 (2011) 1831–1837.

[30] H. Wang, G. Zheng, W. Li, X. Wang, D. Sun, Direct-writing organic three-dimensional nanofibrous structure, Appl. Phys. A 102 (2011). 457–46.

[31] C. Chang, V.H. Tran, J. Wang, Y.-K. Fuh, L. Lin, Direct-write piezoelectric polymeric nanogenerator with high energy conversion efficiency, Nano Lett 10 (2010) 726–731.

[32] F.-L. Zhou, P.L. Hubbard, S.J. Eichhorn, G.J.M. Parker, Jet deposition in near-field electrospinning of patterned polycaprolactone and sugar-polycaprolactone core–shell fibers, Polymer 52 (2011) 3603–3610.

[33] Y. Zhang, X. He, J. Li, Z. Miao, F. Huang, Fabrication and ethanol-sensing properties of micro gas sensor based on electrospun SnO_2 nanofibers, Sens. Actuat. B-Chem. 132 (1) (2008) 67–73.

[34] M. Rinaldo, F. Ruggieri, L. Lozzi, S. Santucci, Well-aligned TiO2 nanofibers grown by near-field-electrospinning, J. Vac. Sci. Technol. B 27 (2009) 1829–1833.

[35] D. Di Camillo, V. Fasano, F. Ruggieri, S. Santucci, L. Lozzi, A. Camposeo, et al., Near-field electrospinning of light-emitting conjugated polymer nanofibers, Nanoscale 5 (2013) 11637–11642.

[36] A.K. El-Aufy, Nanofibers and Nanocomposites Poly (3,4-ethylene dioxythiophene)/Poly (styrene sulfonate) by Electrospinning, PhD dissertation, Drexel Unoversity, USA, 2004.

[37] S.J. Chronakis, S. Grapenson, A. Jakob, Conductive polypyrrole nanofibers via electrospinning: electrical and morphological properties, Polymer 47 (5) (2006) 1597–1603.

[38] I.D. Norris, M.M. Shaker, F.K. Ko, A.G. MacDiarmid, Electrostatic fabrication of ultrafine conducting fibers: polyaniline/polyethylene oxide blends, Synthetic Metals 114 (2000) 109–114.

[39] A.G. MacDiarmid, W.E. Jones, I.D. Norris, J. Gao, A.T. Johnson, N.J. Pinto, et al., Electrostatically-generated nanofibers of electronic polymers, Synth. Met. 119 (2001) 27–30.

[40] E. Tavakkola, H. Tavanaia, A. Abdolmalekib, M. Morsheda, Production of conductive electrospun polypyrrole/poly (vinyl pyrrolidone) nanofibers, Synth. Met. 231 (2017) 95–106.

[41] Y.W. Ju, J.H. Park, H.R. Jung, W.J. Lee, Electrochemical properties of polypyrrole/sulfonted SEBS composite nanofibers prepared by electrospinning, Electrochim. Acta 52 (14) (2007) 4841–4847.

[42] S. Cetiner, F. Kalaoglu, H. Karakas, A.S. Sarac, Electrospun nanofibers of polypyrrole-poly (acrylonitrile-co-vinyl acetate), Textile Res. J. 80 (17) (2010) 1784–1992.

[43] M. Yanilmaz, F. Kalaoglu, H. Karakas, A.S. Sarac, Preparation and characterization of electrospun polyurethane-polypyrrole nanofibers and films, Journal of Applied Polymer Science 125 (5) (2012) 4100–4108.

[44] Andrea Camposeo, Luana Persano, Dario Pisignano, Light-emitting electrospun nanofibers for nanophotonics and optoelectronics, Macromol. Mater. Eng. 298 (5) (2013) 487–503.

[45] I.M. Alarifi, A. Alharbi, W.S. Khan, R. Asmatulu, Carbonized electrospun PAN nanofibers as highly sensitive sensors in SHM of composite structures, J. Appl. Polym. Sci. (2015). Available from: https://doi.org/10.1002/app.43235.

[46] I.M. Alarifi, W.S. Khan, A.K.M.S. Rahman, Y. Kostogorova-Beller, R. Asmatulu, Synthesis, analysis and simulation of carbonized electrospun nanofibers infused carbon prepreg composites for improved mechanical and thermal properties, Fibers Polym. 17 (2016) 1449–1455.

[47] I.M. Alarifi, A. Alharbi, W.S. Khan, A.K.M.S. Rahman, R. Asmatulu, Mechanical and thermal properties of carbonized PAN nanofibers cohesively attached to surface of carbon fiber reinforced composites, Macromol. Symp. 365 (2016) 140−150.

[48] I.M. Alarifi, A. Alharbi, W.S. Khan, A. Swindle, R. Asmatulu, Thermal, electrical and surface properties of electrospun polyacrylonitrile nanofibers for structural health monitoring, Materials 8 (2015) 7017−7031.

[49] I.M. Alarifi, A. Alharbi, O. Alsaiari, R. Asmatulu, Training the engineering students on nanofiber-based SHM systems, Trans. Tech. STEM Educ. 1 (2016) 59−67.

[50] J. Ma, Q. Zhang, A. Mayo, Z. Ni, H. Yi, Y. Chen, et al., Thermal conductivity of electrospun polyethylene nanofibers, Nanoscale 7 (40) (2015) 16899−16908.

[51] X. Wang, V. Ho, R.A. Seqalman, D.G. Cahill, Thermal conductivity of high-modulus polymer fibers, Macromolecules 46 (12) (2013) 4937−4943.

[52] V. Singh, T.L. Bougher, A. Weathers, Y. Cai, K. Bi, M.T. Pettes, et al., High thermal conductivity of chain-oriented amorphous polythiophene, Nat. Nanotechnol. 9 (2014) 384−390.

[53] Z. Zhong, M.C. Wingert, J. Stzalka, H.H. Wang, T. Sun, R. Chen, et al., Structure-induced enhancement of thermal conductivities in electrospun polymer nanofibers, Nanoscale 6 (14) (2014) 8283−8291.

[54] C. Canetta, S. Guo, A. Narayanaswamy, Measuring bthermal conductivity of polystyrene nanowires using the dual-cantilever technique, Rev. Sci. Instrum. 85 (10) (2014) 104901.

[55] O. Ero-Phillips, M. Jenkis, A. Stamboulis, Tailoring crystallinity of electrospun plla fibers by control of electrospinning parameters, Polymer 4 (3) (2012) 1331−1348.

[56] H. Shao, X. Zhang, S. Liu, F. Chen, J. Xu, Y. Feng, Preparation of pure nanofibers via electrospinning, Mater. Lett. 65 (2011) 1775−1777.

[57] W. Ponhan, S. Maensiri, Fabrication and magnetic properties of electrospun copper ferrite ($CuFe_2O_4$) nanofibers, Solid State Sci. 1 (2009) 479−484.

[58] N.A.M. Barakat, B. Kim, H.Y. Kim, Production of smooth and pure nickel metal nanofibers by the electrospinning technique: nanofibers possess splendid magnetic properties, J. Phys. Chem. C 113 (2) (2009) 531−536.

[59] X. Chen, K.M. Unruh, C. Ni, B. Ali, Z. Sun, Q. Lu, et al., Fabrication, formation mechanism, and magnetic properties of metal oxide nanotubes via electrospinning and thermal treatment, J. Phys. Chem. C 115 (2011) 373−378.

[60] P. Gupta, R. Asmatulu, G. Wilkes, R.O. Claus, Superparamagnetic flexible substrates based on submicron electrospun Estane® fibers containing MnZnFe-Ni nanoparticles, J. Appl. Polym. Sci. 100 (2006) 4935−4942.

[61] P. Gupta, R. Asmatulu, R. Claus, G. Wilkes "Superparamagnetic Flexible Substrates-based on Submicron Electrospun Estane® Fibers Containing MnZnFe-Ni Nanoparticles," Materials Research Society Symposium Proceedings (MRS 2005 Spring Meeting, San Francisco), Vol. 877E, S5.7.1−S5.7.6.

[62] W.S. Khan, R. Asmatulu "Magnetic Properties of Electrospun PVP and PAN Nanocomposite Fibers Associated with NiZn-Ferrite Nanoparticles," SAMPE Fall Technical Conference, Fort Worth, TX, October 17−20, 2011, 9p.

Characterization of electrospun nanofibers

13

CHAPTER OUTLINE

13.1 STRUCTURAL CHARACTERIZATION OF ELECTROSPUN NANOFIBERS

13.1.1 ELECTROSPINNING NANOFIBERS

Electrospinning is a process of using the DC electric field between a syringe tip (mainly metal wires immersed into the polymeric solution) and metallic collector screens (e.g., Cu, Al, Ni, Zn, stainless steel, and other conductive alloys), where the negative polarity is applied to create a potential difference for a better electrospinning process and electric charge grounding takes place. This high potential difference usually overcomes the surface tension of the polymeric solutions to produce nanofibers [1−5]. During the electrospinning process, polymeric solution is pushed out of the syringe tip a small distance (called the jet length—usually a few cm long) through the elongation process, and then bending instability takes

Synthesis and Applications of Electrospun Nanofibers. DOI: https://doi.org/10.1016/B978-0-12-813914-1.00013-4

place for plastic deformations. Prior to the nanofibers reaching the collector surface, a nanofiber in one fiber line or multiple fibers based on the applied voltage will be stretching quite a bit, where the fibers will undergo a lot of elongation to reduce the fibers from micron size to nanosize. Most of the solvents in the polymeric solutions are evaporated during the traveling so that nanofibers are not merging on the collector screen to form thin films instead of nanofiber films [6–9]. Fig. 13.1 shows a schematic illustration of the electrospinning process from a polymeric solution to nanofiber formations [10].

A number of different parameters, including voltage, collector distance, pump speed, humidity, conductivity, temperature, and viscosity of the polymeric solution affect the electrospinning process, so these parameters must be controlled precisely to achieve good nanofibers in micron, submicron, and nanoscale ranges with various surface properties. During the electrospinning process, Maxwell stress tensor equations are considered to calculate the amount of stresses and critical voltage values. Maxwell stress tensors in the electrospinning process can be defined as follows [11–13]:

$$\sigma_{ij} = \varepsilon V_i V_j \frac{1}{\mu} B_i B_j - \frac{1}{2}\left(\varepsilon V^2 + \frac{1}{\mu_0} B^2\right)\delta_{ij} \tag{13.1}$$

where ε is the permittivity of the solution, V is applied voltage to the spinneret (spinning voltage), B is the magnetic part and dividing by H^2. H is the distance between the spinneret and the collector screen. Eq. (13.1) can be further reduced by neglecting some of the small variables that have low impacts on the electrospinning process and fiber formations [8]:

$$\sigma = (\varepsilon V^2)/H^2 \tag{13.2}$$

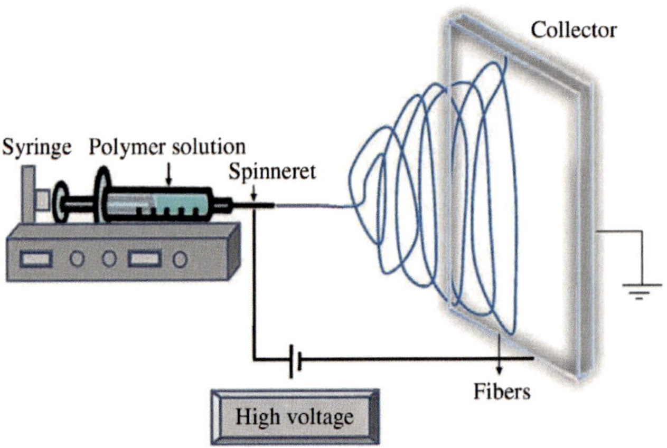

Schematic illustration of the electrospinning process [10].

The critical spinning voltage V_c can be obtained by balancing the Maxwell stress and spinneret (capillary stress), which has to overcome the surface tension of the electrospinning solution to start the electrospinning process [8].

$$V_c = \sqrt{\frac{\gamma H^2}{\varepsilon r}} \tag{13.3}$$

This is basically the balanced equation that is required to start the electrospinning process at different DC voltages (8−15 kV) for different polymers and solutions. As can be seen from Eq. (13.3), collector distance, surface tension of the polymeric solution, and relative permittivity of the solvent/solution are the driving parameters for the critical electrospinning voltages. A number of different parameters, including hydrostatic pressure, inertia, and viscoelastic forces can also be analyzed during the calculations owing to the possible positive or negative impact of the forces acting on the solutions of polymers and selected solvents. However, their impact is negligible compared to the high electrical field. At the right conditions of the process and system parameters, the polymeric solutions will have the mutual charge repulsion to initiate nanofiber formation [8−13].

13.1.2 XRD CHARACTERIZATION

The characterization of nanofibers, and other nanomaterials and nanostructures, has been primarily based on selected surface analysis techniques. Conventional characterization methods are mostly developed for bulk materials, but some of them still can be used to characterize nanoscale materials. Structural characterizations of nanofibers typically include X-ray diffraction (XRD), scanning electron microscope (SEM), transmission electron microscope (TEM), scanning probe microscope (SPM)—two major members of scanning tunneling microscope (STM) and atomic force microscope (AFM), and other various electron microscopes. In addition to them, there are also physical and chemical characterization methods that are utilized in many areas of nanotechnology, including laser-based optical spectroscopy, electron spectroscopy, ionic spectroscopy, Raman spectroscopy, Fourier-transform infrared spectroscopy (FTIR), differential scanning calorimetry (DSC), thermal gravimetric analysis (TGA), X-ray photoelectron spectroscopy (XPS), UV−vis spectroscopy, and vibrating sample magnetometer (VSM) [13−16].

XRD is an analytical technique mainly used for the structural characterization of a crystalline or semicrystalline material and can provide information about the unit cell dimensions and properties. The samples to be characterized by the XRD method are finely ground (powder), homogenized, or average bulk composition before analyzing. Bragg's law is the major law for the XRD analysis of solid substrates to make precise quantification of test results performed on the samples to find out the crystal structures of the specimens. One can consider lattice planes that are separated by a distance d_{hkl} (interplanar distance), as is seen in Fig. 13.2 [17−19]. In Bragg law (Eq. 13.4), the scattered X-rays from two parallel planes

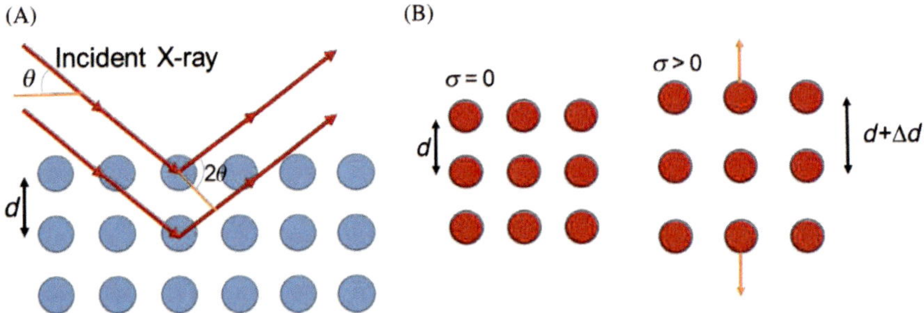

FIGURE 13.2

Images showing (A) the incoming X-ray beam in and reradiating a small portion of its intensity, and (B) strain formation into the unit cell structure when a stress is applied [17–19].

interact with each other to create constructive interference for the structural analysis [16,17,20,21].

$$n\lambda = 2d\sin\theta \tag{13.4}$$

in which n is the number of the wavelength, λ is the wavelength of X-rays, θ is the angle of diffracted X-rays, and d is the spacing between atomic planes. In order to calculate the strain values of the crystalline structure when a stress is applied, the following equation may be used.

$$\varepsilon = \frac{d - d_o}{d_o} = \frac{\Delta d}{d} \tag{13.5}$$

If there is no inhomogeneous strain, the crystal size or thickness can be estimated using the peak width following Scherrer's formula [17]:

$$D = \frac{K\lambda}{B\cos\theta_B} \tag{13.6}$$

where D is the thickness (diameter) of crystallite, K is the constant dependent on crystallite shape (usually 0.89), λ is the X-ray wavelength, B is the full width at half maximum (FWHM) or integral breadth, and θ_B is the diffraction (Bragg) angle [17].

In addition to the conventional XRD method, small-angle X-ray scattering (SAXS) is another analytical X-ray technique for the structural characterization of solid and fluid materials in small sizes (e.g., molecular and nanometer ranges). In this technique, electron density fluctuation on the order of 10 nm or larger can be adequate to create considerable scattered X-ray intensities at an angle of less than 5 degree ($2\theta < 5$ degree). SAXS is used to explore structural details in the order of 0.5–50 nm range for various nanomaterials, including [17]:

- Polymeric nanofilms and nanofibers;
- Particle sizing of suspended nanoparticles;

- Life science and biotechnology entities (e.g., proteins, viruses, enzymes, and DNA complexes);
- Surface area of catalyst per volume;
- Micro- and nanoemulsions;
- Hydrogel complexes;
- Liquid crystals (properties between solid and liquid).

Asmatulu et al. reported about the integration of graphene and C_{60} into TiO_2 electrospun nanofibers for enhanced energy conversion efficiencies [18]. TiO_2 nanofibers associated with C_{60} and graphene nanoscale inclusions at various concentrations (0, 1, 2, 4, and 8 wt.%) were produced using poly(vinyl acetate), dimethylformamide, and titanium (IV) isopropoxide precursor solutions. At the beginning, the produced nanofibers were in organic phases, which were heat treated at 300 °C for 2 h in an oven (ambient oxygen), and then further heated up to 500 °C in an Ar atmosphere for an additional 12 h to crystallize the TiO_2 nanofibers after removing organic parts of the nanofibers. In order to eliminate the decomposition of the graphene and C_{60} in the TiO_2 nanofibers, the temperature was kept constant at 500 °C under Ar atmosphere. XRD studies were conducted on the samples to determine whether the graphene, C_{60}, and TiO_2 nanofibers are in crystalline forms [19]. Fig. 13.3 shows the XRD peaks of plain TiO_2, TiO_2

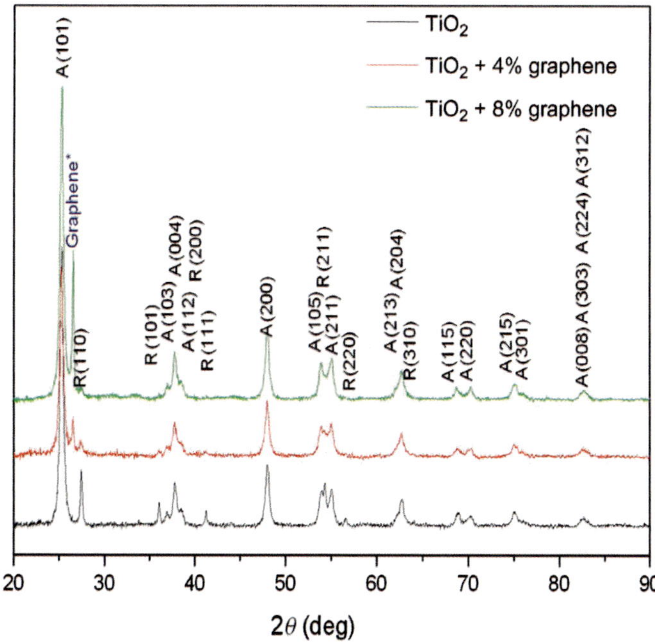

FIGURE 13.3

XRD peaks showing plain TiO_2, and TiO_2 with 4 and 8 wt.% graphene into the TiO_2 nanofibers [18,19].

with 4 and 8 wt.% graphene into the TiO_2 nanofibers [18,19]. The XRD peak indicated that the TiO_2 nanofibers were all crystalline, and single-layer graphene nanoflakes were presented into the TiO_2 nanofibers. From the X-ray diffraction patterns, it is clear that both the anatase and rutile phase of TiO_2 are also present, but the amount of rutile is relatively low.

Mohammed et al. reported the superhydrophobic electrospun PAN nanofibers for gas diffusion layers of proton exchange membrane (PEM) fuel cells for cathodic water management applications [22]. This study was mainly focused on the gas diffusion layers (GDLs) fabricated using electrospinning with different concentrations of hydrophobic and hydrophilic agents in proton exchange membrane fuel cells (PEMFCs) after the carbonization process. The objective of the research was to investigate the cathodic water management efficiency of carbonized and surface-treated PAN nanofibers. After the carbonization tests, the XRD method was used to characterize the crystallinity of the prepared nanofibers, as shown in Fig. 13.4 [22]. As can be seen, the XRD strong diffraction peak at $2\theta = 25$ appears to be the ideal carbonized fibers. The d spacing of the XRD peak was calculated to be about 3.56 A$^\circ$ with a reflection of (100) plane corresponding to the α phase crystals. The surfaces of carbon nanofibers were preferentially hydrophobized using Ultra-Ever Dry solutions (for top and bottom coatings) and Krytox 157 FSH Oil, and hydrophilic hydrophilized using potassium permanganate ($KMnO_4$) and sulfuric acid (H_2SO_4). The water condensation tests conducted on the surfaces of the functionalized GDL indicated that the proposed approach significantly improved water management in the fuel cell [22,23].

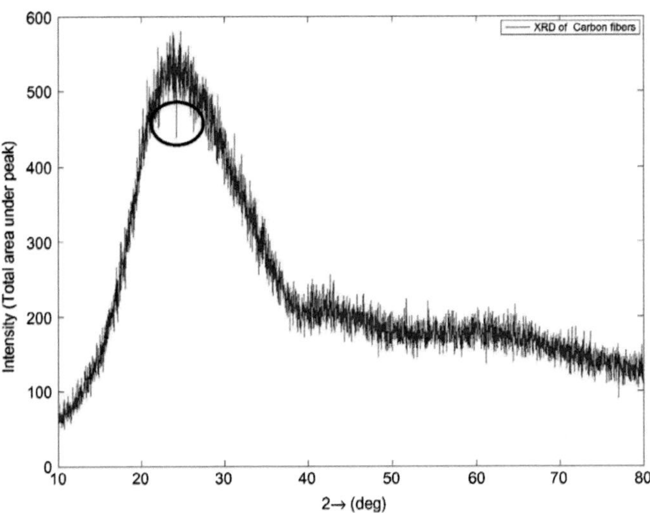

FIGURE 13.4

XRD studies conducted on the crystallinity PAN nanofibers developed for the GDL of fuel cells.

13.1.3 SEM CHARACTERIZATION

SEM is an electron microscope that is capable of generating high-resolution images of the surfaces of bulk and nanomaterials at nanoscale ranges. Based on the images created by SEM, those images have characteristics of two-dimensional and three-dimensional appearances, which are useful for judging the surface structures of the specimens. Depending of the types of SEM and their applications, the beam size can be adjustable to 10 nm or below. Backscattered and field emission SEMs are commonly used SEM techniques by industries and scientific communities to analyze the surface morphologies of substrates. The backscattered SEM primarily scatters off the nucleus of an atom, and then the backscatter electron detector provides good compositional contrast due to the fact that the larger the nucleus of the atom, the more the electron backscattering is. Field emission SEM generally provides ultra-high-resolution images down to 1 nm and is one of the most demanding electron microscopes for many scientific and technological applications. SEMs also have energy-dispersive X-ray spectroscopy (EDS or EDAX) capabilities to find out the elemental distributions of the samples' top surfaces [17].

Srikant investigated the regeneration behaviors of neuroglial cells on artificially made electrospun PCL scaffolds embedded with highly conductive carbon-based nanomaterials [24]. It was stated that nerve damage could result in a major decrease in the quality of life and represents one of the major public health concerns; thus, tissue engineering became the most promising approach to reinstate the nervous system back to health. The objective of the study was to investigate the effects of nanoscale scaffold designs with electrospun nanofibers incorporated with graphene, carbon nanotubes (CNTs), and C_{60} to promote astrocyte cell culture. Fig. 13.5 shows SEM images of nanoscaffolds fabricated by the electrospinning process: PCL-only nanofibers, PCL-CNT, PCL-graphene, and PCL-fullerene nanocomposite fibers [24]. The diameters of the prepared nanocomposite fiber scaffolds were in the ranges of 100−400 nm, which can easily mimic the extracellular matrix for nerve damage treatments. SEM studies further confirmed the neural cell attachments on the surface of nanofiber-based scaffolds [24]. The carbonaceous nanomaterials (e.g., carbon fibers, graphene, CNTs, and C_{60}) are highly conductive materials, so these nanomaterials may enhance the signal that the brain sends to different parts of the body when needed.

13.1.4 TEM CHARACTERIZATION

TEM is an electron microscope that is capable of generating near atomic level images on both bulk materials and nanomaterials. In the TEM method, highly focused electron beams pass through ultra-thin nanosize structures of organic and inorganic materials or small biological specimens, such as viruses, bacteria, DNA, and so on [17]. After passing electrons through the thin specimen, the electrons strike a fluorescent screen and create an image from the top to bottom

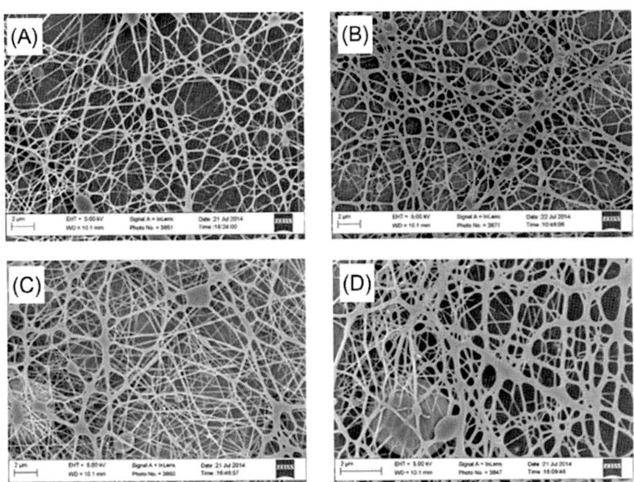

FIGURE 13.5

SEM images showing nanoscaffolds fabricated by electrospinning methods: (A) PCL only nanofibers, (B) PCL-CNT, (C) PCL-graphene, and (D) PCL-fullerene nanocomposite fibers [24].

surface—inside and out of the object. There are imaging and diffraction modes of the TEM equipment based on the applications, structures, and types of materials. TEM also has EDS/EDAX capabilities to find out the elemental distributions of the sample surfaces. Prior to the TEM analysis, the larger sample is milled or ion beam etched to reduce the size to thin enough (~ 50 nm) for electrons to pass through specimens. The beam size can be adjusted down to 0.5 nm. Electrons are accelerated much the same way as in SEM, but travel through the specimen. Fig. 13.6 shows the diffraction patterns of crystalline, semicrystalline, and amorphous materials [17−19,22]. In addition to conventionally used TEM, the other TEM, called high-resolution TEM (HRTEM) was also developed to create images at atomic scales. This will determine the crystallographic properties of bulk and nanomaterials, epitaxial growth mechanisms, lattice mismatches, defect formations, as well as atomic planes and orders, boundaries, and phase transitions [17−19,22].

Asmatulu et al. synthesized and analyzed the electrospun TiO_2 nanofibers incorporated with nanoscale graphene and C_{60} for improved dye-sensitized solar cell (DSSC) efficiencies [18]. Electrospun TiO_2 nanofibers were produced using poly(vinyl acetate), dimethylformamide, and titanium (IV) isopropoxide precursor solutions, stabilized at 300 °C for 2 h, and then heat-treated at 500 °C in Ar gas for 12 h to crystalize the nanocomposite fibers [19]. Fig. 13.7 shows TEM images of electrospun TiO_2 nanofibers incorporated with C_{60} at different magnifications [18,19]. TEM diffraction pattern and images indicated that all the nanofibers were in crystalline forms and graphene nanoflakes were well-integrated into the prepared

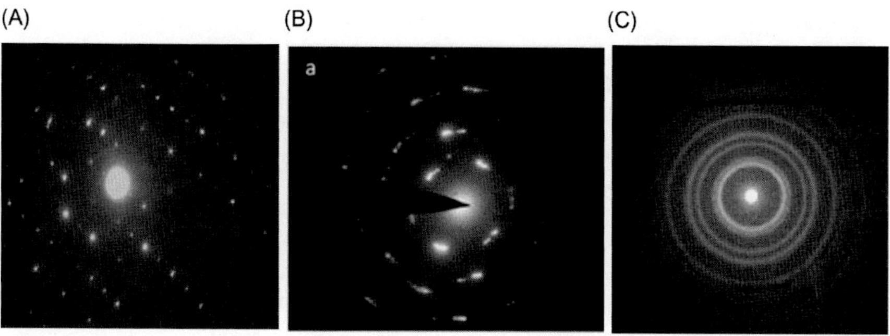

FIGURE 13.6

TEM diffraction patterns of (A) crystalline, (B) semicrystalline, and (C) amorphous materials [17].

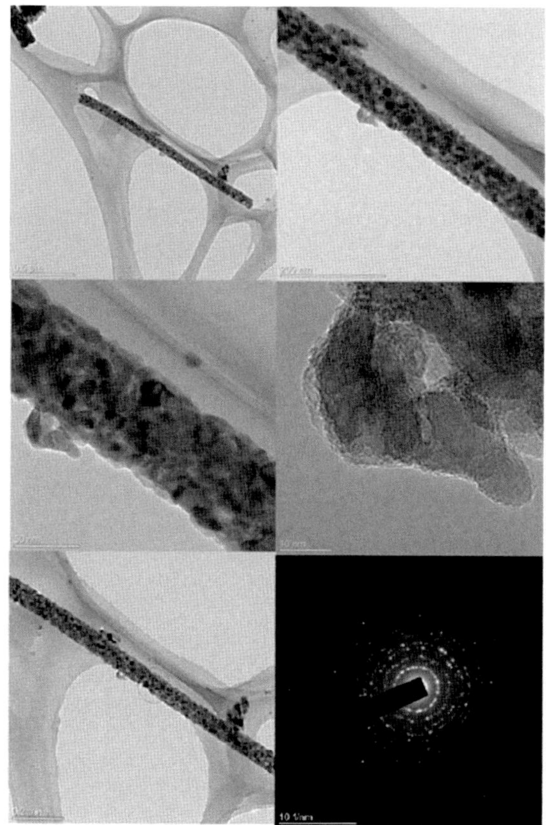

FIGURE 13.7

TEM images of electrospun TiO_2 nanofibers incorporated with 4 wt.% graphene at different magnifications and diffraction patterns [18,19].

electrospun nanofibers. It is assumed that this process will likely increase the efficiencies of the DSSCs. In the same study, C_{60} nanoscale inclusions were also integrated into the electrospun TiO_2 nanofibers, and analyzed by TEM equipment.

13.2 AFM CHARACTERIZATION

AFM is a high-resolution scanning probe microscope with a resolution of fractions of Angstrom, which has more than 1000 times higher resolution when compared to the classical optical microscope. AFM contains a micro- and nanoscale cantilever with a silicon or silicon nitrite sharp tip (probe) at its end that is used to scan the specimen surface at the nanoscale level. CNTs can be attached to the tip of the cantilever to make a sharper tip to increase the image resolution of the surfaces at the angstrom level. However, the CNT-based AFM tip is highly fragile, and the operator must be extra careful during the mounting and surface scanning. Depending on the requirements, applied forces between the tip and surface can be measured for the detection of mechanical contact force, magnetic, van der Waals interaction, electrostatic, capillary, chemical, and steric forces [17]. Usually, three different AFM modes can be employed in an AFM unit, including contact mode, noncontact mode, and tapping mode. Fig. 13.8 shows the different modes of AFM, including contact mode, noncontact mode, and tapping mode [25].

The contact mode in which the tip of the cantilever scans the sample in close contact with the surface is the most commonly used method for surface force measurements. In the noncontact mode, the tip usually hovers about 5−15 nm above the surface of the substrate. There are some difficulties with the contact mode AFM because of the higher wear rate and failure of the tips during scanning. On the other hand, tapping mode AFM deals with the problems associated

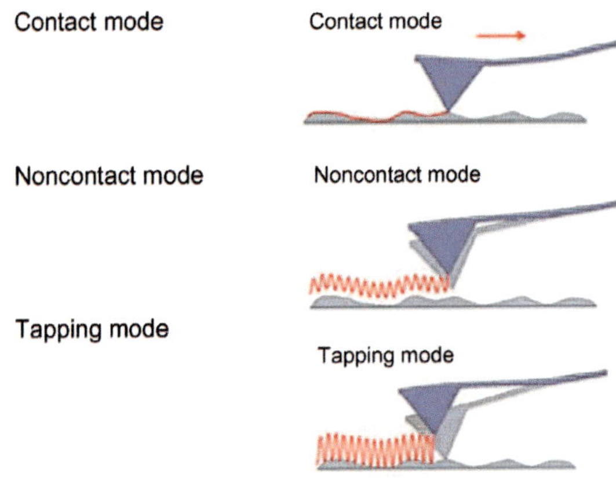

FIGURE 13.8

Different modes of AFM, including contact mode, noncontact mode, and tapping mode [25].

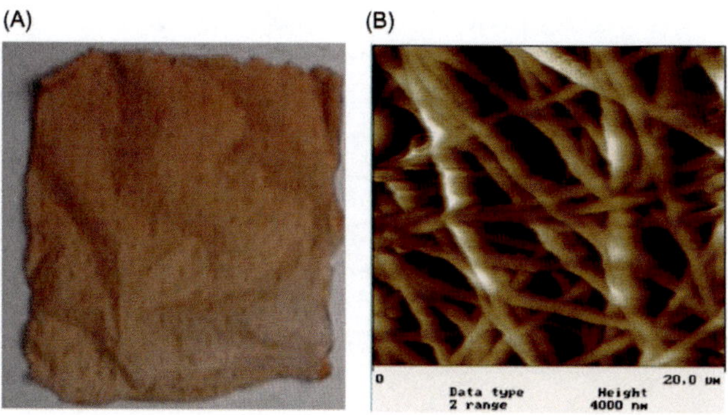

FIGURE 13.9

(A) Photograph and (B) AFM image of the electrospun Estane nanofibers incorporated with 26 wt.% MnZnFe/Ni magnetic nanoparticles obtained at 18 kV voltage, 3 mL h^{-1} pump speed, and 15 cm separation distance.

with friction, tip tackling with the rough surfaces, electrostatic forces, adhesion, and other difficulties by the tapping process. Fig. 13.9 shows photographs and an AFM image of electrospun Estane nanofibers incorporated with 26 wt.% MnZnFe/Ni magnetic nanoparticles obtained at 18 kV voltage, 3 mL h^{-1} pump speed, and 15 cm separation distance [13,17].

13.3 OPTICAL SPECTROSCOPY

13.3.1 FTIR CHARACTERIZATION

FTIR spectroscopy is a method of analyzing the composition of organic and inorganic materials (solid, liquid, or gas) based on the chemical bonds that have different characteristics of energy levels and bond starching capabilities. The primary goal of FTIR spectroscopy is to measure how well the samples absorb different wavelengths of light, usually between 3700 and 500 nm. The light energy that has not been absorbed during the IR exposing can be detected and displayed on a graph (spectrum) that has a series of peaks, each of which represents particular chemical bond energies, enabling a conservation to identify the chemical structures of the substrates [17]. Based on the size, types and applications, the focused beam area is around 40 × 40 μm^2 up to 6 × 6 mm^2. Table 13.1 gives the important FTIR absorption peaks for different functional groups, their wavelengths, and intensities [26].

Studies in electrospun nanofibers have been substantially growing in many industries, such as environmental health and safety, biomedical, energy, sensor, separation science, defense, and so on [27−32]. Alharbi et al. investigated the highly hydrophilic electrospun polyacrylonitrile (PAN)/polyvinylpyrrolidone

Table 13.1 The Important FTIR Absorption Peaks for Different Functional Groups, Their Wavelengths, and Intensities [26]

Functional Groups	Wavelengths (cm^{-1})	Intensity
Water OH tretch	3700–3100	Strong
Alcohol OH stretch	3600–3200	Strong
Carboxylic acid OH stretch	3600–2500	Strong
N–H stretch	3500–3350	Strong
$\equiv C - H$ stretch	~3300	Strong
$= C$–H stretch	3100–3000	Weak
–C–H stretch	2950–2840	Weak
–C–H aldehydic	2900–2800	Variable
$C \equiv N$ stretch	~2250	Strong
$C \equiv C$ stretch	2260–2100	Variable
$C = O$ aldehyde	1740–1720	Strong
$C = O$ anhydride	1840–1800, 1780–1740	Weak, strong
$C = O$ ester	1750–1720	Strong
$C = O$ ketone	1745–1715	Strong
$C = O$ amide	1700–1500	Strong
$C = C$ alkene	1680–1600	Weak
$C = C$ aromatic	1600–1400	Weak
CH_2 bend	1480–1440	Medium
CH_3 bend	1465–1440, 1390–1365	Medium
C–O–C stretch	1250–1050 several	Strong
C–OH stretch	1200–1020	Strong
NO_2 stretch	1600–1500 and 1400–1300	Strong
C–F	1400–1000	Strong
C–Cl	800–600	Strong
C–Br	750–500	Strong
C–I	~500	Strong

(PVP) nanofibers associated with an antibiotic (gentamicin sulfate) as a filtration membrane for water treatments [32]. In these studies, powder forms of PAN and PVP were mixed together and dissolved in dimethylformamide (DMF), and then different weight percentages of gentamicin sulfate powder were added to the prepared solution before the electrospinning process. Gentamicin was added for antibacterial and elimination of biofouling purposes, whereas PVP was added to make the surface of the membrane hydrophilic to improve the filtration rate and efficiency. FTIR studies conducted on the PAN and PAN with PVP nanofibers, and the test results are shown in Fig. 13.10 [32]. FTIR analysis is considered to be a useful tool for determining the chemical interactions of PAN and other incluions in the nanofibers.

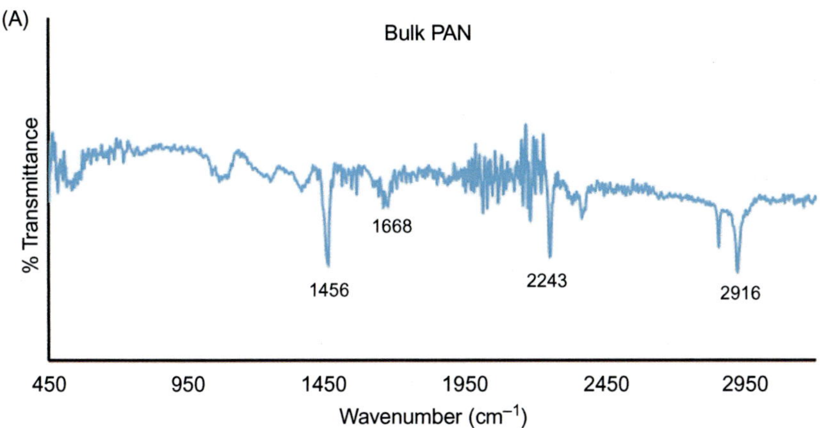

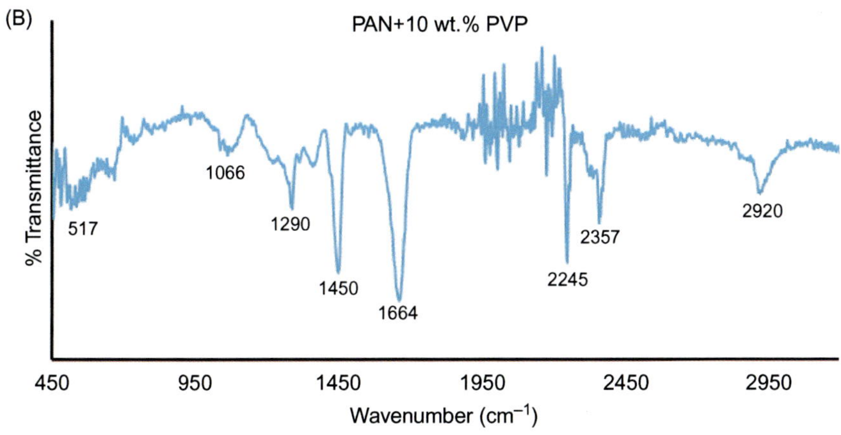

FIGURE 13.10

FTIR spectra of (A) bulk PAN electrospun nanofibers, and (B) PAN with 10 wt.% PVP nanofibers.

The FTIR spectra of electrospun PAN fibers have different peaks related to the existence of CH_2, $C\equiv N$, $C=O$, $C-O$, and $C-H$ bonds (Fig. 13.10A). A peak at $2916\ cm^{-1}$ is mainly because of the $C-H$ bonds, such as CH, CH_2, and CH_3. Another peak at $2243\ cm^{-1}$ may be the indication of nitrile ($C\equiv N$) bonds which is usually seen into the PAN molecular chains/structures. The spectra corresponding to $1668\ cm^{-1}$ are attributed to the cyclic $C=O$ bond for the methyl acrylate comonomer, while the peak at $1456\ cm^{-1}$ is mainly due to the CH bond [32].

From Fig. 13.10B, FTIR spectra of PAN with 10 wt.% PVP nanofibers can be seen. The band incorporated with the pyrrolidone ($C=O$ group) is located at $1664\ cm^{-1}$. Also, the vibrational band of $1698\ cm^{-1}$ represents $C=O$ stretching of the PVP polymer film, while the absorption peak at $2920\ cm^{-1}$ shows the C$-$H asymmetric stretching of the CH_2. The other bonded stretching at a wavelength of $1664\ cm^{-1}$ can be attributed to the vibrational band of $C=O$, which also suggests that some of the H-bonding carbonyl groups may exist in PVP structures. The other two wavelengths at 1290 and $1450\ cm^{-1}$ are attributed to the C$-$N stretching vibration and C$-$H bending vibration of PVP, respectively [32].

13.3.2 RAMAN SPECTROSCOPY

Raman spectroscopy is a spectroscopic technique to identify the vibrational, rotational, and other low-frequency modes and provides a structural fingerprint of molecules in a substrate. In some senses and applications, Raman spectroscopy results may be more reliable when compared to FTIR test results. This spectroscopy is used to probe molecular and lattice vibrations which couple to the electric field of laser light (Ar, Kr, etc.) through the polarization tensor [17]. The information in this spectroscopy can be used to determine molecular and crystal symmetry, as well as bond strengths of the substrates. This technique can be applied for the phase identification purposes, structural disorders, and defects (D = defect bond and G = graphitic bond) characterizations (e.g., carbon fibers, CNTs, and graphene). Atomic shifts in the wavelengths, which are mainly caused by stress and defects, can be identified. It is expected that the D band is low while the G band is high for better carbon nanofibers [17].

Alharbi et al. studied the electrospun PAN and PVP nanofibers incorporated with different inclusions as a filtration medium for water treatments [32]. Fig. 13.10 depicts a typical Raman spectra of PAN and PAN + 10 wt.%PVP nanofibers. The wavelength was in the range of 160 to $3160\ cm^{-1}$, where various Raman spectroscopy peaks superimposed on a broad feature over almost the entire region. It was noticed that the entire spectra are primarily due to the second-order Raman scattering, and only sharp peaks can be the first-order Raman peaks. The Raman peak of $2990\ cm^{-1}$ is mainly attributed to the first-order Raman scattering. This peak is very sensitive to the crystalline structure and microstructure of the selected nanofibers. This peak clearly specifies that the PAN nanofiber sample can be highly crystalline material at this condition [32].

Fig. 13.11B indicates that no major broadening peaks of Raman spectroscopy were observed, either in pure PAN nanofibers or PAN fibers with different wt.% of PVP. This result indicates that there might not be a major breakdown of the long-range order of crystallinity in both nanofiber samples. In fact, Raman spectra clearly show the long-range order of crystallinity in the samples, which may be happening during the electrospinning of PAN nanofibers. As can be seen from Fig. 13.11A, B, the intensity of the Raman band corresponding to the C$-$N vibration band at 719 and $785\ cm^{-1}$, and C$-$C stretching vibration at 901 and $897\ cm^{-1}$ are mainly because of the presence of PVP in PAN nanofibers. The other bands at $1109\ cm^{-1}$ are owing to the C$-$N stretching of pure PVP. Overall,

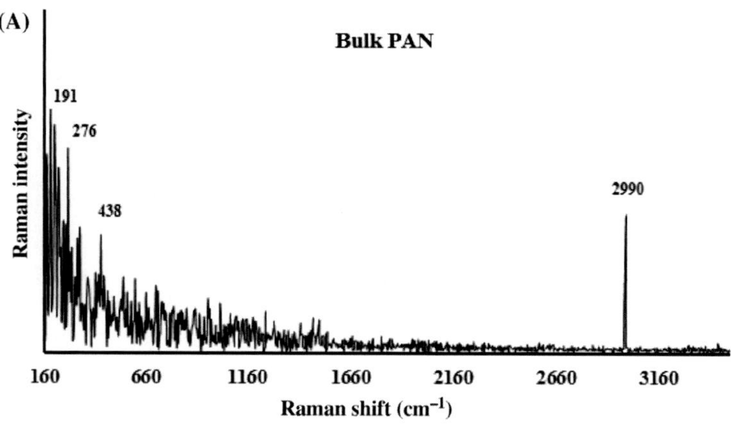

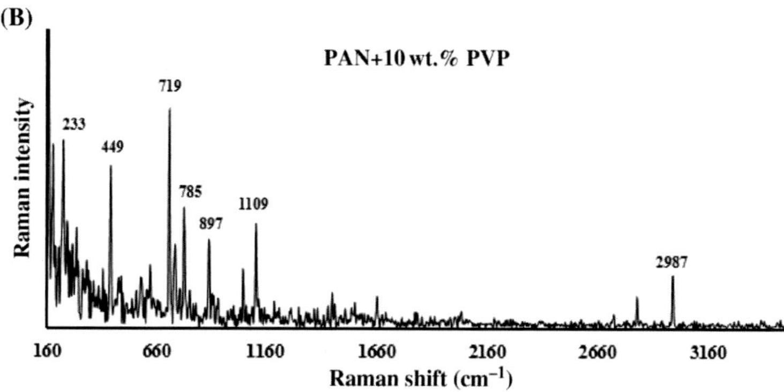

FIGURE 13.11

Raman spectra of (A) bulk PAN nanofibers and (B) PAN with 10 wt.% PVP nanofibers.

the Raman spectroscopy results indicate that PVP was well integrated into the PAN nanofibers [32].

13.4 OTHER CHARACTERIZATIONS OF NANOFIBERS

13.4.1 DSC CHARACTERIZATION

DSC consists of the quantification of heat released in a chemical process, either a reaction or a conformational alteration of the organic molecules during the heat variations (usually between 0 and 400°C although other temperatures are possible). Through the DSC process, one can determine some parameters, such as the heat of reaction ($\Delta_r H$), which is a part of the enthalpic changes related to the process of a chemical reaction. If the value of $\Delta_r H$ is a negative, the process is called "exothermic," which generally releases heat; however, if the value of $\Delta_r H$ is a positive, the process is called "endothermic," which mainly requires heat input to initiate the chemical reactions [17].

Alarifi et al. reported about the electrical, thermal, and surface hydrophobic properties of electrospun PAN nanofibers for the purpose of the structural health monitoring of composite aircraft and wind blades [33–35]. The study mainly talked about the fabrication of PAN-based electrospun nanofibers, stabilization, and carbonization in order to remove all noncarbonaceous materials prior to the structural health monitoring. A number of characterization techniques, such as DSC, water contact angle, thermogravimetric analysis (TGA), and FTIR were utilized to determine the surface, thermal, and other chemical and physicochemical properties of the carbonized electrospun PAN nanofibers. Fig. 13.12 shows the DSC thermogram of the PAN fibers for heat flow, nonreversible heat flow, and reversible heat flow [33].

As can be seen from Fig. 13.12, the glass transition temperature of the PAN nanofibers is approximately 104°C. It was detected that there was no melting of the materials as indicated by the green line of the DSC test results; nonetheless, it could be seen with modulated DSC (modulate 1.5°C/60 S) as specified by the blue line in the curves. From the DCS test results, PAN nanofibers had a broad exothermal peak at 312.4°C, which may be due to the cyclization process. Onset

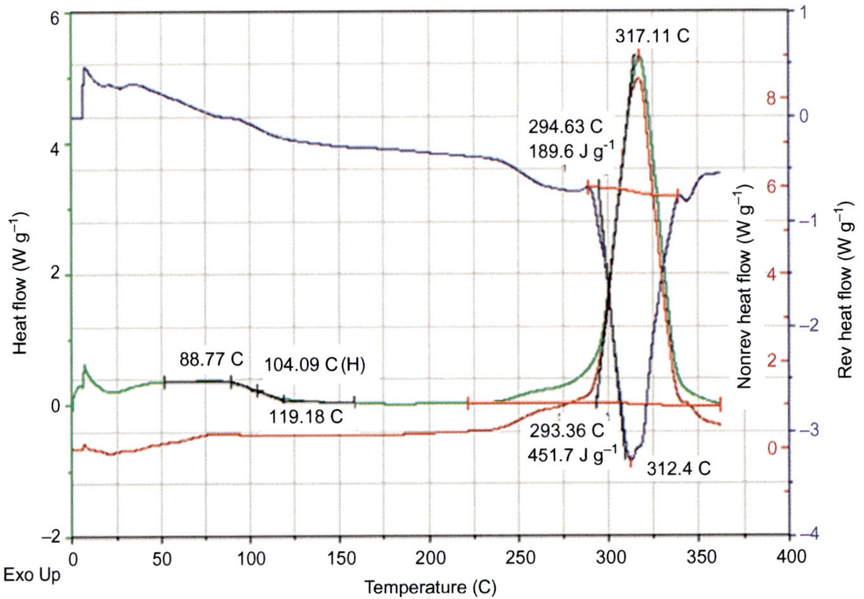

FIGURE 13.12

DSC thermogram of the PAN nanofibers for heat flow, nonreversible heat flow, and reversible heat flow [33].

temperature was about 294.6°C, and the thermal reaction was then completed at about 330°C. The peak broadening in the exothermic region and its corresponding temperature can be related to the stabilization time. When the stabilization time was enhanced, the cyclization peak is usually broadened, and the cyclization temperature is noticeably improved. It was stated that electrospun PAN nanofibers were cyclized only through the free radical mechanism and provided a peak at 312.4°C. The cyclization of nitrile groups in PAN polymer is extremely exothermic, leading to the fragmentation of polymer chains because of the rapid heat building-up in the substrates.

Khan et al. investigated the thermal properties of electrospun PAN nanofibers incorporated with various graphene nanoplatelets and multiwalled carbon nanotubes (MWCNTs) using the DSC method [36]. The DSC method was employed to find out the glass transition temperature (T_g), melting temperature (T_m), and heat flow of the polymeric nanocomposite fibers. These thermal properties are very important properties throughout the thermal processing of polymers, applications, fire retardancy, and their storage [36,37]. The DSC test results showed that the pure PAN nanofibers provided a T_g of 104.1°C; nonetheless, in the presence of 2 and 4 wt.% of graphene nanoflake inclusions in the same PAN fibers, the T_g values were increased to 105.1°C and 105.8°C, respectively. Similar studies conducted on PAN nanofibers associated with 2 and 4 wt.% of MWCNTs provided T_g values of 105.1°C and 108.2°C. The increase in nanoscale inclusions beyond 4 wt.% made some slight reductions in the T_g values of the PAN nanocomposite fibers, which may be attributed to the clustering effects of nanoscale inclusions at higher concentrations [36].

13.4.2 TGA CHARACTERIZATION

TGA is a thermal analysis method to measure physical and chemical properties of materials (mainly organic and flammable) under various temperature and time variances. This method usually provides information about phase transitions, desorption, absorption, as well as thermal decomposition, chemisorptions, and solid−gas reactions (e.g., oxidation or reduction) of various materials. TGA works mainly based on the characteristics of selected materials that exhibit either mass loss or gain due to the oxidation, decomposition, or loss of volatiles (e.g., moisture and volatile organic compounds). During many experimental studies, DSC and TGA test results are usually compared to each other to analyze the thermal behaviors of materials at the same time [17]. The sample capacity of TGA is about 200 mg (ideal sample weight in the range of 10−50 mg), with a balance sensitivity of 0.1 microgram, and it has a temperature range from ambient to 1500°C. TGA samples can be in the form of fiber, powder, film, gel, or small pieces, so the interior temperature of the sample remains close to the measured gas temperature during the experimentation [17].

Jabbarnia et al. investigated the mechanical, thermal, and electrical properties of polyvinylidene fluoride (PVdF) and polyvinylpyrrolidone (PVP) nanofibrous membranes prepared for the supercapacitor applications [38,39]. Carbon black nanoparticles ($\sim$50 nm) with different weight percentages (0, 0.25, 0.5, 1, 2, and 4 wt.%) were dispersed in *N,N*-dimethylacetamide (DMAC) and acetone prior to the electrospinning processes at 25 KV DC voltage, 2 mL pump speed, and 25 cm tip-to-collector distances. The capillary tube (syringe needle) diameter (inside diameter) was about 0.5 mm. Fig. 13.13 shows the SEM images of electrospun PVDF/PVP fibers incorporated with 1, 2, and 4 wt.% carbon black particles [38]. The diameters of the PVDF/PVP nanocomposite fibers were between 50 and 250 nm. Some thermal properties of nanofibrous membranes were analyzed using TGA, in which weight losses were determined as a function of temperature and time. During the TGA studies, the specimens were heated from 50°C to 1000°C in a nitrogen atmosphere at a heating rate of 10°C min^{-1}. Fig. 13.14 shows the TGA thermogram of PVDF/PVP nanofibers incorporated with 0.25, 0.5, 1, 2, and 4 wt.% carbon black particles [38]. The test results revealed that most of the nanofibers underwent single-step degradation between approximately 440°C and 450°C. Addition of carbon black into PVDF/PVP nanofibers at higher concentrations provided some initial stability. Overall, all the test results indicated that many of the physical properties of the nanocomposite separators were

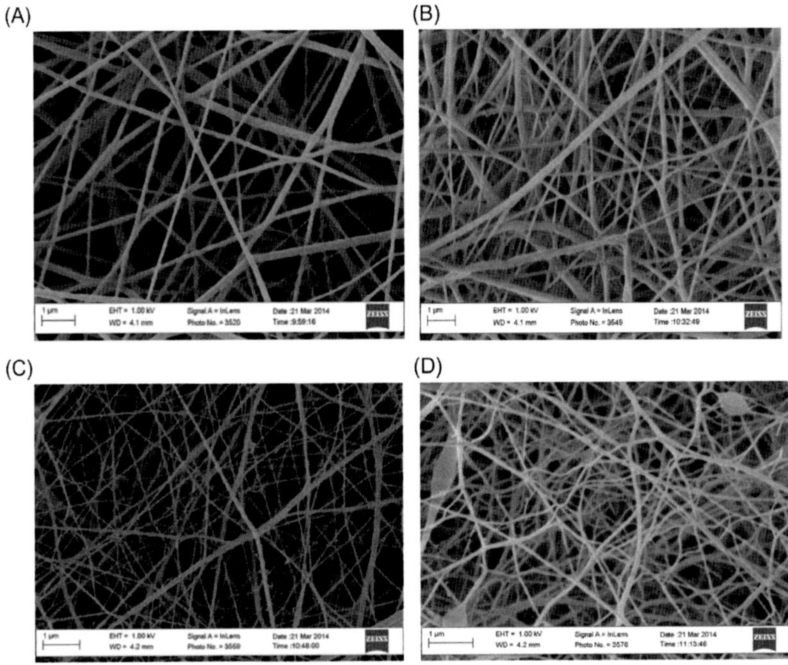

FIGURE 13.13

SEM images of (A) electrospun PVDF/PVP fibers incorporated with (B) 1 wt.%, (C) 2 wt.%, and (D) 4 wt.% carbon black particles developed for supercapacitor separators.

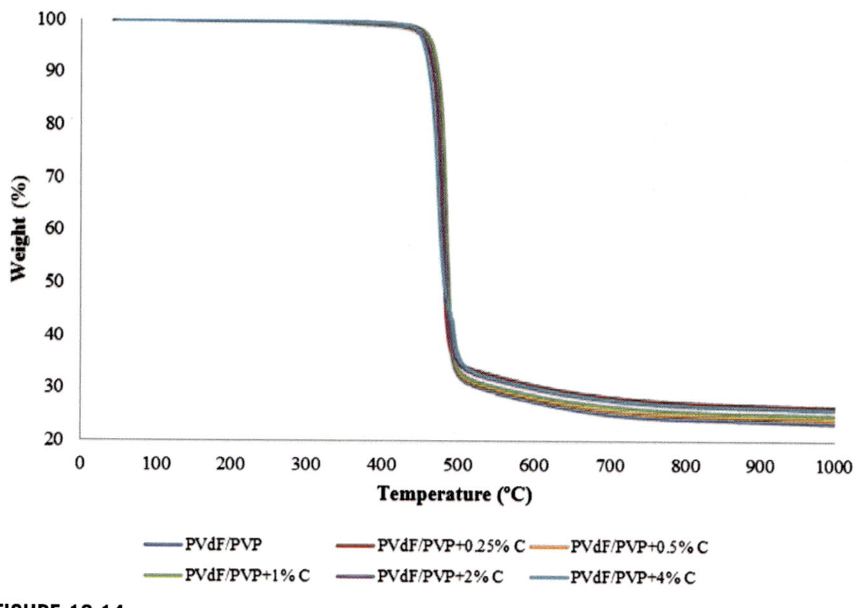

FIGURE 13.14

TGA thermogram of PVDF/PVP nanofibers incorporated with 0.25, 0.5, 1, 2, and 4 wt.% carbon black particles.

considerably improved as a function of carbon black inclusions in the polymeric structures. This study may be useful for the applications of nanofiber-based supercapacitor separators and other energy storage devices.

13.4.3 XPS CHARACTERIZATION

X-ray photoelectron spectroscopy (XPS) (also called "ESCA") measures the chemical composition and elemental distribution of the top 0–10 nm of solid surfaces (e.g., organic, inorganic, and biological). The source of soft X-ray stimulates the ejection of photoelectrons, and then its kinetic energy is measured by the electrostatic analyzer during the experiment. Any changes in the level of energy can result in chemically shifted valence states of the atoms from which the electrons are ejected from the source. It is a surface-sensitive quantitative spectroscopic technique and mainly provides chemical information about the sample surface. XPS beam size can be 1 μm and up to 1 mm, depending on the application and type of XPS unit [17]. In addition to the compositional characteristics of the materials, XPS can also determine the empirical formula of pure materials, chemical or electronic state of each element in the surface, uniformity of elemental composition across the surface, elements that contaminate the surface, and uniformity of elemental composition as a function of ion beam etching.

Xiong et al. reported about the flexible membranes of MoS_2/C nanofibers by electrospinning as binder-free anodes for high-performance sodium-ion battery applications [40]. MoS_2 is a uniform 1-D layered material, similar to graphene and boron nitrate-layered structures. The lithium-ion battery has been considered one of the main power sources for a number of electronic, communication, and other portable devices, and nowadays it is the only promising technology for many applications in electric vehicles and green energy storage due to the higher energy density and long cycle life compared to several other batteries. Highly flexible nanofiber mats with MoS_2/carbon nanofibers were produced using the electrospinning method, and it was determined that MoS_2 nanosheets were uniformly embedded into the interconnected carbon nanofibers (~ 150 nm). XPS analysis conducted on MoS_2/carbon fiber materials showed that Mo, S, C, and O elements were present in the MoS_2/carbon nanofiber films. EDS (or EDX) results also confirmed the existence of those elements in the electrospun nanofibers. The high-resolution XPS spectra of Mo is at 3d while S is at 2p. The XPS peaks of 232.8 and 229.5 eV can be because of the Mo 3d3/2 and Mo 3d5/2. The other peak at 236.0 eV is indexed to Mo61 3d5/2 of MoO_3, which is mainly related to the surface oxidation of the nanofiber samples. Furthermore, the XPS peaks at 162.4 and 163.5 eV in S 2p spectra are also typical characteristics of the S22 of MoS_2 [40].

13.4.4 UV−VIS SPECTROSCOPY

Ultraviolet visible spectroscopy (UV−vis) is a simple, fast, and affordable method of measuring the spectra of micro- and nanoscale samples between 190 and 1100 nm wavelengths. It is primarily used in nearly all analytical fields, such as solubility of pigments in water and organic solvents, characterization of substances, quantitative determination of substances in solution, and dissolution kinetics of organic and inorganic materials. Generally, two cuvettes (reference and target sample) with liquid or air are used for comparison to quantify the unknown material's behavior [17,41]. For example, if white light passes through or is reflected by the substance with different colors, the characteristic portion of the mixed wavelengths is absorbed; thus, the remaining light can be considered about the complementary color to the wavelength(s) absorbed during the spectroscopy test.

Ding et al. studied the influence of anatase-rutile mixed phase of TiO_2 nanofibers and ZnO blocking layer on DSSC photoanodes for the fabrication of a solar energy conversion device [42]. It was assumed that high efficiency in DSSC was expected using the 1D TiO_2 nanostructures due to the effective electron transport; however, the test results showed that DSSCs provided fairly low efficiencies when compared to the nanoparticle-based TiO_2 materials because of the low dye adsorption, longer length, and smooth surfaces of the nanofibers. In order to address these issues, the authors proposed a simple methodology of using a thick layer of TiO_2 electrospun nanofiber films as photoanodes to produce high energy conversion efficiency. For the improved the DSSC performance, anatase-rutile mixed phase of TiO_2 nanofibers was created by high-temperature sintering (500°C), and applying a thin layer of ZnO films via the atomic layer deposition technique. The thicknesses of mixed-phase (~ 15.6 wt.% rutile) TiO_2 nanofiber and ZnO film were 40 μm and 15 nm, respectively. The photoelectric conversion efficiency and short-circuit current values were found to be 8.01% and 17.3 mA cm^{-2}. The UV−vis absorption spectra conducted on sensitized TiO_2 nanofiber films showed that when the sintering temperatures were increased from 500°C to 600°C, the amount of light absorption was considerably increased due to the crystallinity increase, which is a good indication of overall increasing DSSC efficiencies [42].

13.4.5 VSM CHARACTERIZATION

VSM is a scientific technique that measures magnetic properties of bulk and nanoscale materials (solid, gel, or liquid) as a function of the magnetic field, temperature, and time. A sample is first magnetized under a uniform magnetic field and then sinusoidally vibrated via piezoelectric materials to measure the magnetic moment of the samples. The changes in resulting magnetic flux induce a noticeable voltage in the sensing coils, which will be proportional to the magnetic moment of the target materials. The most common VSM measurement method is the hysteresis loop method to determine magnetic properties at ambient temperature [2,17].

Khan et al. studied the magnetic properties of electrospun PVP and PAN nanocomposite fibers integrated with NiZn-ferrite ($Ni_{0.6}Zn_{0.4}Fe_2O_4$) nanoscale inclusions [2]. The electrospun nanocomposite fibers at different NiZn-ferrite nanoparticle (22 nm in diameter) concentrations were produced using 2.5 mL h^{-1} pump speed, 25 KV DC voltage, and 25 cm distance. $Ni_{0.6}Zn_{0.4}Fe_2O_4$ nanoparticles were initially prepared by the coprecipitation of $NiSO_4$, $ZnSO_4$, and $Fe_2(SO_4)_3$ according to the following equation:

$$0.6\,NiSO_4 + 0.4\,ZnSO_4 + Fe_2(SO_4)_3 + 8\,NaOH = >0.6\,Ni\,(OH)_2 + 0.4\,Zn(OH)_2$$
$$+ 2Fe(OH)_3 + 4\,Na_2(SO_4) = > Ni_{0.6}Zn_{0.4}Fe_2O_4$$

Fig. 13.15 shows SEM images of the electrospun nanocomposite PAN fibers as a function of magnetite concentration [2]. The diameters of the nanocomposite fibers were in the ranges of 150−300 nm. The magnetic properties of the nanocomposite fibers were investigated using the VSM method. Fig. 13.16 shows the magnetic moment of PAN nanocomposite fibers as a function of applied fields [2]. The VSM test results indicated that all the prepared nanocomposite fibers had superparamagnetic behaviors, and the magnetic saturation values of the magnetic nanocomposite fibers improved by increasing the NiZn-ferrite concentrations in the nanocomposite fibers [13−15]. These superparamagnetic nanofibers can also be used for biomedical, energy, filtration and separation, sensor, structural health monitoring, and defense applications.

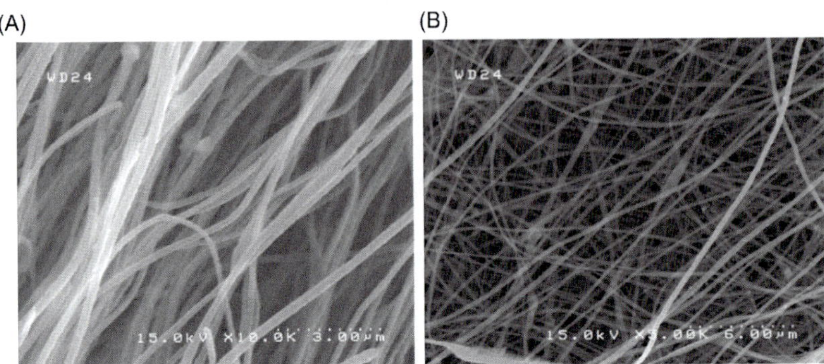

FIGURE 13.15

SEM images of the electrospun nanocomposite PAN fibers with (A) 4 wt.% $Ni_{0.6}Zn_{0.4}Fe_2O_4$ and (B) 16 wt.% $Ni_{0.6}Zn_{0.4}Fe_2O_4$ nanoparticles [2].

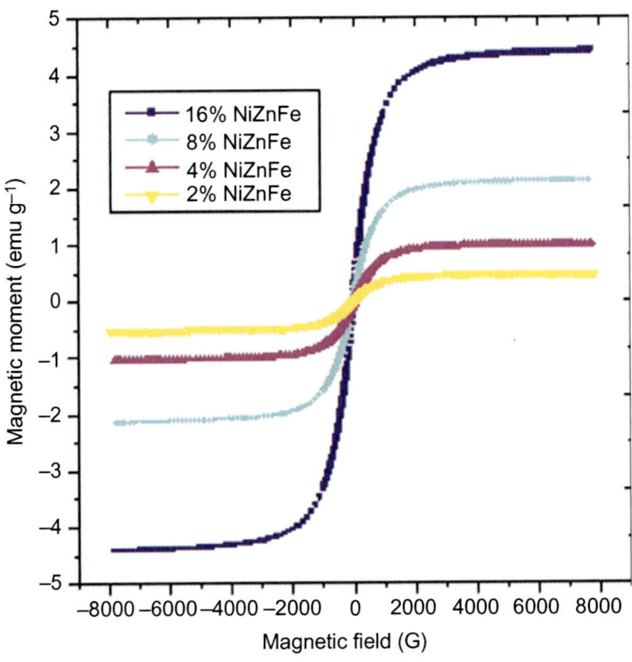

FIGURE 13.16

The magnetic moment of PAN nanocomposite fibers as a function of applied fields [2].

13.5 CONCLUSIONS

In this chapter, three major topics were investigated in detail. In the first part, the structural characterization of electrospun nanofibers was analyzed using different characterization methods (XRD, SEM, TEM, and AFM). Prior to this step, some information regarding electrospinning and jet formation were summarized. In the second part, optical spectroscopy measurement techniques (FTIR and Raman spectroscopy) were explained in terms of nanofiber characterizations and properties. In the last part of the chapter, the other characterization techniques for electrospun nanofibers, such as DSC, TGA, XPS, UV−vis, and VSM were evaluated. These studies indicated that many of the conventionally used characterization techniques can be used for various nanofibers. The characterizations of electrospun nanofibers (in organic and inorganic forms) are vitally important for different industrial applications of the nanofibers.

REFERENCES

[1] S. Cavaliere, Electrospinning for Advanced Energy and Environmental Applications, CRC Press, Boca Raton, FL, 2015.

[2] W.S. Khan, and R. Asmatulu "Magnetic Properties of Electrospun PVP and PAN Nanocomposite Fibers Associated with NiZn-Ferrite Nanoparticles," SAMPE Fall Technical Conference, Fort Worth, TX, October 17−20, 2011, 9p.

[3] S. Ramakrishna, K. Fujihara, W.E. Teo, T.C. Lim, Z. Ma, An Introduction to Electrospinning and Nanofibers, World Scientific, London, UK, 2005.

[4] Y. Gogotsi, Nanomaterials Handbook, CRC Press, Boston, 2006.

[5] R. Andrews, D. Jacques, A.M. Rao, T. Rantell, F.Y. Chen, J. Chen, et al., Nanotube composite carbon fibers, Appl. Phys. Lett. 75 (1999) 1329−1331.

[6] X.L. Xie, Y.W. Mai, X.P. Zhou, Dispersion and alignment of carbon nanotubes in polymer matrix: a review, Mater. Sci. Eng. R Rep. 49 (2005) 89−112.

[7] A. Theron, E. Zussman, A.L. Yarin, Electrostatic field-assisted alignment of electrospun nanofibers, Nanotechnology 12 (2001) 384−390.

[8] L.Y. Yeo, J.R. Friend, Electrospinning carbon nanotube polymer composite nanofibers, J. Exp. Nanosci. 1 (2) (2006) 177−209. Available from: https://doi.org/10.1080/17458080600670015.

[9] J.L. Ferreira, S. Gomes, C. Henriques, J.P. Borges, J.C. Silva, Electrospinning polycaprolactone dissolved in glacial acetic acid: Fiber production, nonwoven characterization, and in vitro evaluation, J. Appl. Polym. Sci. 131 (22) (2014). Available from: https://doi.org/10.1002/app.41068.

[10] R. Pignatello, "Advances in biomaterials science and biomedical applications," Rijeka, Croatia: InTech, 2013, doi:10.5772/54125 (Biofabrication of tissue scaffolds).

[11] D. Reneker, A. Yarin, H. Fong, S. Koombhongse, Bending instability of electrically charged; liquid jets of polymer solutions in electrospinning, J. Appl. Phys. 87 (2000) 4531−4557.

[12] A. Yarin, S. Koombhongse, D. Reneker, Bending instability in electrospinning of nanofibers, J. Appl. Phys. 89 (2001) 3018−3026.

[13] P. Gupta, R. Asmatulu, G. Wilkes, R.O. Claus, Superparamagnetic flexible substrates based on submicron electrospun Estane® fibers containing mnznfe-ni nanoparticles, J. Appl. Polym. Sci. 100 (2006) 4935−4942.

[14] G. Cao, Y. Wang, Nanostructures and Nanomaterials: Synthesis, Properties, and Applications, Second ed., World Scientific, London, UK, 2011.

[15] B. Rogers, J. Adams, S. Pennathur, Nanotechnology Understanding Small Systems, Second ed., CRC Press, Boca Raton, FL, 2011.

[16] R.K. Goyal, Nanomaterials and Nanocomposites: Synthesis, Properties, Characterization and Applications, CRC Press, Boca Raton, FL, 2017.

[17] R. Asmatulu, Introduction to Nanotechnology, Class Text Notes, Wichita State University, 2017.

[18] R. Asmatulu, M.A. Shinde, A. Alharbi, I.M. Alarifi, Integrating graphene and C_{60} into TiO_2 nanofibers via electrospinning process for the enhanced energy conversion efficiencies, Macromol. Symp. 365 (2016) 128−139.

[19] M. Shinde, "Synthesis and Analysis of Electrospun TiO_2 Nanofibers Incorporated with Nanoscale Inclusions for Improved DSSC Efficiencies," M.S. Thesis, Wichita State University, December, 2013.

[20] R. Asmatulu, August Nanotechnology Safety, Elsevier, Amsterdam, The Nederland, 2013.

[21] N. Nuraje, R. Asmatulu, G. Mul, November Green Photo-Active Nanomaterials: Sustainable Energy and Environmental Remediation, RSC Publishing, Cambridge, England, 2015.

[22] S. Mohammad, M.N. Uddin, G. Hwang, R. Asmatulu "Superhydrophobic PAN Nanofibers for Gas Diffusion Layers of Proton Exchange Membrane Fuel Cells for Cathodic Water Management," International Journal of Hydrogen Energy, 2017. http://dx.doi.org/10.1016/j.ijhydene.2017.07.229.

[23] S. Mohammad, "Investigating Superhydrophobic Behaviors of Carbonized PAN Nanofibers on Gas Diffusion Layers of PEM Fuel Cells," M.S. Thesis, Wichita State University, November 23, 2015.

[24] M. Srikant, "Investigating the Regeneration Behaviors of Neuroglial Cells on Artificially Made Electrospun PCL Scaffolds Embedded with Conductive Nanomaterials," Ph.D. Dissertation, Wichita State University, July 10, 2015.

[25] http://slideplayer.com/slide/9702681/ (accessed 10.12.18).

[26] http://www.chem.ucla.edu/∼bacher/General/30BL/IR/ir.html (accessed 11.12.18).

[27] Y. Li, M. Ceylan, B. Sherstha, H. Wang, Q.R. Lu, R. Asmatulu, et al., Nanofibers support oligodendrocyte precursor cell growth and function as a neuron-free model for myelination study, Biomacromolecules 15 (2014) 319−326.

[28] A. Alharbi, I.M. Alarifi, W.S. Khan, A. Swindle, R. Asmatulu, Synthesis and characterization of electrospun polyacrylonitrile/graphene nanofibers embedded with $SrTiO_3/NiO$ nanoparticles for water splitting, J. Nanosci. Nanotechnol. 17 (2017) 1−9.

[29] K.B. Mahat, I.M. Alarifi, A. Alharbi, R. Asmatulu, Effects of UV light on mechanical properties of carbon fiber reinforced PPS thermoplastic composites, Macromol. Symp. 365 (2016) 157−168.

[30] M.I. Alarifi, A. Alharbi, W.S. Khan, A.K.M.S. Rahman, R. Asmatulu, Mechanical and thermal properties of carbonized PAN nanofibers cohesively attached to surface of carbon fiber reinforced composites, Macromol. Symp. 365 (2016) 140−150.

[31] A. Alharbi, I.M. Alarifi, W.S. Khan, R. Asmatulu, Synthesis and analysis of electrospun $SrTiO_3$ nanofibers with NiO_x nanoparticles shells as photocatalysts for water splitting, Macromol. Symp. 365 (2016) 246−257.

[32] A. Alharbi, I.M. Alarifi, W.S. Khan, R. Asmatulu, Highly hydrophilic electrospun polyacrylonitrile / polyvinypyrrolidone nanofibers incorporated with gentamicin as filter medium for dam water and wastewater treatment, J. Membr. Sep. Technol. 5 (2016) 38−56.

[33] I.M. Alarifi, A. Alharbi, W.S. Khan, A. Swindle, R. Asmatulu, Thermal, electrical and surface properties of electrospun polyacrylonitrile nanofibers for structural health monitoring, Materials 8 (2015) 7017−7031.

[34] I.M. Alarifi, A. Alharbi, W.S. Khan, R. Asmatulu, Carbonized electrospun PAN nanofibers as highly sensitive sensors in SHM of composite structures, J. Appl. Polym. Sci. (2015). Available from: http://dx.doi.org/10.1002/a. 43235.

[35] I.M. Alarifi, A. Alharbi, O. Alsaiari, R. Asmatulu, Training the engineering students on nanofiber-based SHM systems, Trans. Tech. STEM Educ. 1 (2016) 59−67.

[36] W.S. Khan, M. Ceylan, A. Jabbarnia, L. Saeednia, R. Asmatulu, Structural investigations of electrospun pan nanofibers incorporated with various nanoscale inclusions, J. Thermal Eng. 3 (2017) 1375−1390.

[37] A. Ghazinezami, W.S. Khan, A. Jabbarnia, R. Asmatulu, Impacts of nanoscale inclusions on fire retardancy, thermal stability, and mechanical properties of polymeric PVC nanocomposites, J. Thermal Eng. 3 (2017) 1308−1318.

[38] A. Jabbarnia, W.S. Khan, A. Ghazinezami, R. Asmatulu, Investigating the thermal, mechanical and electrochemical properties of pvdf/pvp nanofibrous membranes for supercapacitor applications, J. Appl. Polym. Sci. (2016). Available from: http://dx. doi.org/10.1002/a. 43707.

[39] A. Jabbarnia, W.S. Khan, A. Ghazinezami, R. Asmatulu, Tuning the ionic and dielectric properties of electrospun PVdF/PVP nanofibers with carbon black nanoparticles for supercapacitor applications, Int. J. Eng. Res. Appl. 6 (2016) 65−73.

[40] X. Xiong, W. Lou, X. Hu, C. Chen, D. Hou, Y. Huang, Flexible membranes of MoS_2/C Nanofibers by electrospinning as binder-free anodes for high-performance sodium-ion batteries, Scientific Reports 5 (2015) 9254.

[41] A. Usta, R. Asmatulu, Synthesis and characterization of electrically sensitive hydrogels incorporated with cancer drugs, J. Pharmaceut. Drug Deliv. Research (2016). Available from: https://doi.org/10.4172/2325-9604.1000146.

[42] J. Ding, Y. Li, H. Hu, L. Bai, S. Zhang, N. Yuan, The influence of anatase-rutile mixed phase and ZnO blocking layer on dye-sensitized solar cells based on TiO_2 nanofiberphotoanodes, Nanoscale Res. Lett. 8 (2013) 9.

Index

Note: Page numbers followed by "*f*" and "*t*" refer to figures and tables, respectively.

N

T

Tapping mode, 266–267, 266*f*
Targeted chemotherapy, 44–45
Taylor cone, 3–4, 18, 20–21, 43
TCO. *See* Transparent conductive oxide (TCO)
TDSs. *See* Total dissolved solids (TDSs)
TEDA. *See* Trietylene diamine (TEDA)
TEM. *See* Transmission electron microscopy (TEM)
Template synthesis, 9
Textiles, electrospun nanofibers for
  fire-retardant fabrics, 78–81
  metamaterials and light and noise sensitivity, 74–78
  protective clothing, 81–82
  superhydrophobic electrospun nanofibers for nonwettable surfaces, 63–74
TGA. *See* Thermogravimetric analysis (TGA)
Thermal gravimetric analysis. *See* Thermogravimetric analysis (TGA)
Thermally conductive electrospun nanofibers, 248–249
Thermochemical hydrogen production process, 157–158
Thermochemical water splitting, 157–158
Thermogravimetric analysis (TGA), 102, 259, 272
  characterization, 273–275
    SEM images, 274*f*
    TGA thermogram of PVDF/PVP nanofibers, 275*f*
3D cancer model with nanofibers, 44–45
Threshold intensity. *See* Critical voltage limit
Tissue engineering, 263
  electrospun nanofibers for, 219–225
    bone tissue engineering, 225–227
    dental growth, 227–230
    wound healing, 215–219
  scaffolds, 220
Titania (TiO$_2$), 155–156, 203–206, 204*t*
  anatase nanofibers, 121–122
  nanofibers, 261–262
Titanium (IV) isopropoxide, 165–166, 261–262, 264–266
Titanium dioxide (TiO$_2$), 155
TMCh. *See* N,N,N-Trimethylchitosan (TMCh)
Toluene, 89, 147–148
Total dissolved solids (TDSs), 141, 143
Toxicity of nanofibers, 203
  carbon-based nanofibers, 206–208
  nanofibers, 199–201
  nanomaterials, 197–199
  protection methods of nanofibers, 208–210
  surface interaction of nanofibers, 201–206

TPP. *See* Triphenyl phosphate (TPP)
Trabeculi, 226
Transitional metal oxide nanostructures, 117–118
Translating spinneret, 32, 32*f*
Transmission electron microscopy (TEM), 18–19, 82, 90, 259
  characterization, 263–266
  diffraction patterns, 265*f*
  images of electrospun TiO$_2$ nanofibers, 265*f*
Transparent conductive oxide (TCO), 120–121
Trichoderma viride, 95–96
Trietylene diamine (TEDA), 147
N,N,N-Trimethylchitosan (TMCh), 219
Triphenyl phosphate (TPP), 80–81

U

Ultra-Ever Dry solutions, 262
Ultrafiltration (UF), 141
Ultrafine fibers, 164, 170
Ultrafine polymer fibers, 136–137
Ultraviolet (UV), 155
  light exposure tests, 102
Ultraviolet visible spectroscopy (UV–Vis spectroscopy), 259, 276–277

V

Vibrating sample magnetometer (VSM), 259
  characterization, 277–278
    magnetic moment of PAN nanocomposite fibers, 278*f*
    SEM images of electrospun nanocomposite PAN fibers, 278*f*
  measurement method, 277
Viscosity, 27–28
Viscous polymer gel electrode, 122
Volatile organic compounds, 147–148
Volcanoes, 146–147
VSM. *See* Vibrating sample magnetometer (VSM)

W

Wastewater, 143
Water
  condensation tests, 262
  contact angle, 272
  filtration/purification, 140–141
  pollution, 140–141
  splitting, 159–160, 164–168
  treatment, 140
  water-removing membrane, 144–146
  water-soluble polyelectrolyte, 246
  water/liquid/oil filtration, 137
Wenzel and Cassie–Baxter model, 144–145

Made in the USA
Las Vegas, NV
12 March 2026

43501876R00183